AF352799

Smart Sensing Technologies
for Agriculture

Smart Sensing Technologies
for Agriculture

Editors

Viacheslav Adamchuk
Kenneth Sudduth
Asim Biswas

MDPI • Basel • Beijing • Wuhan • Barcelona • Belgrade • Manchester • Tokyo • Cluj • Tianjin

Editors

Viacheslav Adamchuk
Ste-Anne-de-Bellevue
Canada

Kenneth Sudduth
USDA-ARS
USA

Asim Biswas
University of Guelph
Canada

Editorial Office
MDPI
St. Alban-Anlage 66
4052 Basel, Switzerland

This is a reprint of articles from the Special Issue published online in the open access journal *Sensors* (ISSN 1424-8220) (available at: https://www.mdpi.com/journal/sensors/special_issues/ Sensing_for_Agriculture).

For citation purposes, cite each article independently as indicated on the article page online and as indicated below:

LastName, A.A.; LastName, B.B.; LastName, C.C. Article Title. *Journal Name* **Year**, *Article Number*, Page Range.

ISBN 978-3-03936-688-0 (Hbk)
ISBN 978-3-03936-689-7 (PDF)

Contents

About the Editors

Viacheslav Adamchuk obtained a mechanical engineering degree from the National Agricultural University of Ukraine in his hometown. Later, he received both M.S. and Ph.D. degrees in Agricultural and Biological Engineering from Purdue University. Shortly after graduation, Dr. Adamchuk began his academic career as a faculty member in the department of Biological Systems Engineering at the University of Nebraska–Lincoln. There, he taught university students, conducted research, and delivered outreach programs relevant to precision agriculture, spatial data management, and education robotics. In 2010, Dr. Adamchuk joined the Department of Bioresource Engineering at McGill University (Chair of the Department since 2018), while retaining his adjunct status at the University of Nebraska–Lincoln. At McGill University, Dr. Adamchuk has formed the Precision Agriculture and Sensor Systems (PASS) research team that has been actively involved in research and outreach activities across Canada and the USA as well as a variety of other global initiatives.

Kenneth Sudduth leads a multidisciplinary team conducting precision agriculture research on topics including field-scale implementation of precision agriculture systems and evaluation of their economic and environmental effects at the USDA Agricultural Research Service's Cropping Systems and Water Quality Research Unit in Columbia, Missouri, USA. Throughout his career, his research has focused on sensors, instrumentation, and data management and interpretation for soil and crop properties important in precision agriculture. This research is documented in over 150 peer-reviewed journal articles (seven of which have received journal paper awards) and one patent. Dr. Sudduth received a B.S. and M.S. from the University of Missouri and a Ph.D. from the University of Illinois, all in Agricultural Engineering. He serves as an associate editor or editorial board member for several journals, holds an adjunct faculty position at the University of Missouri, and is Past-President of the International Society of Precision Agriculture. He has received over 25 international invitations to present his research and four major awards from three national/international scientific societies.

Asim Biswas is an Associate Professor at the University of Guelph, Canada, and an Adjunct Professor at McGill University, Canada. He is also a Visiting Professor at Jiangxi University of Finance and Economics, China, and a Jinshan Scholar Professor at Jiangsu University, China. Previously, he worked as Assistant Professor at McGill University, Canada, and Environmental Research Scientist at the Commonwealth Scientific and Industrial Research Organisation, Australia. Dr. Biswas completed all three degrees in soil science: a Ph.D. from the University of Saskatchewan; Canada; an M.Sc. from the University of Agricultural Sciences-Bangalore, India; and a B.Sc. from Bidhan Chandra Agricultural University, West Bengal, India. Dr. Biswas's research program on sustainable soil management focuses on providing solutions to increase the productivity and resilience of land-based agri-food production systems. His team works towards developing a more sustainable agri-food system to meet the challenges of a changing climate but also to do so in an environmentally sustainable manner. The core of his research is the design and development of sensors for fast, inexpensive, and accurate characterization of soil properties. He teaches courses related to traditional soil science and data-driven and sensor-based agriculture. He leads and is a member of various committees of international organizations including the International Union of Soil Sciences, International Society of Precision Agriculture, Soil Science Society of America, American Society of Agronomy, and Canadian Society of Soil Science. He is an Associate Editor of five journals and has guest edited another five Special Issues from different journals.

Preface to "Smart Sensing Technologies for Agriculture"

In recent years, agriculture has been transformed from a relatively conservative set of cultural practices to a dynamic industry producing food and other biomaterials in a highly competitive market. Unpredictable weather patterns and public concerns pertaining to the sustainability of food supply and the environment motivate farmers and other agri-businesses to consider various technological advancements. Thus, developments in precision agriculture and related scientific research lead to new means of managing crops and livestock in a more efficient and environmentally friendly manner. Many emerging practices rely on new sensors and new sensor applications to make this aim feasible and economically viable. This Special Issue presents 14 papers discussing cutting-edge research on smart sensing technologies applied to diverse agricultural challenges. Whereas some sensor systems generate data essential for decision support, others become part of closed-loop control systems that automate specific operations and processes.

These presented works rely on geophysical, radiometric, potentiometric, 3D scanning, machine vision, and other sensing principles to characterize soil, plants, and animals as well as their functionality in various production conditions. We hope that you enjoy reading this Special Issue and that you may consider contributing to further developments in smart sensing technologies for agriculture.

Viacheslav Adamchuk, Kenneth Sudduth, Asim Biswas
Editors

 *sensors*

Article

Three-Dimensional Mapping of Clay and Cation Exchange Capacity of Sandy and Infertile Soil Using EM38 and Inversion Software

Tibet Khongnawang [1,2], Ehsan Zare [1], Dongxue Zhao [1], Pranee Srihabun [1,2] and John Triantafilis [1,*]

1   School of Biological, Earth and Environmental Sciences, Faculty of Science, UNSW Sydney, Kensington, NSW 2052, Australia
2   Land Development Regional Office 5, Land Development Department, Khon Kaen 40000, Thailand
*   Correspondence: j.triantafilis@unsw.edu.au

Received: 15 June 2019; Accepted: 2 September 2019; Published: 12 September 2019

Abstract: Most cultivated upland areas of northeast Thailand are characterized by sandy and infertile soils, which are difficult to improve agriculturally. Information about the clay (%) and cation exchange capacity (CEC—cmol(+)/kg) are required. Because it is expensive to analyse these soil properties, electromagnetic (EM) induction instruments are increasingly being used. This is because the measured apparent soil electrical conductivity (EC_a—mS/m), can often be correlated directly with measured topsoil (0–0.3 m), subsurface (0.3–0.6 m) and subsoil (0.6–0.9 m) clay and CEC. In this study, we explore the potential to use this approach and considering a linear regression (LR) between EM38 acquired EC_a in horizontal (EC_{ah}) and vertical (EC_{av}) modes of operation and the soil properties at each of these depths. We compare this approach with a universal LR relationship developed between calculated true electrical conductivity (σ—mS/m) and laboratory measured clay and CEC at various depths. We estimate σ by inverting EC_{ah} and EC_{av} data, using a quasi-3D inversion algorithm (EM4Soil). The best LR between EC_a and soil properties was between EC_{ah} and subsoil clay ($R^2 = 0.43$) and subsoil CEC ($R^2 = 0.56$). We concluded these LR were unsatisfactory to predict clay or CEC at any of the three depths, however. In comparison, we found that a universal LR could be established between σ with clay ($R^2 = 0.65$) and CEC ($R^2 = 0.68$). The LR model validation was tested using a leave-one-out-cross-validation. The results indicated that the universal LR between σ and clay at any depth was precise (RMSE = 2.17), unbiased (ME = 0.27) with good concordance (Lin's = 0.78). Similarly, satisfactory results were obtained by the LR between σ and CEC (Lin's = 0.80). We conclude that in a field where a direct LR relationship between clay or CEC and EC_a cannot be established, can still potentially be mapped by developing a LR between estimates of σ with clay or CEC if they all vary with depth.

Keywords: Three-dimensional mapping; quasi-3D inversion algorithm; cation exchange capacity; clay content; sandy infertile soil

---

1. Introduction

Most cultivated upland areas of northeast Thailand are being used for cash crops (e.g., sugarcane) [1]. However, the soil is sandy and infertile, and they are difficult to improve agriculturally without information about clay and cation exchange capacity (CEC—cmol(+)/kg). In terms of clay, knowledge is important because it is an indication of the capacity of soil to hold moisture and potential to store exchangeable cations [2,3]. Knowledge of the CEC is also necessary because it is a measure of nutrient availability and how well soil pH is buffered against acidification [4] as well as an index of the shrink–swell potential of soil [5]. Therefore, information about the spatial distribution of clay and CEC are required. This is particularly the case in Khon Kaen Province, where poor water holding capacity leads to deep drainage and in some cases rising water tables and soil salinization. In addition, clay (Table 1) and CEC (Table 2) data provides a farmer with information from which fertilizer recommendations can be made.

However, the conventional ways of measuring these soil properties are costly and time-consuming owing to the soil sampling and laboratory analysis. Nevertheless, much research has shown that if many soil samples can be collected, clay and CEC can be mapped using classical geostatistical methods [6,7]. Among the first to map topsoil (0–0.3m) clay and CEC in this way where [8], who used punctual kriging of soil sample locations at the field scale. More recently, [9] predicted topsoil (0–0.15 m) and subsurface (0.3–0.5 m) CEC using various types of kriging (i.e., ordinary) across a large area of North Dakota, USA. Similarly, [10] used additive and modified log-ratio transformation of soil particle size fraction (psf) using ordinary kriging. They then compared this to the untransformed psf data using various kriging techniques (i.e., compositional ordinary- and ordinary-kriging) to predict the topsoil (0–0.1 m) clay, across a very large area in south-eastern Australia. However, a major disadvantage of such geostatistical approaches is that many samples (>100) are required, which need to be spatially correlated and variable [11,12] to yield good results.

To add value to the limited soil data, pedotransfer functions can be used to predict one soil property from another [13,14]. However, to account for short scale variation, easier to acquire ancillary data, which are directly related to clay or CEC are increasingly being used. One of the most widespread are electromagnetic (EM) instruments (i.e., EM38 and EM34), because they measure apparent electrical conductivity (EC_a—mS/m). [15] were among the first to identify a linear regression (LR) between EM34 EC_a and average (0–15 m) clay ($R^2 = 0.73$). [16] developed a LR between EM38 EC_a and average (0–1.5 m) clay ($R^2 = 0.77$) to map clay across a cotton field (244 ha). [17] similarly found a good LR ($R^2 = 0.76$) and mapped clay across different fields. In their comprehensive review, [18] demonstrated many other LR of variable strength ($R^2 = 0.01$–0.94). In terms of CEC, [19] found a LR between topsoil (0–0.2 m) CEC and EC_a, while [20] found a strong LR ($R^2 = 0.74$) between an EM38 and topsoil (0–0.3 m) CEC across various fields. [21] showed how a LR between an EM38 and average (0–0.2 m) CEC ($R^2 = 0.81$) could then be used to map CEC, while [12] established a separate LR to map different topsoil (0–0.075, 0.075–0.15 and 0.15–0.3 m) CEC across a field in Missouri, USA. Again, [18] provided another example of LR between EC_a and CEC (0.50–0.76 m).

Given the sandy and infertile nature of soil in northeast Thailand, chemical and compost fertiliser application guidelines [22,23] have been developed. For example, if clay (%) is known and is small (<15%), the chemical fertiliser rates for nitrogen (N), phosphorus (P_2O_5) and potassium (K_2O) would be 113, 38 and 113 kg/ha, respectively. Alternatively, a compost fertiliser rate of 25 t/ha is suggested. This is similarly the case for CEC. In this research our interest is seeing if we can assist farmers with applying these guidelines by developing digital soil maps (DSM). The first aim is to see if we can develop a LR relationship between EM38 EC_a directly with either topsoil (0–0.3 m), subsurface (0.3–0.6 m) or subsoil (0.6–0.9 m) clay and CEC. We compare this approach with a universal LR we develop between the calculated true electrical conductivity (σ—mS/m) and laboratory measured clay and CEC at various depths, because of recent success in mapping salinity [24] and moisture [25] by inverting EC_a data. While a similar approach was used to map CEC in 3-dimensions by [26], they used a Veris-3100 instrument. Herein, we validate the universal LR using a leave-one-out-cross-validation, considering accuracy, bias and Lin's concordance.

2. Materials and Methods

2.1. Study Area

The study site (Lat 16°11′40.79″ N and Lon 102°43′54.46″ E) is located in the Ban Haet district, Khon Kaen (Thailand). It is situated a short distance to the west of Ban Haet village and located approximately 40 km south of Khon Kaen. The area is approximately 6 ha (150 m × 400 m), with the dominant soil type being an acid sandy loam to sandy Alfisols, described by Land Development Department of Thailand (scale of 1:25,000). The current land use is rain-fed sugarcane farming [1]. The topography across the site is flat to relatively flat with a slope of 0–1%.

Table 1. Chemical and compost fertilizer application guidelines based on clay content for sugarcane in Thailand [22,23].

Clay (%)	Chemical Fertilizer Rates (kg/ha)			Compost Fertilizer Rates (t/ha)
	N	P_2O_5	K_2O	
<15	113	38	113	25
15–18	113	38	75	19
18–35	75	19	75	18
>35	72	38	38	18

The climate is tropical savanna [27]. The mean annual precipitation is around 1,100 mm with the average minimum and maximum temperatures of 18.7 and 35.2 °C, respectively [28]. However, the area has three distinct seasons. The dry-season occurs between mid-February to mid-May with the hottest temperatures in April (43.9 °C) with some rainfall (224.4 mm). Conversely, the rainy season is between May to October with average temperatures typified by July (24.4 °C) which also has the most rainfall (1104 mm). The winter season is between mid-October to mid-February, with maximum temperatures in December (24.2 °C) with limited rainfall (76.3 mm) [28].

Table 2. Liming application guidelines for sugarcane in Thailand when pH less than 5.0 [29].

CEC (cmol(+)/kg)	Lime Application (t/ha)
<4	1.25
4–8	2.5
8–16	4
>16	5

2.2. Data Collection and Interpolation

The instrument used to collect EC_a data was an EM38 [30]. The instrument consists of a transmitter and receiver coil located at either end and spaced 1.0 m apart. The depth of exploration depends on coil configuration. In the horizontal mode, the EM38 measures EC_{ah} and within a theoretical depth of 0–0.75 m. In the vertical mode, the EM38 measures EC_{av} and within a theoretical depth of 0–1.5 m.

In terms of collecting EM38 EC_a data in these two modes, 17 parallel transects were defined and spaced approximately 10 m apart in essentially an east–west orientation. Figure 1b shows the spatial distribution of these transects, which were of unequal length. The survey was conducted on 17 January 2018. In all, 467 measurement sites were visited to measure the EM38 EC_{ah} and EC_{av}. All EC_a measurements were georeferenced using a Garmin Etrex Legend G [31] submeter GPS.

2.3. Soil Sampling and Laboratory Analysis

To determine if a direct linear relationship LR between EC_a or σ could be developed with topsoil (0–0.3 m), subsurface (0.3–0.6 m) and subsoil (0.6–0.9 m) clay or CEC, 46 soil sampling locations were selected. The sampling points were selected according to two criteria, as suggested by [21]. Firstly, locations with small, intermediate and large EC_a were selected; and secondly, samples were spaced evenly across the field. The samples were collected on the 25 January 2018. Figure 1c shows the location of the 46 sampling locations.

The soil samples were air-dried, ground and passed through a 2-mm sieve. Laboratory analysis involved determination of soil particle size fractions (e.g., clay—%) based on Hydrometer method [32]. The cation exchange capacity (CEC—cmol(+)/kg) was also determined based on ammonium saturation method. Regarding this, soil samples were saturated by NH4OAc pH 7.0 and rinsed with NH_4^+ using NaCl (Na^+). Distillation apparatus and titrate method was used to determine CEC [33].

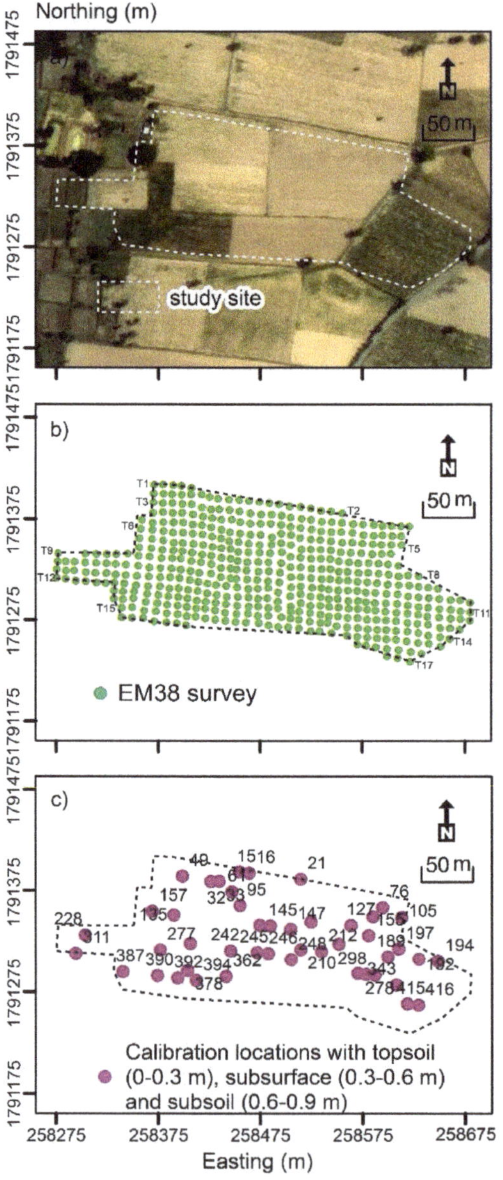

Figure 1. (**a**) Air-photo of study site; (**b**) EM38 survey transects (i.e., 17) and, (**c**) calibration locations (46) where topsoil (0–0.3 m), subsurface (0.3–0.6 m) and subsoil (0.6–0.9 m) samples.

2.4. Quasi-3D Inversion of EM38

The EM4Soil (v304) inversion software package [34] was used to invert EM38 EC_{ah} and EC_{av} to calculate true electrical conductivity (σ—mS/m) and develop electromagnetic conductivity images (EMCI). The quasi-3D inversion algorithm was used in this study. In brief, quasi-3D is a 1-dimensional spatial constrained technique and a forward modelling approach. It assumes that below each EC_a measurement location, an estimate of the 1-dimensional variation of σ is constrained because the EM4Soil software considers neighboring locations where the EM38 EC_a was measured [35].

The raw EC_a data was first gridded using a nearest neighbor technique onto a grid spacing of 10×10 m using the gridding tool available in EM4Soil. The initial model of σ was set equal to 10 mS/m with the maximum number of iterations equal to 10. A homogeneous five-layer initial model was also considered with depths to the top of each layer being 0, 0.3, 0.6, 0.9 and 1.05 m. The same depths would be used by the EM4Soil software to estimate true σ at these depths and for 3D prediction.

To identify the best possible LR between σ and measured clay and/or CEC, there are a number of other parameters which need to be considered, including selection of a forward model (S1, S2), inversion algorithm (cumulative function (CF) and full solution (FS)), and damping factor (λ). With respect to the inversion algorithm there are two variations (S1 and S2) of Occam's regularization [36]. The S2 algorithm constrains the model response (of σ) to be around a reference model. It produces therefore smoother results than that of S1.

Theoretically, the CF model is based on the EC_a cumulative response and is used to convert depth profile conductivity to σ [37] considering the condition of low induction numbers. The FS model is based on the Maxwell equations [38] and is not limited to the small induction number condition. Therefore, the FS can improve models calculated from EC_a data acquired over highly conductive soils (i.e., >100 mS/m). The damping factor (λ) was progressively increased with smaller increments initially and in large increments thereafter. The λ values used in this study were 0.07, 0.3, 0.6 and 0.9 to balance between rough and smooth EMCIs.

2.5. Validation and Comparison with LR

We use a simple linear regression (LR) model which is in the form of:

$$Y = a + bX + \varepsilon \tag{1}$$

where Y is vector of the target property (i.e., clay and CEC) and X is a vector of a predictor while ε is the model's residual.

Figure 2 shows the flow chart of the two different approaches we undertook. First, we looked to see if six independent MLR models that can be developed between the EC_a (i.e., EC_{ah} and EC_{av}) data and clay and CEC in either the topsoil (0–0.3 m), subsurface (0.3–0.6 m) and subsoil (0.6–0.9 m). Secondly, we look to see if a satisfactory LR model can be developed between true electrical conductivity (σ—mS/m) as inverted from EM38 EC_{ah} and EC_{av}, and clay (%) and CEC (cmol(+)/kg) at all depths using a universal LR, which is applicable at any depth.

To determine the robustness of the calibration and the prediction of either clay and CEC, we tested the final DSM using a leave-one-out-cross-validation procedure. This was carried out 46 times and involved the removal of each of the 46 soil sample locations one at a time and from all three increment depths, including topsoil, subsurface and the subsoil.

The accuracy assessment of prediction was examined using root mean square error (RMSE), whereby the closer the RMSE to zero the more accurate the prediction. The prediction bias was estimated by calculating mean error (ME). Again, the closer to zero then the less biased the prediction.

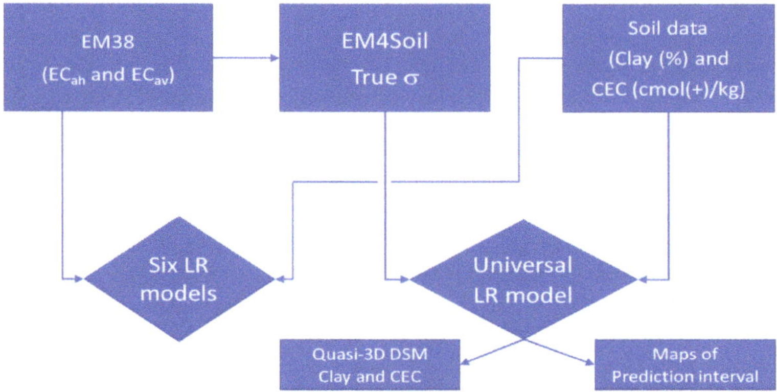

Figure 2. Flow chart of two different approaches to establish a linear regression (LR), between (i) EM38 EC$_a$ (mS/m) in horizontal (EC$_{ah}$) or vertical (EC$_{av}$) and three different depths of clay (%) or CEC (cmol(+)/kg) data (i.e., topsoil (0–0.3 m), subsurface (0.3–0.6 m) and subsoil (0.6–0.9m), and (ii) true electrical conductivity (σ—mS/m) inverted from EM38 EC$_{ah}$ and EC$_{av}$ with clay (%) or CEC (cmol(+)/kg) using universal LR at any depth.

The Lin's concordance correlation coefficient (ρ_c) was also calculated to assess how close the LR model is to the 1:1 relationship overall. This is because Lin's concordance correlation coefficient [39] measures degree of agreement between two variables. The Lin's concordance correlation coefficient is determined from a sample as follows:

$$\rho_c = \frac{2S_{XY}}{S_X^2 + S_Y^2 + \left(\overline{X} - \overline{Y}\right)^2} \tag{2}$$

where $\overline{X}$ and $\overline{Y}$ are means for the two variables (which in our case are the measured and predicted clay or CEC in the topsoil, subsurface or subsoil and S_X^2 and S_Y^2 are the corresponding variances and

$$S_{XY} = \frac{1}{n} \sum_{i-1}^{n} \left(X_i - \overline{X}\right)\left(Y_i - \overline{Y}\right) \tag{3}$$

2.6. Prediction Interval (PI)

To evaluate the uncertainty, the 95% prediction interval (PI) was used to compute the data between the measured and predicted clay and CEC across the study field and in the topsoil, subsurface and subsoil. The PI represents the frequency of possible confidence intervals that contain the true value of the prediction. A broad PI suggests a larger confidence in prediction [40].

3. Results and Discussion

3.1. Preliminary EC$_{ah}$ and EC$_{av}$ Data Analysis

Table 3 shows the summary statistics of the 467 EC$_a$ measured sites during the EM38 survey. The mean EC$_{ah}$ (0–0.75 m) was 23.1 mS/m with a minimum of 14 mS/m and maximum of 35 mS/m. The median (23) was close to the mean, with the EC$_{ah}$ slightly positively skewed (0.2) with a coefficient of variation (CV) of 19.4%. In comparison, the EC$_{av}$ (0–1.5 m) had a larger mean (28.8 mS/m) with a minimum of 18 mS/m and maximum of 45 mS/m. The median EC$_{av}$ was again slightly larger (29.0 mS/m) than the mean with the skewness positive again (0.3) and CV slightly smaller (20.5%).

Table 3. Summary statistics of apparent electrical conductivity (EC_a mS/m) measured by an EM38 instrument for the entire survey area and at the 46 calibration points.

			EC_a (mS/m)				
Data Source	**n**	**Min**	**Mean**	**Median**	**Max**	**Skewness**	**CV (%)**
Survey data							
EC_{ah}	467	14	23.1	23	35	0.2	19.4
EC_{av}	467	18	28.8	29	45	0.3	20.5
Calibration data							
EC_{ah}	46	15	22.3	22	33	0.4	19.5
EC_{av}	46	19	27.5	27.5	42	0.67	19.5

Similarly, the simple statistics of the EC_a data at the 46 calibration locations were relatively close to the surveyed data. The mean EC_{ah} was 22.3 mS/m with a minimum of 15 mS/m and maximum of 33 mS/m. The median was close to the mean (22), with the EC_{ah} slightly positively skewed (0.4) and with a coefficient of variation (CV) of 19.5%. In comparison, the EC_{av} had a larger mean (27.5 mS/m) with a minimum of 19 mS/m and maximum of 42 mS/m. The median EC_{av} was the same value (27.5 mS/m) to mean with the skewness positive again (0.67) and CV slightly smaller (19.5%).

Figure 3a shows the interpolated digital elevation model (DEM). The highest elevation was in the east end of study field (169 m). The elevation gradually decreased toward the south end of the study field where it was lowest (161 m). Figure 3b shows the interpolated contour plot of measured EC_{ah}. The study field was characterized by intermediate-small (15–25 mS/m) EC_{ah} in the northern half. Whereas, intermediate-small to intermediate EC_{ah} (25–35 mS/m) defines the southern.

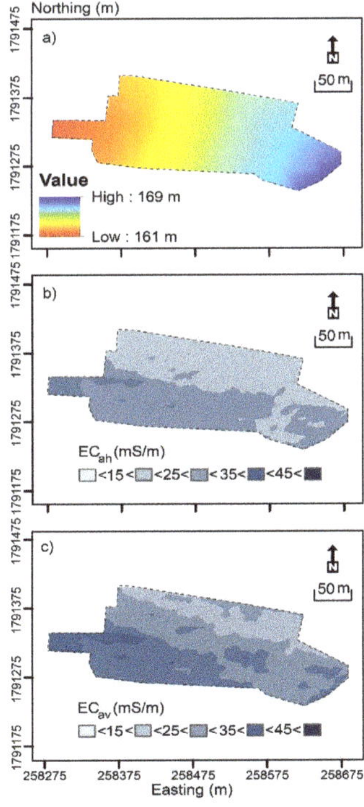

Figure 3. Contour plot of (**a**) elevation (m), and apparent electrical conductivity (EC_a – mS/m) of EM38 measured in (**b**) horizontal (EC_{ah}; 0–0.75 m), and (**c**) vertical (EC_{av}; 0–1.5 m) modes.

Figure 3c shows the contour plot of measured EC_{av}. Again, the study field was characterized by intermediate-small EC_{av} in the northern half which started from the east through the north west corner and intermediate-large EC_{av} (35–45 mS/m) in the west. From Figure 3b,c and Table 3, we surmise that the subsurface and subsoil were likely to be more conductive than the topsoil.

3.2. Preliminary Clay and CEC Data Analysis

Table 4 shows the summary statistics of measured clay (%) at the 46 sample locations. The mean topsoil (0–0.3 m) clay was 11.9% with a minimum of 9.4% and maximum of 16.8%. The median was similar (12%) and the skewness and CV were 0.80 and 12.2, respectively. In the subsurface (0.3–0.6 m), the mean, minimum and maximum were all slightly larger (15.6%, 10.6% and 20.6%, respectively). The median was again similar (15.3%) to the mean, with the clay being positively skewed (0.1). In comparison, the subsoil (0.6–0.9 m) mean of 19.1%, minimum (14.1%) and a maximum (23.4%) was larger again. The median was 19% with a positive skewness (0.1).

Table 4 also shows the summary statistics of measured CEC (cmol(+)/kg) at the sample locations. The mean CEC in the topsoil was 3.3 cmol(+)/kg with a minimum of 2.3 cmol(+)/kg and maximum of 5 cmol(+)/kg. The median was slightly smaller (3.2 cmol(+)/kg) than mean and the skewness and CV were 1.3 and 14.8, respectively. The subsurface mean CEC was larger (4.1 cmol(+)/kg) than the topsoil, with the minimum (2.5 cmol(+)/kg) and maximum (6.3 cmol(+)/kg) also larger. Again, the median was 4 cmol(+)/kg with the skewness and CV were being 0.6 and 18.5. As with the subsoil clay, subsoil CEC had a larger mean (4.9 cmol(+)/kg) minimum (3.5 cmol(+)/kg) and maximum (6.7 cmol(+)/kg). The median was smaller (4.7 cmol(+)/kg) than mean with a skewness and CV of 0.7 and 15.1 respectively.

Table 4. Summary statistics of measured clay (%) and CEC (cmol(+)/kg) at the 46 calibration locations.

Property/Depth	n	Min	Mean	Median	Max	Skewness	CV (%)
clay (%)							
topsoil (0–0.3 m)	46	9.4	11.9	12	16.8	0.8	12.2
subsurface (0.3–0.6 m)	46	10.6	15.6	15.3	20.6	0.1	17
subsoil (0.6–0.9 m)	46	14.1	19.1	19	23.4	0.1	11.2
CEC (cmol(+)/kg)							
topsoil (0–0.3 m)	46	2.3	3.3	3.2	5	1.3	14.8
subsurface (0.3–0.6 m)	46	2.5	4.1	4	6.3	0.6	18.5
subsoil (0.6–0.9 m)	46	3.5	4.9	4.7	6.7	0.7	15.1

3.3. Spatial Distribution of Clay and CEC Data

Figure 4 shows the contour plot of measured clay (%) and CEC (cmol(+)/kg). Figure 4a shows measured clay in the topsoil (0–0.3 m), which was characterised by small clay (< 15%) across the field. Figure 4b shows measured clay in the subsurface (0.3–0.6 m), which was characterised by a slightly larger clay varying between intermediate-small (15–18%) in the middle of the field and intermediate (18–21%) clay along the southern margin and in the west.

Figure 4c shows measured clay of the subsoil (0.6–0.9 m). It was characterised in the northern half predominantly by intermediate-small clay. In the west and along the southern margin, clay was intermediate to intermediate-large (21–24%). Clearly, clay increases with depth on average.

Figure 4d shows measured CEC (cmol(+)/kg) from the topsoil. Most of the field was characterised by small (<3.8 cmol(+)/kg) to intermediate-small CEC (3.8–4.5 cmol(+)/kg), except the small area in the west where it was intermediate-large (5.3–6 cmol(+)/kg).

Figure 4e shows measured CEC in the subsurface, which was generally larger and in accord with the areas where clay was also large. Figure 4f shows measured CEC in the subsoil. As with the clay, as shown in Table 5, the CEC increased with depth on average. For the most part, larger clay and CEC were in accord with the increasing EC_{ah} and EC_{av} from north to south as shown in Figure 3b,c, respectively.

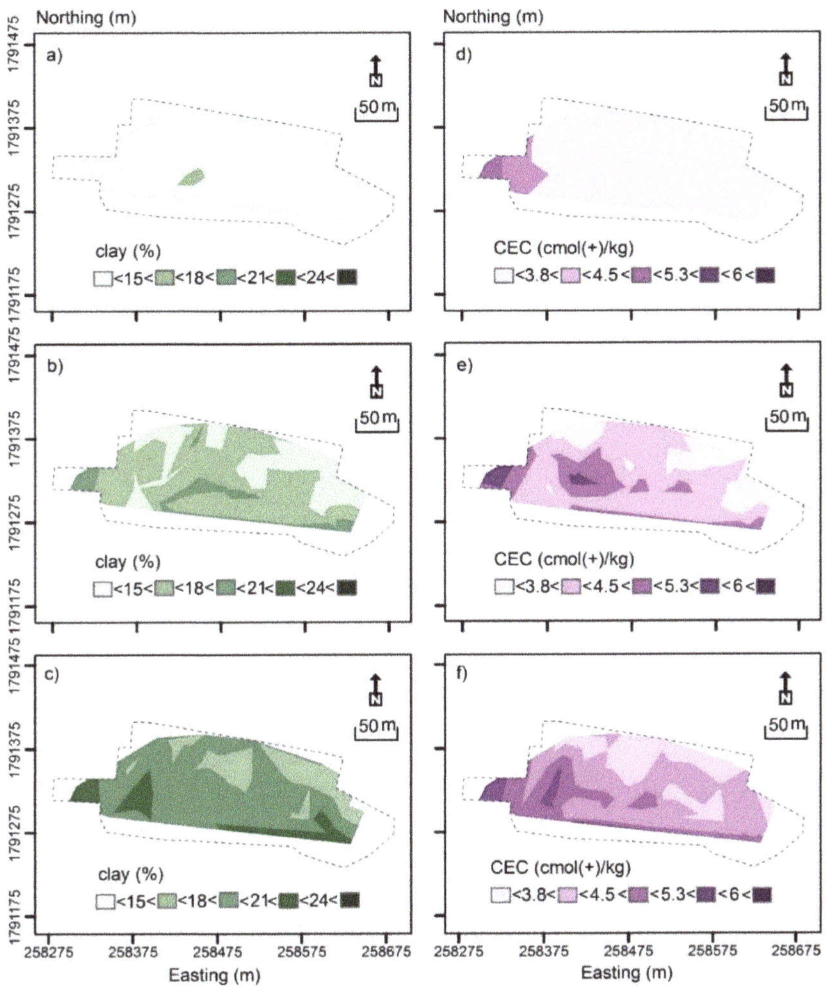

Figure 4. Contour plots of measured (**a**) topsoil (0–0.3 m), (**b**) subsurface (0.3–0.6 m) and (**c**) subsoil (0.6–0.9 m) clay (%) and (**d**) topsoil, (**e**) subsurface and (**f**) subsoil cation exchange capacity (CEC—cmol(+)/kg).

3.4. Linear Regression of Clay and CEC of Individual Depth Increment and ECa

Figure 5 shows the LR between measured clay and CEC versus EC_a. Figure 5a shows EC_{ah} and topsoil (0–0.3 m) clay was small ($R^2 = 0.21$). Slightly better correlations were achieved between EC_{ah} and subsurface (0.3–0.6 m) and subsoil (0.6–0.9 m) clay ($R^2 = 0.33$ and 0.43, respectively). Figure 5b shows EC_{av} and topsoil clay was also small (0.19), with similarly poor correlations achieved between EC_{av} and subsurface ($R^2 = 0.3$) and subsoil ($R^2 = 0.42$). Figure 5c shows that equivalent results were achieved between EC_{ah} and CEC, however the correlations were larger in the topsoil (0.50), subsurface ($R^2 = 0.47$) and subsoil clay ($R^2 = 0.56$) as compared to clay. Figure 5d shows again the same trend of correlation, with the linear regression between EC_{av} and measured CEC equivalent to topsoil ($R^2 = 0.51$), subsurface ($R^2 = 0.47$) and subsoil ($R^2 = 0.56$) CEC. We conclude that there was no satisfactory correlation between the measured soil properties and EC_{ah} and EC_{av} and with increasing depths and therefore no valid LR calibrations to predict these soil properties across our study field. We attribute this to the small CV and subtle differences in clay and CEC across the field.

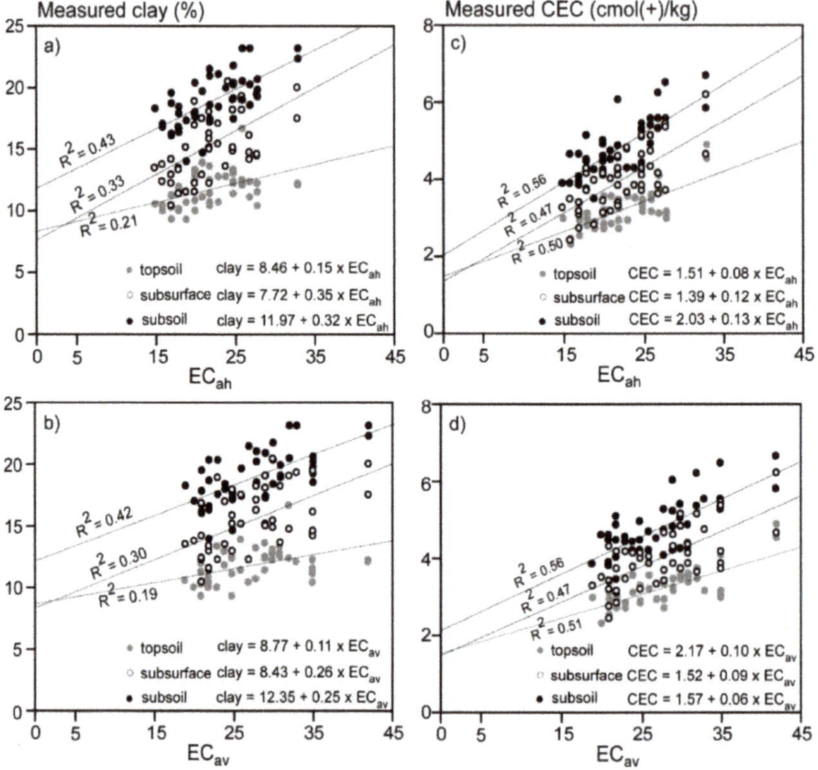

Figure 5. Plots of linear regression (LR) between apparent electrical conductivity (EC_a—mS/m) measured in (**a**) horizontal (EC_{ah}; 0–0.75 m) and (**b**) vertical (EC_{av}; 0–1.5 m) modes of EM38 and measured clay (%) in the topsoil (0–0.3 m), subsurface (0.3–0.6 m) and subsoil (0.6–0.9) and plots of LR between (**c**) EC_{ah} and (**d**) EC_{av} and measured cation exchange capacity (CEC—cmol(+)/kg) in the topsoil, subsurface and subsoil.

3.5. Linear Regression of Clay for All Depth Increment and σ

To determine if a better coefficient of determination (R^2) could be obtained between clay and σ, we inverted the EM38 EC_{ah} and EC_{av} using EM4Soil. Table 5 shows the R^2 between estimated σ (mS/m) obtained from EM4Soil (quasi-3D) and measured clay at all depths (i.e., topsoil, subsurface and subsoil). Table 5 shows that with increase in the damping factor (i.e., λ), the correlation between σ and clay decreases, regardless of algorithm (S1 or S2) or forward model (CF or FS). The best coefficient of determination ($R^2 = 0.65$) was when S1 algorithm and FS forward model were used with a λ value of 0.07. The σ values calculated using these parameters were selected to establish a linear regression (LR) between σ and clay. This R^2 was a little smaller to that achieved by [38] who developed a LR ($R^2 = 0.74$) between σ and clay along a single transect in a large irrigated area. We attribute this to the fact that we predicted σ using a quasi-3D inversion, compared to the quasi-2D used along the transect. The effect was greater smoothing herein, as a larger number of neighbours were used to estimate σ.

Table 5. Coefficient of determination (R^2) achieved between the measured clay and CEC with the estimated true electrical conductivity (σ) generated by inverting EM38 apparent electrical conductivity (EC_a—mS/m) using the EM4Soil quasi-3D model, cumulative function (CF) or full solution (FS), algorithms S1 or S2, and various damping factors (λ).

Clay	λ	S1, CF	S1, FS	S2, CF	S2, FS
	0.07	0.631	0.648	0.596	0.616
	0.3	0.625	0.643	0.537	0.559
	0.6	0.622	0.643	0.465	0.487
	0.9	0.619	0.637	0.409	0.430
CEC					
	0.07	0.666	0.674	0.656	0.666
	0.3	0.666	0.674	0.641	0.655
	0.6	0.667	0.672	0.599	0.617
	0.9	0.668	0.676	0.559	0.577

Figure 6a shows the LR between σ (S1, FS, and $\lambda = 0.07$) and measured clay (%). The LR was made using the bivariate fit tool [41] and include fitted lines, associated fitted confidence intervals (confidence limits for the mean value) and individual confidence intervals (confidence limits for individual predicted values). The LR developed was of the form: clay $= 6.04 + 0.50\sigma$.

Table 6 shows the summary statistics, which indicates the LR was statistically significant (<0.0001*). We note that in Figure 6a topsoil (0–0.3 m) σ was smallest (8–22 mS/m), however, subsurface (0.3–0.6 m) σ was intermediate (11–25 mS/m) with subsoil (0.6–0.9 m) σ largest (17–37 mS/m). The increasing σ was a function of EC_{av} being larger than EC_{ah}, as shown in Figure 3b,c, respectively. The reason for the increasing σ was due to increasing clay with depth.

Table 6. Summary statistics of the linear regression (LR) model established between the calculated true electrical conductivity (σ) and measured clay (%) and CEC (cmol(+)/kg). The σ was estimated using the full solution (FS), S1 algorithm and a damping factor (λ) of 0.07 and 0.9 respectively.

Clay	Parameter	Estimate	SE	t-ratio	Prob > |t|	R^2
	Intercept	6.04	0.63	9.62	<0.0001*	0.65
	0.07, S1, FS	0.50	0.03	15.84	<0.0001*	
CEC						
	Intercept	1.46	0.16	9.09	<0.0001*	0.68
	0.9, S1, FS	0.13	0.01	16.85	<0.0001*	

Figure 6b shows the leave-one-out cross-validation applied for clay by removing each of the 46 sampling locations one at a time. Specifically, the topsoil, subsurface and subsoil samples of one of the sample locations. It was apparent that predicted agreed with measured clay. The prediction precision was good as indicated by a small RMSE (1.78%). The bias of prediction was also good since the ME was close to zero (0.25%). The large Lin's concordance (0.78) between measured and predicted clay signifies good agreement, with the coefficient of determination strong ($R^2 = 0.64$).

3.6. Linear Regression of CEC for All Depth Increments and σ

Using the same approach, we used for clay, we wanted to determine if a satisfactory LR could be established between CEC and σ estimated from the inversion of EC_{ah} and EC_{av} EM38 data and using EM4Soil. Table 5 also shows the coefficient of determination (R^2) between estimated σ (mS/m) and CEC at all depths (i.e., topsoil, subsurface and subsoil). As was the case for clay, when λ increased the coefficient between σ and CEC decreased, regardless of algorithm (S1 or S2) or forward model (CF or FS). However, this was the not the case for S1, where the best R^2 (0.68) overall was achieved when the FS forward model was used with a λ value of 0.9. While this coefficient was equivalent to that achieved for clay, [39] managed to develop a LR (R^2 = 0.89) between σ and CEC in a small field in Spain. We attribute their success to the fact the EC_a data, while collected on 12 m transect spacings, was collected continuously (Veris-3100) and gridded onto a 5 × 5 m grid. In addition, the range in CEC (i.e. range) was much larger in Spain (1–23 cmol(+)/kg) than in this field (2.3–6.7 cmol(+)/kg).

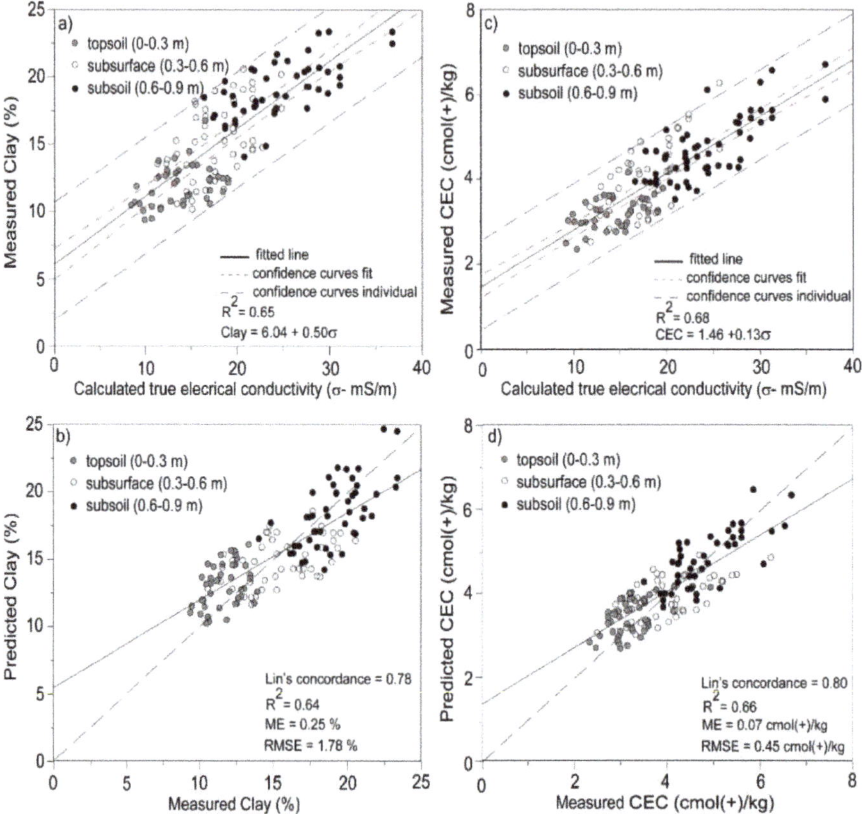

Figure 6. Plots of measured (**a**) clay (%) and (**c**) cation exchange capacity (CEC—cmol(+)/kg) versus true conductivity (σ—mS/m) using λ of 0.07 and 0.9 and S1 inversion algorithm with full solution, respectively; and, plots of measured versus predicted (**b**) clay and (**d**) CEC using leave-one-out cross-validation, for topsoil (0–0.3 m), subsurface (0.3–0.6 m) and subsoil (0.6–0.9 m).

Figure 6c shows the LR between σ and measured CEC (cmol(+)/kg) using these parameters, including the fitted lines, confidence curves and individual confidence curves. The LR was CEC = 1.46 + 0.13σ, with the summary statistics (Table 6) indicating it was significant (<0.0001*). We note that topsoil (0–0.3 m) σ was small (1.80–4.25 mS/m), with subsurface (0.3–0.6 m) intermediate σ (2.2–5.0 mS/m) and subsoil (0.6–0.9 m) σ largest (3.8–7.5 mS/m). Again, increasing σ was a function of the EC_{av} (Figure 3c) being larger than EC_{ah} (Figure 3b). As would therefore be expected, the estimates of subsoil σ were larger than topsoil σ.

Figure 6d shows the leave-one-out cross-validation applied by removing each of the 46 sampling locations one at a time. Again, the topsoil, subsurface and subsoil samples of one of the sample locations was removed and this process was repeated once for each sample location. Predicted CEC was in good concordance (Lin's = 0.80) with measured CEC, with prediction precision (RMSE = 0.53 cmol(+)/kg) and bias (ME = 0.07 cmol(+)/kg), being small and close to zero, respectively. The coefficient of determination was also strong (R^2 = 0.66).

3.7. Digital Soil Maps of Clay and CEC

Figure 7 shows predicted clay and CEC across the study area as 3D models. Figure 7a was generated after applying the LR (clay = 6.04 + 0.50σ) developed between σ and clay shown in Figure 6a. The predictions were then made using the quasi-3D model generated estimates of σ estimated from passing the gridded EC_a data collected from the EM38 EC_a across the rest of the area and through EM4Soil and the quasi-3D algorithm. Figure 7b was generated after applying the LR (CEC = 1.46 + 0.13σ) developed between σ and CEC shown in Figure 6b, with predictions applied to the quasi-3D modelled EM38 EC_a data as described above for clay.

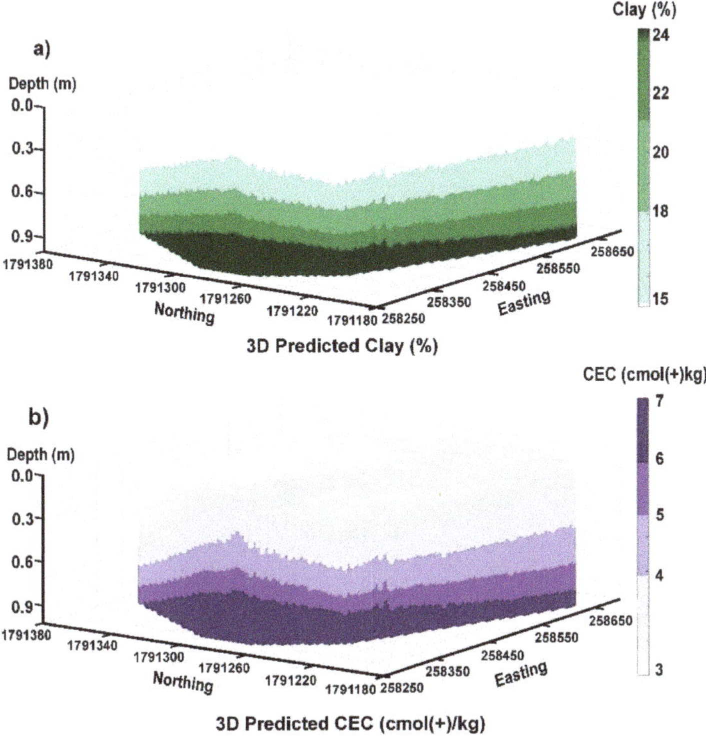

Figure 7. Predicted (**a**) clay (%) and (**b**) CEC (cmol(+)/kg) generated from inversion of EM38 apparent electrical conductivity (EC_a—mS/m) using EM4Soil and S1 inversion algorithm, full-solution (FS) with damping factor (λ) = 0.07 (clay) and 0.9 (CEC). Note: Calculated true electrical conductivity (σ—mS/m) used to predict clay and CEC from linear regression in Figure 6a,b, respectively.

Figure 8 shows the Digital Soil Mapping (DSM) of predicted clay and CEC at the three sampled depths generated from the LR developed between σ (mS/m) and clay and or CEC. Figure 8a shows predicted topsoil (0–0.3 m) clay. The DSM was like the contour plot generated from the 46 sampling points (Figure 4a) alone. Figure 8b shows the DSM for the subsurface (0.3–0.6 m). The northern half was characterised by small (<15%) clay, while the southern half by intermediate-small (15–18%) clay. These predictions appear less consistent with measured subsurface clay (Figure 4b). Figure 8c shows predicted clay for the subsoil (0.6–0.9 m).

Figure 8d shows the DSM of predicted topsoil (0–0.3 m) CEC. The predicted CEC was consistent with measured CEC, which was small (<3.8 cmol(+)/kg) across most of the field. However, the same could not be said about the small parcel of land along the western margin. Here, there was a band of measured CEC, which was intermediate-small (3.8–4.5 cmol(+)/kg) and intermediate (4.5–5.3 cmol(+)/kg). Predicted CEC was small, however, as was predicted across the entire field.

There were similar inconsistencies in the subsurface (0.3–0.6 m) predicted CEC as shown in Figure 8e. Here the field was divided into the northern half, which was characterised by predicted CEC which was small, whereas in the southern half it was intermediate-small, with small pockets of intermediate CEC. This was not the case for measured CEC (Figure 4e). These results were consistent with the difference between measured and predicted CEC. Similar results were evident for subsoil (0.6–0.9 m) CEC (Figure 8f) and measured CEC (Figure 4f).

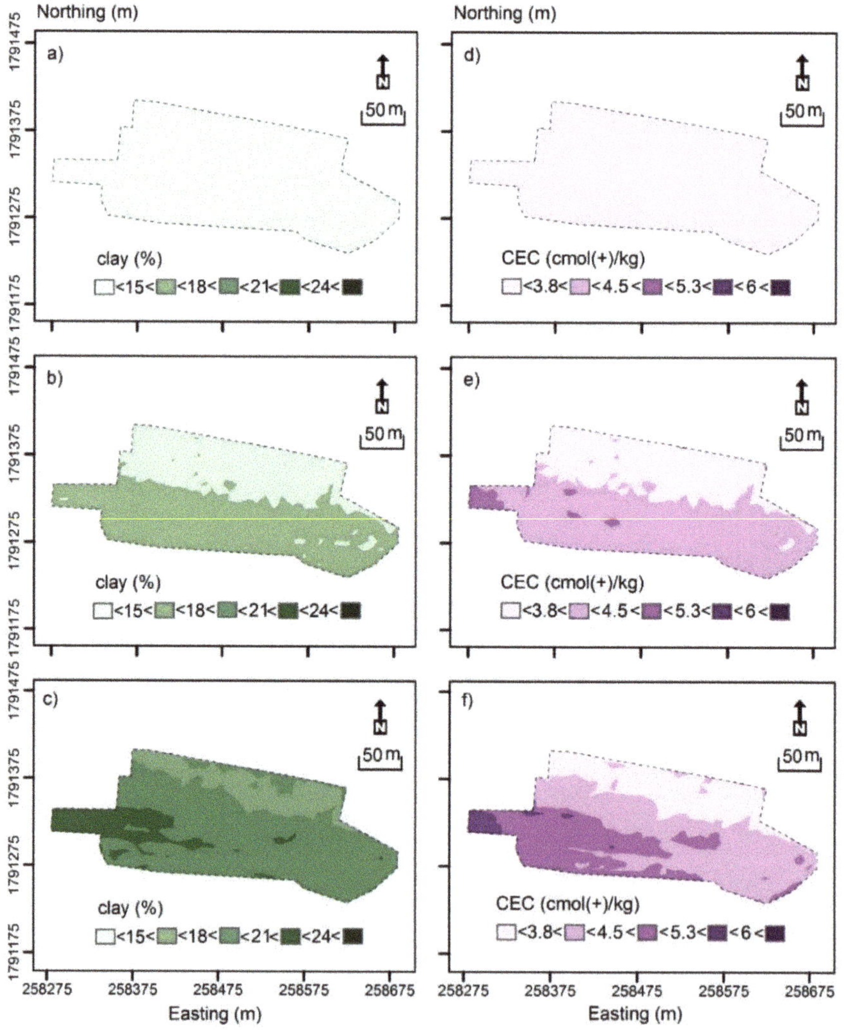

Figure 8. Spatial distribution of predicted clay (%) at depths of (**a**) topsoil (0–0.3 m), (**b**) subsurface (0.3–0.6 m) and (**c**) subsoil (0.6–0.9 m) generated using linear regression (LR) model (Figure 5a), and cation exchange capacity (CEC—cmol(+)/kg) in (**d**) topsoil, (**e**) subsurface and (**f**) subsoil.

3.8. Mapping the Prediction Interval (PI) of Predicted Clay and CEC

To better understand the uncertainty in the predicted DSM of clay and CEC, we mapped the prediction interval (PI). Figure 9a shows the PI for predicted topsoil (0–0.3 m) clay. It indicates that in the southern half, the PI was small (1.0%), whereas the northern half was intermediate-small (1.0–1.5%). Figure 9b shows PI calculated for the subsurface (0.3–0.6 m) clay. Here, PI was generally small across the whole field. Of most interest was the wider PI (>1.5%) in the subsoil (0.6–0.9 m) clay shown in Figure 9c. This was particularly the case for the southern half of the field and in the west.

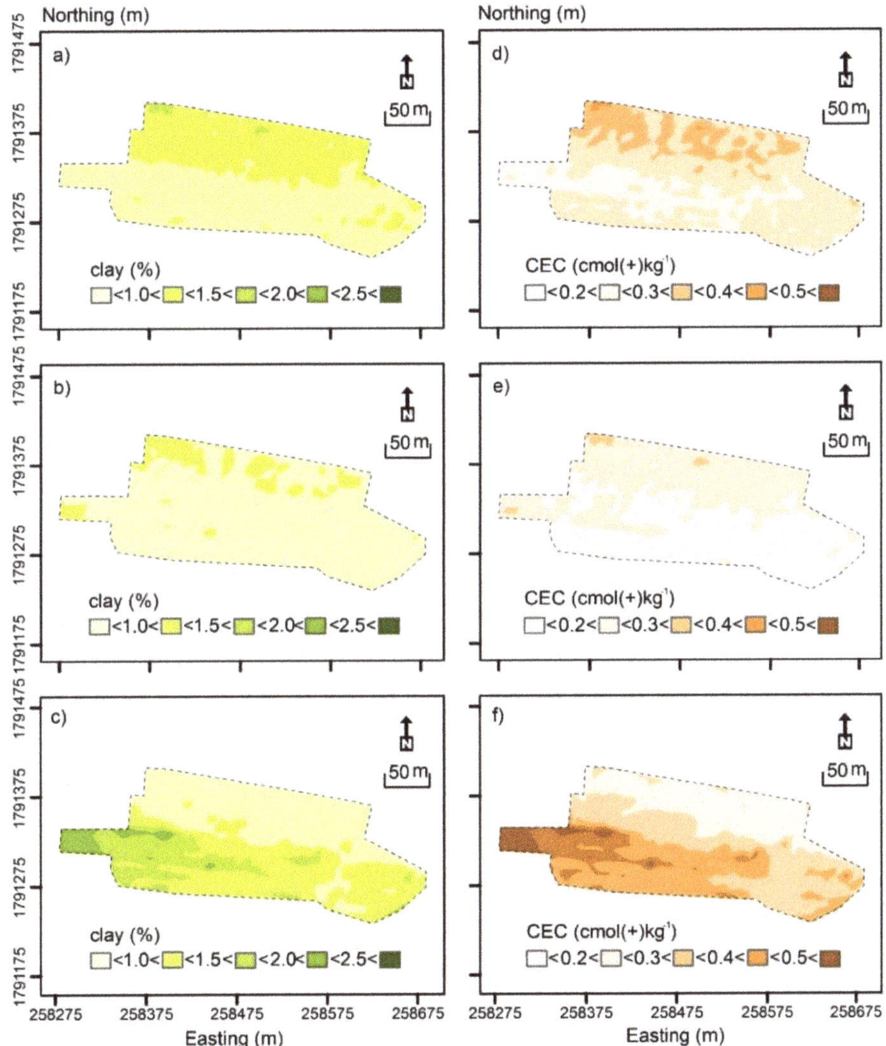

Figure 9. Contour plots of the prediction interval (PI) of the predictions made for clay (%) at different depths including (**a**) topsoil (0–0.3 m), (**b**) subsurface (0.3–0.6 m) and (**c**) subsoil (0.6–0.9 m); and cation exchange capacity (CEC—cmol(+)/kg) in the (**d**) topsoil, (**e**) subsurface and (**f**) subsoil.

Figure 9d shows the PI for predicted topsoil (0–0.3 m) CEC. It showed equivalent patterns to clay. However, there were more classes of PI, with more uncertainty evident in the southern half, where

PI was small and intermediate-small (0.2–0.3 cmol(+)/kg). Figure 9e shows the PI calculated for the subsurface (0.3–0.6 m) CEC. Here the PI was generally small (<0.2 cmol(+)/kg) in the southern half. Of most interest was the much wider PI (>0.5 cmol(+)/kg) values for subsoil (0.6–0.9 m) CEC shown in Figure 9f. This was particularly the case for the southern half and in the western margin.

We attribute these PI differences to various factors. Generally, the larger PI were associated with predicted clay and CEC in the subsoil, with the next largest PI associated with the topsoil. With respect to the topsoil, we attribute the larger PI for being a function of EC_{ah} and EC_{av} having a much larger theoretical depth of measurement of 0–0.75 and 0–1.5 m, respectively, compared with estimating σ to a depth of 0–0.3 m. Our ability to resolve topsoil σ, and hence predict clay or CEC, was therefore poor. To improve the estimation of σ and perhaps reduce the PI, we could collect EC_a at various heights. This was the approach of [42], who showed that in combination with EM31 EC_{ah} and EC_{av}, EM38 EC_{ah} and EC_{av} at a height of 0.6 m was optimal to make a LR between σ and CEC to predict CEC at 0.3 m increments and to a depth of 2.0 m along a single transect.

Herein, short scale variation was also problematic, owing to the fact that we collected EM38 EC_a data on an approximate 10 × 10 m grid. To better account for some of the short scale variation, and reduce the PI, the approach of [26] would be appropriate. This is because they collected EC_a from a DUALEM-21 instrument which was mobilized and coupled with a GPS and data logging capabilities. This will allow more detailed and closely spaced EC_a data to be collected. We believe a similar approach here, instead of the grid (10 × 10 m) we used, will reduce smoothing and improve prediction accuracy.

3.9. Soil Improvement Guidelines Based on Clay and CEC Maps

In spite of the uncertainty, the topsoil (0–0.3 m) and subsurface (0.3–0.6 m) clay DSM (Figure 8a and b, respectively) are valuable in terms of practical soil improvement. In the areas where clay was small (<15%), that is in the topsoil and the subsurface in northern half of the area, chemical and compost fertiliser application rates can be suggested according to the guidelines given by [22] and [23], respectively. Specifically, as shown in Table 1, for the areas with small (<15%) clay, chemical fertiliser rate for nitrogen (N), phosphorus (P_2O_5) and potassium (K_2O) can be recommended at 113 (kg/ha), 38 (kg/ha) and 113 (kg/ha), respectively. Alternatively, compost fertiliser at the rate of 25 t/Ha can be suggested. In the southern half of the study area, the rate could be reduced, owing to the slightly larger subsurface clay (15–18%) and specifically for chemical fertiliser for N (113 kg/ha), P_2O_5 (38 kg/ha) and K_2O (75 kg/ha), while a compost fertiliser (19 t/ha) could also be considered.

With respect to the topsoil CEC map (Figure 8d) the small (<3.8 cmol(+)/kg) CEC are equally informative. Overall, the small (<10 cmol(+)/kg) CEC indicates very poor soil fertility and practically no shrink–swell potential [19]. In situations like this, the subsoil is sometimes more resilient than the topsoil and consideration can be given to invert the profile using a moldboard plough. However, here this was not the case, because even though the subsoil (0.6–0.9 m) CEC was larger (i.e., intermediate-small; <4.5 cmol(+)/kg) it was still below this threshold.

On the other hand, the topsoil map does provide information which could be useful in assisting with the rate of lime application for the purpose of fertilisation. Specifically, and given this part of Northeast Thailand is a sugarcane growing area, the lime application guidelines indicated in Table 2 could be used. This is despite the fact that the guidelines were developed for application in similarly sandy soil and in tropical climates of North Queensland [29]. Given these guidelines and based on the small (<3.8 cmol(+)/kg) topsoil CEC, a rate of 1.25 t/h of lime should be applied across the entire area.

4. Conclusions and Discussion

We were unable to develop any satisfactory linear regression (LR) between EC_{ah} and EC_{av} with measured topsoil (0–0.3 m), subsurface (0.3–0.6 m) and subsoil (0.6–0.9 m) clay (%) or CEC (cmol(+)/kg). We attribute this to the small variation in EC_a as well as clay and CEC across the study field and at these three depths. However, the estimates of true electrical conductivity (σ—mS/m) generated by inverting

EC_{ah} and EC_{av} and using a quasi-3D algorithm (EM4Soil), enabled the development of a universal LR calibration for both clay and CEC and which included the capability to predict both soil properties in the topsoil, subsurface and subsoil. For clay we found the S1 inversion algorithm with full-solution (FS) and using a damping factor (λ) = 0.07 was optimal (R^2 = 0.65) with the LR expressed as follows: clay (%) = 6.04 + 0.50σ. For CEC the S1 inversion algorithm, full-solution (FS) and a damping factor (λ) = 0.9 was optimal (R^2 = 0.68) and could be estimated as follows: CEC (cmol(+)/kg) = 1.46 + 0.13σ.

We were able to predict and subsequently map the spatial distribution of clay and CEC in the topsoil, subsurface, and subsoil. Subsequently, the uncertainty of these maps was assessed using the prediction interval (PI). We attribute the larger PI in the topsoil to be a function of EC_{ah} and EC_{av} having a theoretical depth of measurement of 0–0.75 and 0–1.5 m, respectively. Given that we were estimating σ to a depth of a topsoil, our ability to do this was not completely satisfactory. This was similarly the case for the larger PI associated with the subsoil depth.

The solution to these problems would be to collect additional EC_a data to better estimate σ. Using our existing EM38, we could collect additional EC_a at various heights or by collecting additional data with an EM31. This was the approach carried out by [43], who showed that in combination with EC_{ah} or EC_{av} of EM31, and EC_{ah} or EC_{av} of EM38 at a height of 0.6 m was optimal to make a LR with CEC at 0.3 m increments and to a depth of 2.0 m along a single transect. Alternatively, EC_a data could be collected using a multiple-coil EM instrument such as a DUALEM-421 as shown by [44–46].

Author Contributions: T.K. conceived, designed and performed the experiments as well as collected ancillary and soil samples; T.K. and E.Z. performed data analysis using EM4Soil software; J.T. and T.K. wrote the paper; T.K. and D.Z. performed methodology, validation and others contributions; P.S. contributed reagents/materials/analysis tools and soil sample analysis.

Acknowledgments: This research was fully financial supported by Land Development Regional Office 5, Land Development Department (LDD), Ministry of Agriculture and Cooperatives, Thailand.

Conflicts of Interest: All the authors declare there is no conflict of interest in this paper and the funders had no role in the design of the study; in the collection, analyses, or interpretation of data; in the writing of the manuscript, or in the decision to publish the results.

References and Notes

1.	FAO. *Management of Tropical Sandy Soils for Sustainable Agriculture "A Holistic Approach for Sustainable Development of Problem Soils in the Tropics" 27th November–2nd December 2005, Khon Kaen Thailand*; FAO: Bangkok, Thailand, 2005; 24p.

2.	Chiu, Y.C.; Huang, L.N.; Uang, C.M.; Huang, J.F. Determination of cation exchange capacity of clay minerals by potentiometric titration using divalent cation electrodes. *Colloid Surf.* **1990**, *46*, 327–337. [CrossRef]

3.	Zhao, D.; Zhao, X.; Khongnawang, T.; Arshad, M.; Triantafilis, J. A Vis-NIR spectral library to predict clay in Australian cotton growing soil. *Soil Sci. Soc. Am. J.* **2018**, *82*, 1347–1357. [CrossRef]

4.	Xu, R.; Zhao, A.; Yuan, J.; Jiang, J. pH buffering capacity of acid soils from tropical and subtropical regions of China as influenced by incorporation of crop straw biochars. *Soils Sediments* **2012**, *12*, 494–502. [CrossRef]

5.	Vaught, R.; Brye, K.R.; Miller, D.M. Relationships among Coefficient of Linear Extensibility and Clay Fractions in Expansive, Stoney Soils. *Soil Sci. Soc. Am. J.* **2006**, *70*, 1983–1990. [CrossRef]

6.	Castrignanò, A.; Giugliarini, L.; Risaliti, R.; Martinelli, N. Study of spatial relationships among some soil physico-chemical properties of a field in central Italy using multivariate geostatistics. *Geoderma* **2000**, *97*, 39–60. [CrossRef]

7.	Asadzadeh, F.; Akbarzadeh, A.; Zolfaghari, A.; Taghizadeh-Mehrjardi, R.; Mehrabanian, M.; Rahimi-Lake, H.; Sabeti-Amirhandeh, M. Study and comparison of some geostatistical methods for mapping cation exchange capacity (CEC) in soils of northern Iran. *Ann. Fac. Eng. Hunedoara* **2012**, *10*, 59–66.

8.	Burgess, T.M.; Webster, R. Optimal interpolation and isarithmic mapping of soil properties, I the semi-variogram and punctual kriging. *J. Soil Sci.* **1980**, *31*, 315–331. [CrossRef]

9.	Wu, J.; Norvell, W.A.; Hopkins, D.G.; Smith, D.B.; Ulmer, M.G.; Welch, R.M. Improved Prediction and Mapping of Soil Copper by Kriging with Auxiliary Data for Cation-Exchange Capacity. *Soil Sci. Soc. Am. J.* **2003**, *67*, 919–927. [CrossRef]

10. Odeh, I.O.A.; McBratney, A.B.; Chittleborough, D.J. Further results on prediction of soil properties from terrain attributes: Heterotopic cokriging and regression-kriging. *Geoderma* **1995**, *67*, 215–226. [CrossRef]

11. Webster, R.; Oliver, M.A. Sample adequately to estimate variograms of soil properties. *Eur. J. Soil Sci.* **1992**, *43*, 177–192. [CrossRef]

12. Jung, W.K.; Kitchen, N.R.; Sudduth, K.A.; Anderson, S.H. Spatial characteristics of claypan soil properties in an agricultural field. *Soil Sci. Soc. Am.* **2006**, *70*, 1387–1397. [CrossRef]

13. Khaledian, Y.; Brevik, E.C.; Pereira, P.; Cerdà, A.; Fattah, M.A.; Tazikeh, H. Modeling soil cation exchange capacity in multiple countries. *Catena* **2017**, *158*, 194–200. [CrossRef]

14. Sulieman, M.; Saeed, I.; Hassaballa, A.; Rodrigo-Comino, J. Modeling cation exchange capacity in multi geochronological-derived alluvium soils: An approach based on soil depth intervals. *Catena* **2017**, *167*, 327–339. [CrossRef]

15. Williams, B.G.; Hoey, D. The use of electromagnetic induction to detect the spatial variability of the salt and clays of soils. *Aust. J. Soil Res.* **1987**, *25*, 21–27. [CrossRef]

16. Triantafilis, J.; Huckel, A.I.; Odeh, I.O.A. Comparison of statistical prediction methods for estimating field-scale clay using different combinations of ancillary variables. *Soil Sci.* **2001**, *66*, 415–427. [CrossRef]

17. Mertens, F.M.; Patzold, S.; Welp, G. Spatial heterogeneity of soil properties and its mapping with apparent electrical conductivity. *Plant Nutr. Soil Sci.* **2008**, *171*, 146–154. [CrossRef]

18. Heil, K.; Schmidhalter, U. The Application of EM38: Determination of Soil Parameters, Selection of Soil Sampling Points and Use in Agriculture and Archaeology. *Sensors* **2017**, *17*, 2540. [CrossRef]

19. Bishop, T.F.A.; McBratney, A.B. A comparison of prediction methods for the creation of fieldextent soil property maps. *Geoderma* **2001**, *103*, 149–160. [CrossRef]

20. Sudduth, K.A.; Kitchen, N.R.; Bollero, G.A.; Bullock, D.G.; Wiebold, W.J. Comparison of Electromagnetic Induction and Direct Sensing of Soil Electrical Conductivity. *Agron. J.* **2003**, *95*, 472–482. [CrossRef]

21. Triantafilis, J.; Lesch, S.M.; La Lau, K.; Buchanan, S.M. Field level digital soil mapping of cation exchange capacity using electromagnetic induction and a hierarchical spatial regression model. *Soil Res.* **2009**, *47*, 651–663. [CrossRef]

22. Paisanchareon, K. *Research and Development on Soil Water and Fertilizers for Sugarcane*; Department of Agriculture, Ministry of Agriculture and Cooperatives: Krung Thep Maha Nakhon, Thailand, 2015.

23. Office of Soil Survey and Land Use Planning. *Miracle of Soils: Soil Groups for Cash Crops Cultivation in Thailand*; Land Development Department, Ministry of Agriculture and Cooperatives: Krung Thep Maha Nakhon, Thailand, 2005.

24. Zare, E.; Huang, J.; Monteiro Santos, F.A.; Triantafilis, J. Mapping salinity in three dimensions using a DUALEM-421 and electromagnetic inversion software. *Soil Sci. Soc. Am. J.* **2015**, *79*, 1729–1740. [CrossRef]

25. Huang, J.; Scudiero, E.; Clary, W.; Corwin, D.L.; Triantafilis, J. Time-lapse monitoring of soil water content using electromagnetic conductivity imaging. *Soil Use Manag.* **2017**, *33*, 191–204. [CrossRef]

26. Koganti, T.; Moral, F.J.; Rebollo, F.J.; Huang, J.; Triantafilis, J. Mapping cation exchange capacity using a Veris-3100 instrument and invVERIS modelling software. *Sci. Total Environ.* **2017**, *599–600*, 2156–2165. [CrossRef]

27. Khedari, J.; Sanprajak, A.; Hirunlarbh, J. Thailand climatic zones. *Renew Energ.* **2002**, *25*, 267–280. [CrossRef]

28. Thai Meteorological Department Website. 2017. Available online: https://www.tmd.go.th/en/ (accessed on 25 April 2017).

29. Sugar Research Australia. Nutrient Management Guidelines for Sugarcane in New South Wales. 2017. Available online: http://www.sugarresearch.com.au/icms_docs/194337_SIX_EASY_STEPS_Nutrient_Guidelines_for_NSW.pdf (accessed on 28 February 2017).

30. Geonics. Ltd. 1745 Meyerside Drive, Unit 8, Mississauga, Ontario L5T 1C6, Canada.

31. Garmin International, Inc. 1200 East 151st Street, Olathe, KS 66062-3426, USA.

32. Beretta, A.N.; Silbermann, A.V.; Paladino, L.; Torres, D.; Bassahun, D.; Musselli, R.; García-Lamohte, A. Soil texture analyses using a hydrometer: Modification of the Bouyoucos method. *Cien. Investig. Agrar.* **2014**, *41*, 263–271. [CrossRef]

33. Chapman, H.D. Cation Exchange Capacity. In *Methods of Soil Analysis*; Black, C.A., Ed.; American Society of Agronomy: Madison, WI, USA, 1965; pp. 891–901.

34. EMTOMO. *EM4Soil v304*; EMTOMO: Lisboa, Portugal, 2017.

35. Monteiro Santos, F.A.; Triantafilis, J.; Bruzgulis, K. A spatially constrained 1D inversion algorithm for quasi-3D conductivity imaging: Application to DUALEM-421 data collected in a riverine plain. *Geophysics* **2011**, *76*, B43–B53. [CrossRef]

36. Constable, S.C.; Parker, R.L.; Constable, C.G. Occam's inversion: A practical algorithm for generating smooth models from electromagnetic sounding data. *Geophysics* **1987**, *52*, 289–300. [CrossRef]

37. McNeill, J.D. *Electromagnetic Terrain Conductivity Measurement at Low Induction Numbers*; Geonics Ltd.: Mississauga, ON, Canada, 1980.

38. Kaufman, A.A.; Keller, G.V. Frequency and Transient Sounding Methods in Geochemistry and Geophysics. *Geophys. J. Int.* **1983**, *77*, 935–937.

39. Lawrence, I.; Lin, K. A concordance correlation coefficient to evaluate reproducibility. *Biometrics* **1989**, *45*, 255–268.

40. Curran-Everett, D. Explorations in statistics: Confidence intervals. *Adv. Physiol. Educ.* **2009**, *33*, 87–90. [CrossRef]

41. SAS Institute. *JMP Version 13*; SAS Institute Inc.: Cary, NC, USA, 2017.

42. Triantafilis, J.; Wong, V.; Monteiro Santos, F.A.; Page, D.; Wege, R. Modeling the electrical conductivity of hydrogeological strata using joint-inversion of loop-loop electromagnetic data. *Geophysics* **2012**, *77*, 285–294. [CrossRef]

43. Triantafilis, J.; Monteiro Santos, F.A. Resolving the spatial distribution of the true electrical conductivity with depth using EM38 and EM31 signal data and a laterally constrained inversion model. *Aust. J. Soil Res.* **2010**, *48*, 434–446. [CrossRef]

44. Zhao, X.; Wang, J.; Zhao, D.; Li, N.; Zare, E.; Triantafilis, J. Digital regolith mapping of clay across the Ashley irrigation area using electromagnetic induction data and inversion modelling. *Geoderma* **2019**, *346*, 18–29. [CrossRef]

45. Davies, G.B.; Huang, J.; Monteiro Santos, F.A.; Triantafilis, J. Modeling coastal salinity in quasi 2D and 3D using a DUALEM-421 and inversion software. *Groundwater* **2015**, *53*, 424–431. [CrossRef]

46. Triantafilis, J.; Roe, J.A.E.; Monteiro Santos, F.A. Detecting a leachate-plume in an aeolian sand landscape using a DUALEM-421 induction probe to measure electrical conductivity followed by inversion modelling. *Soil Use Manag* **2011**, *27*, 357–366. [CrossRef]

Article

Real-Time Electrical Resistivity Measurement and Mapping Platform of the Soils with an Autonomous Robot for Precision Farming Applications

İlker Ünal *, Önder Kabaş and Salih Sözer

Department of Machine, Technical Science Vocational School, Akdeniz University, 07070 Antalya, Turkey;
okabas@akdeniz.edu.tr (Ö.K.); sozer@akdeniz.edu.tr (S.S.)
* Correspondence: ilkerunal@akdeniz.edu.tr; Tel.: +90-506-928-9903

Received: 30 October 2019; Accepted: 2 December 2019; Published: 1 January 2020

Abstract: Soil electrical resistivity (ER) is an important indicator to indirectly determine soil physical and chemical properties such as moisture, salinity, porosity, organic matter level, bulk density, and soil texture. In this study, real-time ER measurement system has been developed with the help of an autonomous robot. The aim of this study is to provide rapid measurement of the ER in large areas using the Wenner four-probe measurement method for precision farming applications. The ER measurement platform consists of the Wenner probes, a y-axis shifter driven by a DC motor through a gear reducer, all installed on a steel-frame that mount to an autonomous robot. An embedded industrial computer and differential global positioning system (DGPS) were used to assist in real-time measuring, recording, mapping, and displaying the ER and the robot position during the field operation. The data acquisition software was codded in Microsoft Visual Basic.NET. Field experiments were carried out in a 1.2 ha farmland soil. ER and DGPS values were stored in Microsoft SQL Server 2005 database, an ordinary Kriging interpolation technique by ArcGIS was used and the average ER values were mapped for the soil depth between 0 and 50 cm. As a result, ER values were observed to be between 30.757 and 70.732 ohm-m. In conclusion, the experimental results showed that the designed system works quite well in the field and the ER measurement platform is a practical tool for providing real-time soil ER measurements.

Keywords: soil; soil electrical resistivity; autonomous robot; real-time measurement; precision farming; mapping

---

1. Introduction

Soil electrical resistivity is an important indicator to indirectly determine the soil properties in the plant production because the suitable soil conditions and water are vital sources for plant root growth and solute transport, including plant nutrients and fertilizers [1]. The knowledge of soil resistivity is a valuable data in determining the composition of soil; such as for example, moisture [2], salinity [3], porosity [4], organic matter level [5], bulk density [6], and soil texture [7].

Soil investigation studies are commonly performed to determine the properties of the soil that involves topsoil and subsoil exploration such as physical mapping, soil sampling, and laboratory testing. Soil sampling and laboratory tests are usually performed to make subsoil investigation. Especially, the borehole method has been widely used to determine the soil properties due to its good data accuracy derived from the direct test method. However, this method has several difficulties and limitations such as high cost, time consuming, and insufficient data for huge farmlands. In this context, geophysical methods offer the chance to overcome some of the problems inherent in more conventional soil investigation techniques for soil structure characterization at larger spatial and temporal scales [8]. Soil is strongly correlated and can be quantified through the geoelectrical properties [9]. Moreover, the soil resistivity is an important property that is a geoelectrical quantity that measures how the soil reduces the electric current flow through it.

The soil composition is one of the most important factors affecting soil properties and has a heterogeneous structure consisting of solid, liquid, and gas phases. The solid and liquid phases are a determinative factor in soil electrical resistivity and behavior of electrical fields [10]. ER measurement was made of at the end of the 19th century by dipping two probes into the soil and measuring the voltage drop between two probes, which impinge a defined current into the soil. In this method, measurement results were incorrect as it intrinsically includes the sum of both soil resistivity and the contact resistivity between the probe and soil [11]. Wenner [12] suggested that the four-probe ER measurement method for minimizing contributions is caused by the soil-probe contact problems. ER measurement has been conducted with the four-probe method in soil studies since 1931 for evaluating soil moisture [13,14] and salinity [15,16] under field conditions. All the soil ER measurements applied in soil science are still based on the standard four-probe method since that time.

In the Wenner four-probe method, the system consists of four probes which are equally spaced (a) from each other to measure apparent soil ER [17]. The outer probes (C1 and C2) are used as the current source (I) for current injection to the soil and the inner probes (P1 and P2) are used as the voltage source to measure voltage difference (V) between the inner probes [12]. Wenner resistance (RW) between inner probes is calculated by dividing voltage by current. A Wenner configuration is shown in Figure 1. The apparent soil ER (ρ_E) with this configuration is calculated by Equation (1):

$$\rho_E = 2 * \pi * a * R_W \tag{1}$$

where: ρ_E = measured apparent soil resistivity (ohm-m), a = probe spacing (m), R_W = Wenner resistance (ohm). Four probes are dipped into the soil to be surveyed to a depth of not more than 1/20 the distance between the probes. The measured apparent soil ER value is average resistivity of the soil at a depth equivalent to the distance "a" between two probes. The measuring depth can be changed by changing the distance between the probes.

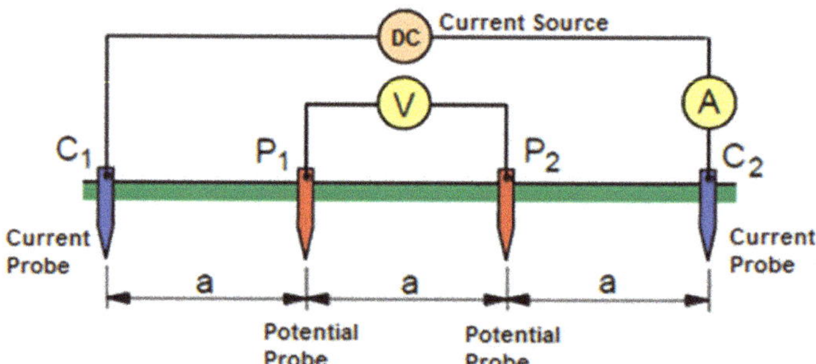

Figure 1. Four-probe Wenner configuration.

Today, agricultural production is carried out in huge farmlands and ER data should be collected fully automated to make a precision assessment of the farmland. Moreover, real-time soil resistivity measurement is important to map spatial farmland heterogeneity for precision farming applications. A useful approach for soil-property investigations is to use proximal soil measurements that combine soil sensors and data analysis methods to obtain high resolution soil data of the huge farmland [18]. Two portable systems are used to measure real time soil resistivity or conductivity of soil in agricultural studies—electrode-soil contact based system and noncontact electromagnetic induction (EM) system. The Veris (Veris Technologies, Inc., Salina, KS, USA) that has six coulter probes arranged in a Wenner method is the most important electrode–soil contact system to use to obtain multiple ER measurements representing different depths in agriculture [19]. The ARP (automatic resistivity profiling, Geocarta, France) is another Wenner based electrode–soil contact system using to acquire and process in real-time both electrical resistivity data and GPS information [20]. The other is the EM38 (Geonics Ltd., Mississauga, ON, Canada) that is the most widely used noncontact EM system in agriculture [21]. The electrode–soil contact-based system has the advantage that it does not require user setup configuration and measures different soil depth [22]. On the other hand, the non-contact EM system is lighter in weight, smaller in size, and thus easier to handle [23].

Real-time soil ER measurement is an important criterion for precision soil survey and can provide continuous measurements to determine temporal variables and soil structure over a huge farmland. The purpose of this study is the development and application of a real-time soil ER measurement system based on the four-probe Wenner method by the help of an autonomous robot for precision farming applications. Finally, for a thorough assessment of a measurement process, field study results are presented to identify the components of variation in the real-time soil ER measurement process.

2. Materials and Methods

The main objective of the designed system is to measure the apparent ER of the soil and map it. The system consists of four main parts:

1. Four-probes Wenner-based measurement platform: It is attached to an autonomous robot and it moves vertically to dip the Wenner probes in the soil. The Wenner probes are attached to the movable platform to measure the apparent ER values of the soil instantaneously.
2. Autonomous robot and steering algorithms: The autonomous robot is a four-wheel robot which steers with four DC motors. The differential steering mechanism is used. It can be steered point-to-point both autonomously and manually [24].
3. Data acquisition system: The system is used to collect data from a digital multimeter and a DGPS receiver on the measurement platform for the storing and mapping process.
4. Software development: The software is used to store data received from electronic instruments and to produce suitable database files to the mapping program.

2.1. Four-Probes Wenner-Based Measurement Platform

The developed measurement system is shown in Figure 2. The measurement platform was made of stainless steel. Some parts of the platform were made of the square steel tube $30 \times 30 \times 3$ mm. The mechanical structure of the measuring system consists of two parts, called the H-shaped carrier grid and the Wenner measurement platform. For vertical movement of the measurement system, an H-shaped carrier grid was constructed by using two 30×30, 910 mm, and by three 30×30, 800 mm square steel tubes. Then, this grid was attached to the autonomous robot. The H-shaped grid held the Wenner measurement system. The H-shaped grid has two steel linear guides adjusted by 30 mm linear rail shaft guide supports and pillow blocks. The length of the linear guides is 910 mm. It was used a linear actuator, made up of a 30×850 mm ball screw, operated by a 24 V—500 W–1440 rpm DC motor which was coupled to a 1:40 reduction gearbox for the vertical movement. The DC motor was mounted on the H-shaped carrier grid. A 740×400 mm rectangular U-shaped sliding platform was attached to the H-shaped carrier grid. A square flanged nut ball screw is mounted on the back of the sliding platform. The ball screw was coupled to this square flange nut. Then, the Wenner measurement platform was connected to this rectangular U-shaped sliding platform.

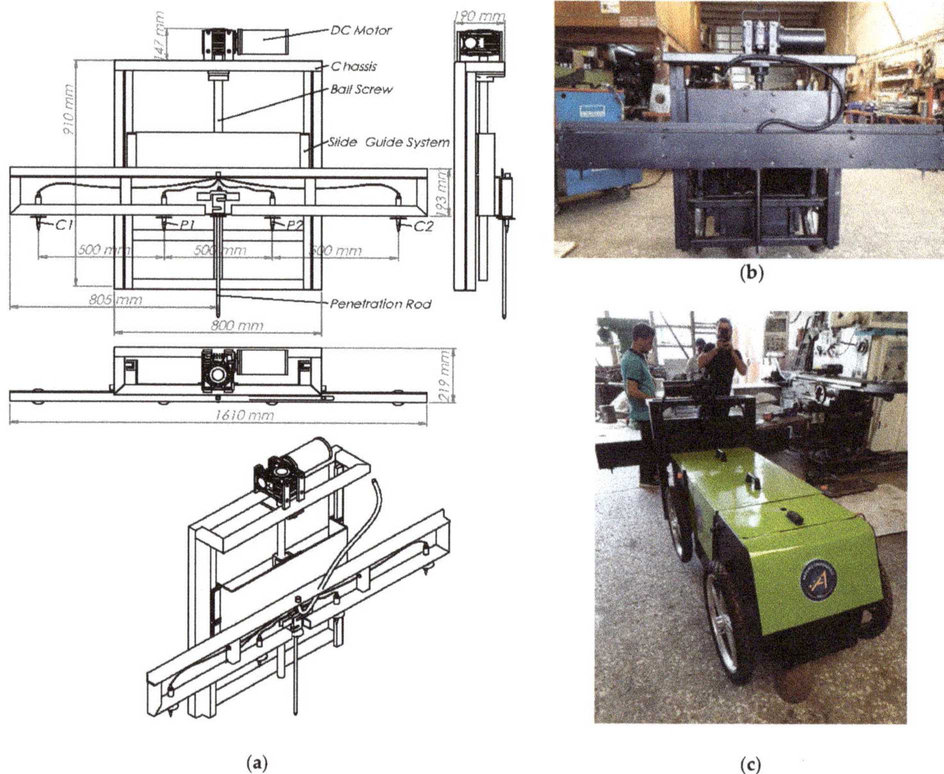

Figure 2. The developed measurement system. (**a**) Full scale technical drawing of the measurement platform; (**b**) full developed measurement platform; (**c**) four-wheel drive agricultural robot, which has the Wenner measurement system.

The Wenner measurement platform has four Wenner probes. Four steel Wenner probes were linearly mounted on the Wenner measurement platform at 500 mm intervals to measure the average apparent ER values of the soil between 0–500 mm. The fiber isolation rings were used to ensure electrical isolation between the platform and the probes. The Wenner probes are 12 mm in diameter and 25 mm long. The probe length must be 1/20 of the distance between the probes. The distance between probes can be changed to measure apparent ER resistivity of the soil at the different depths. Each probe was wired with the insulated single core cable. The cables of the C1 and C2 probes are connected directly to a 24-volt battery to inject electric current into the soil. The cables of the P1 and P2 probes are connected to a multimeter to measure the potential difference between the probes in the soil. The system is also capable of measuring soil penetration resistance.

2.2. Autonomous Robot and Steering Algorithms

The four-wheel drive agricultural robot which is used in this study can be steered both autonomously and manually. Four 2.50 × 17 rubber wheels were chosen to steer the robot in field conditions. It has a differential steering mechanism. In this system, the speed difference between the right and left wheels of the robot can be created. To make the mobile robot steer in a straight line, speeds of the all wheels must be the same. If the speeds of the right and left wheels are different, the robot rotates to the slow wheels side. When the right and left wheels are rotated opposite each other, the mobile robot can be able to rotate 360 degrees where it is. The robot is powered by four

24 V—0.25 kW—1440 rpm DC motors which were coupled to a 1:10 reduction gearbox. Each wheel of the robot was independently coupled to motor-gearbox assemblies mounted on the robot chassis. In this way, the torque generated by the motors can be transmitted completely to the wheels. The robot's weight is approximately 150 kg with batteries and the measurement system and the maximum speed is 20 km/h. Two RoboteQ FDC3260 three-channel DC motor control units (Roboteq Inc., Scottsdale, AZ, USA) were used to steer the robot by varying the speed and direction of the motors. The two 12 V-90 Ah rechargeable maintenance-free sealed batteries were used as the power source of the robot and other equipment's. Moreover, two batteries were connected in the series to provide 24 V for the DC motors.

In order to operate the mobile robot both manually and autonomously, the navigation program was codded in Visual Studio.NET 2015 using Visual Basic.NET language. This program was firstly coded for two-wheel drive robots by the author in 2015 [24], and rearranged to the four-wheel drive robots for this study. However, the navigation algorithm is the same. The flowchart for autonomous drive of the mobile robot is given in Figure 3. The flowchart of the quadrant control mechanism is given in Figure 4.

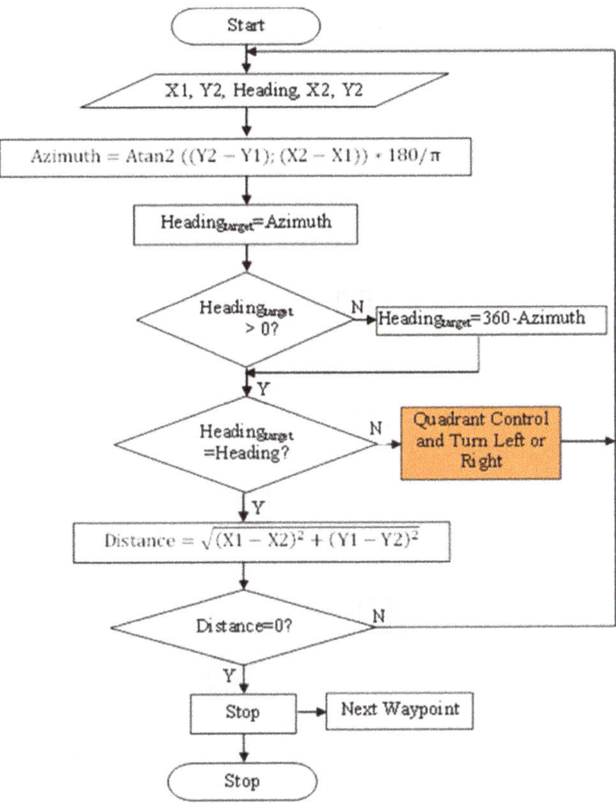

Figure 3. The flowchart for autonomous drive of the mobile robot.

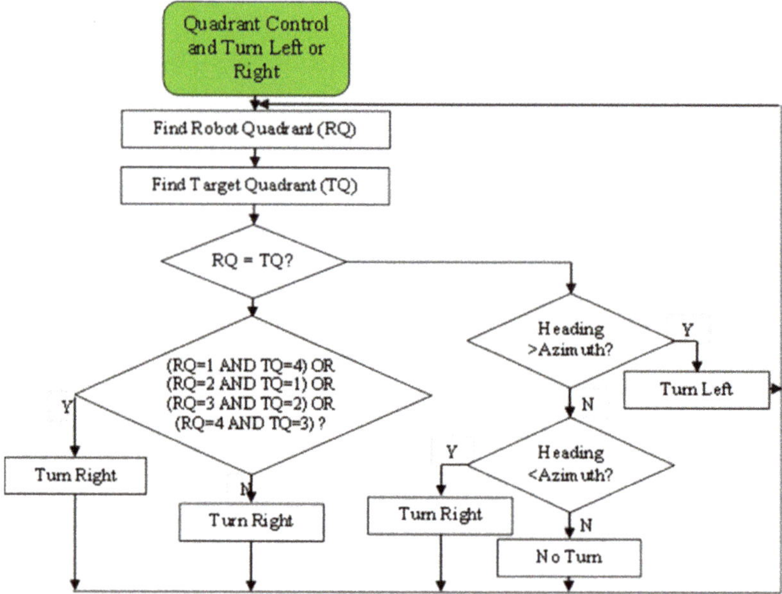

Figure 4. The flowchart of the quadrant control mechanism.

In autonomous guidance algorithm, the angle difference is calculated using heading and azimuth angles. In this way, the robot can be steered to the desired direction. After that, the distance between the robot location and the target location is calculated. Lastly, when the heading angle is equal to the azimuth angle and the distance is equal to zero, the robot arrives at the desired location. In quadrant control, the compass dial is used to determine the quadrant in which the current point was located. The calculated azimuth angle helped in determining the quadrant in which the target point was located [24].

2.3. Data Acquisition System

The data acquisition system is described for two processes. The first is the autonomous steer system of the robot and the second is the measurement system of the apparent ER of the soil. In this study, Lilliput PC-700 industrial all-in-one touchscreen computer (Zhangzhou Lilliput Electronic Technology Co., Fujian, China) was used to manage and communicate with each other all of the electronic-based equipment placed on the robot. In the autonomous steer system, a Promark 500 RTK-GPS (Magellan Co., Santa Clara, CA, USA) receiver was used to collect geographical data. These data were used to determine the geographic location (latitude, longitude, speed, time, etc.) of the robot. In addition, the RTK-GPS receiver was used to determine the measurement location of the apparent ER of the soil for mapping. The Honeywell HMR3200 (Honeywell International Inc., Charlotte, NC, USA), which is a digital compass, was used to collect the precision heading angle of the robot for navigation software. The Protek 506 handheld digital multimeter (MCS Test Equipment Ltd., Denbighshire, UK) was used to measure voltage between P1 and P2 Wenner probes. Two RoboteQ FDC3260 three-channel DC motor control unit were used to control robot steering and also movement of the Wenner measurement platform. The RS232 protocol was used for connecting the industrial computer and other electronic devices.

2.4. Software Development

A program was developed using Microsoft Visual Basic.NET 2015 programming language to steer the robot both autonomously and manually and measure apparent ER values of the soil. It was used to control the robot and measuring system, monitor the telemetry data, store all the data in the database, and prepare the suitable file for ArcGIS mapping software. The program consists of two parts: Navigation software (Figure 5) and soil resistivity measurement (Figure 6).

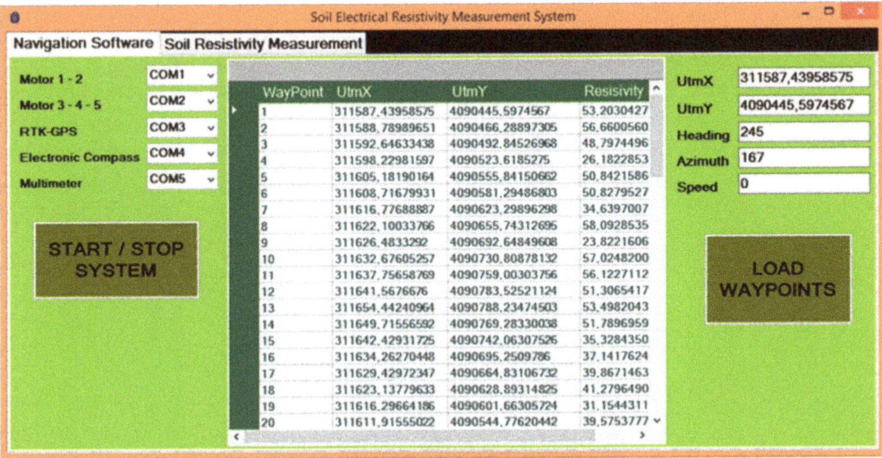

Figure 5. Navigation software.

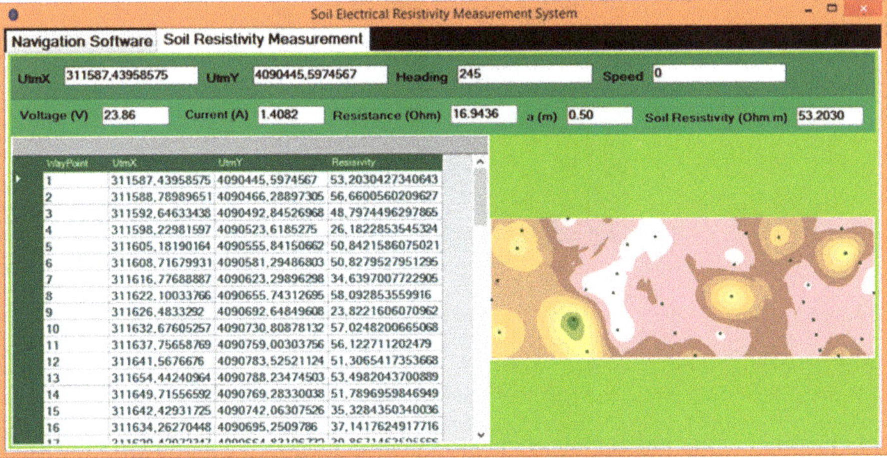

Figure 6. Soil resistivity measurement.

In the navigation software, the waypoint file can be loaded from database into the program. In this way, the waypoints can be used to steer the robot from point to point for autonomous guidance. Each waypoint includes a longitude (X2) and latitude (Y2) value which is the location of the target point to be measured. There are two important angles for robot navigation: Robot's heading angle and azimuth angle of the target point. Robot's heading angle is taken from the HMR3200 electronic compass by the navigation software. Additionally, the azimuth angle is continuously calculated by the navigation software. Moreover, the distance between robot position (X1, Y1) and target position (X2, Y2) is calculated by the software. These calculations were shown in Figure 3.

In the soil resistivity measurement part, the current applied to the C1 and C2 probes and the voltage collected from probes P1 and P2 is monitored. Additionally, the Wenner resistance (Rw) is calculated instantly during the measurement using the obtained current and the measured voltage values. All data are stored instantly to the SQL Server 2005 database. The ArcObjects SDK 10 Microsoft .NET Framework was used to prepare and display the soil ER maps using ArcMap interface in ArcGIS.

2.5. Experimental Field and Data Collection

The field experiments were carried out at the Batı Akdeniz Agricultural Research Institute, in Aksu, Antalya, Turkey (36°56′34.46″ N and 30°53′04.10″ E). The experimental field has an area of 1.2 ha and an elevation of 35 m above the sea level. In this field, the corn silage was harvested on July 25, 2019. The soil type is silty-clay, having a dark brown color, consists of 18% sand, 40% silt, and 42% clay. The organic matter content was 1.4%. Soil bulk density was 1.29 g/cm^3, the water content was 6.8%, and the average soil penetration resistance value was 1.62 MPa between 0 and 20 cm. The experimental field was shown in Figure 7.

Figure 7. The experimental field.

During the study within the experimental field, the robot was autonomously steered to 72 different geographical points and the average apparent soil ER values were collected for 0–50 cm depth. In this system, autonomous stop-and-go measurement method was used to measure the apparent soil ER values in large farmlands. The stop-and-go method is all about stopping the robot when taking the measurement. In this method, the agricultural robot goes to the first measurement point and stops, takes the measurement, and goes to the next measurement point. In this procedure, the digital map with high-resolution content of the experimental field was transferred into the ArcGIS 10.5 mapping software to determine the measurement points in large farmlands. In this way, a total 72 different GPS waypoints were randomly determined for autonomously steering the robot to the measurement point. All waypoints were stored to the database. After that, the agricultural robot was steered point-to-point to measure apparent soil ER value. All measured data were stored into the SQL Server 2005 database by the soil resistivity measurement software.

3. Results

In this study, the data obtained from all measured points were imported into Microsoft SQL Server 2005 database and mapped using ArcGIS 10.5 mapping software. In ArcGIS, ordinary Kriging interpolation was used to generate the contour map which makes a prediction of the apparent soil ER values in other parts of the experimental field for sampling.

The selection of the type of Kriging interpolation to use depends on the characteristics of the spatial data. Soil properties can spatially differ from point to point. As with most soil physical properties, soil is not also homogeneous in terms of electrical resistivity. In this regard, ordinary Kriging interpolation was used in this study. Simple Kriging is based on the theory of stationarity. This means that the mean and variance remain constant and are known in all locations. On the other hand, ordinary Kriging is a spatial estimation method and a linear geostatistical method that assumes that the mean may vary in the study area and does not remain constant. Universal Kriging is used to estimate spatial means when the data have a strong trend. This means that the trend is scale dependent. The apparent soil ER data may display trends over small geographic areas but at the scale of the huge farmlands, there is no trend that can be modeled by simple functions. Simple and universal Kriging interpolations were not chosen for this study because of these reasons.

A summary of the method used for Kriging interpolation is given in Table 1. The histogram of the apparent soil ER values is given in Figure 8. As can be seen in Figure 8, the minimum value of the soil ER is 30.757 ohm-m and the maximum value is 70.732 ohm-m. The skewness and Kurtosis values were observed as −0.14091 and 1.7091, respectively. Due to the skewness value being between −1 and −0.5, the data are reasonably skewed. This would mean that the sample data for the apparent soil ER are approximately symmetric. The Kurtosis value is low (<3). This means that the data are slightly platykurtic, the lack of outliers are in data, and the extreme values are less than that of the normal distribution.

Table 1. The field study data.

Method	
Name	Kriging
Type	Ordinary
Output Type	Prediction
Neighbors to Include	5
Include at Least	2
Sector Type	Four and 45 degree
Major Semiaxis	21.669411886777
Minor Semiaxis	21.669411886777
Angle	0
Variogram	Semivariogram
Number of Lags	12
Lag Size	2.429475299987
Model Type	Stable
Parameter	1.131640625
Range	21.669411886777
Anisotropy	No
Partial Sill	164.681994538028

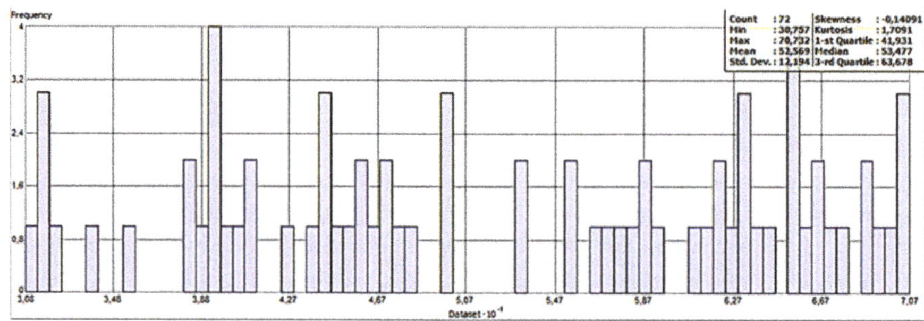

Figure 8. The histogram of the apparent soil ER (Electrical Resistivity) values.

The normal QQ plot graph was used to show the quantiles of the difference between the predicted and measured values and the corresponding quantiles from a standard normal distribution. As can be seen in Figure 9, the errors appear to be normally distributed even though there is a slight, possibly curved trend in the plot. Prediction errors of the ordinary Kriging method are given in Table 2.

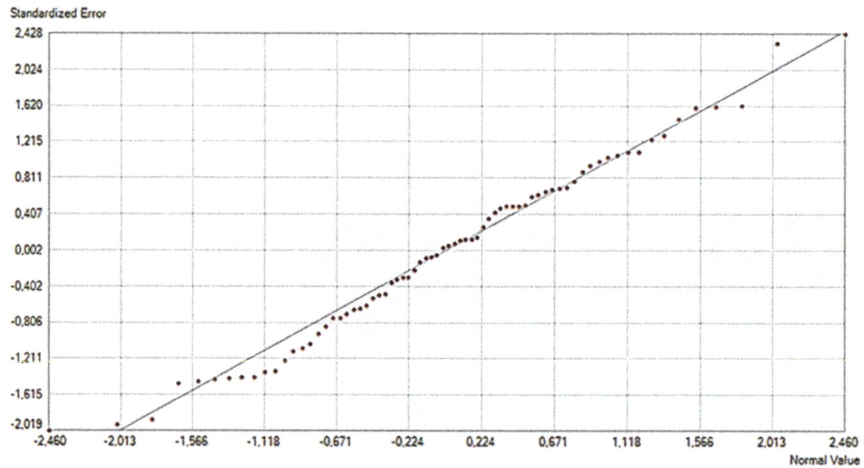

Figure 9. The normal QQ (Quantile – Quantile) plot graph of the standardized error.

Table 2. Prediction errors of the ordinary Kriging method.

Prediction Errors (ohm-m)	
Number of Samples	72
Mean	-0.2602
Root Mean Square	13.1404
Mean Standardized	0.01682
Root Mean Square Standardized	1.02197
Average Standard Error	12.7975
Regression Function (Predicted)	$-0.0133 * x + 52.8556$
Regression Function (Error)	$-1.0133 * x + 52.8556$
Regression Function (Standardized Error)	$-0.0747 * x + 3.89899$

In order to obtain the interpolation of the soil apparent ER values, the map of soil apparent ER on the experimental field was interpolated by using the ordinary Kriging approach. The interpolation map is given in Figure 10. Moreover, the Voronoi map of the study is given in Figure 11. It is observed that the ER values on the left side of the map are higher than the right side when the map is examined visually. This is also clearly seen on the Voronoi map of the study. If the measured points are close to each other, apparent soil ER values are approximately homogeneous. However, when the distance between the points increases, homogeneity decreases.

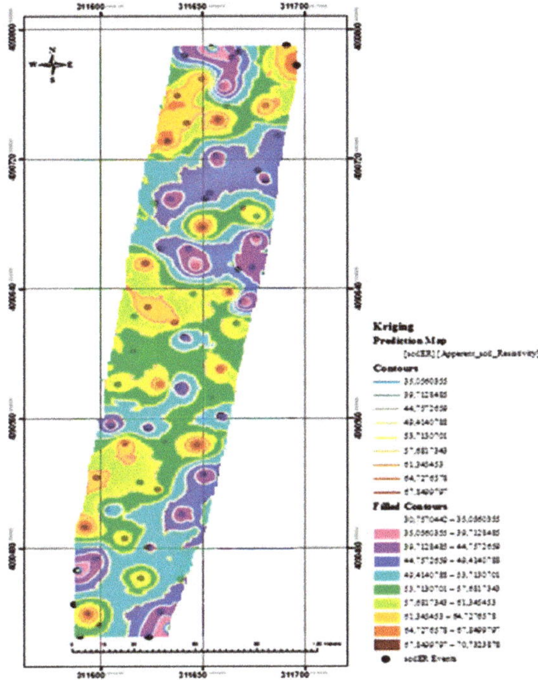

Figure 10. The interpolation map of the soil ER.

Figure 11. Voronoi map of the study.

4. Discussion

The apparent soil ER values depend on several parameters such as size of the soil, porosity, and water content. Hunt [25] indicated that the electrical resistivity varies from 1.5 ohm-m and below for wet clay soils to more than 2400 ohm-m for massive and hard bedrocks (Table 3).

Table 3. ER values of the different soil types.

Materials	Resistivity (ohm-m)
Clayey soils: Wet to moist	1.5–3.0
Silty clay and silty soils: Wet to moist	3–15
Silty and sandy soils: Moist to dry	15–150
Bedrock: Well fractured to slightly fractured with moist soil-filled cracks	150–300
Sand and gravel with silt	About 300
Sand and gravel with silt layers	300–2400
Bedrock: Slightly fractured with dry soil-filled cracks	300–2400
Sand and gravel deposits: Coarse and dry	>2400
Bedrock: Massive and hard	>2400

In the literature, there are not many studies measuring the apparent ER of the soil by the using mobile Wenner platform. However, there have been studies autonomously determining soil electrical resistivity of different soil types about soil science. Giao et al. [26] measured the electric resistivity of over 50 clay soil samples collected worldwide in the laboratory. Researchers have also measured the electric resistivity of over 50 soil samples taken from different locations in South Korea. As a result, they said that the sandy soil has a resistivity of above 10 ohm-m, the silty soil has a resistivity from 5 to 10 ohm-m. Juandia and Syahril [27] measured soil resistivity on 25 points across the study area using the Schlumberger configuration. Soil type was silty-sand. They reported that the average soil resistivity varied from 33 to 40.5 ohm-m. Rossi et al. [28] investigated the potential use of a direct current (DC) continuous resistivity profiling on-the-go sensor in precision viticulture. The authors used an automatic on-the-go DC recording resistivity meter (ARP, automatic resistivity profiling. Geocarta, Paris, France) in the three soil layers (V1 = 0–0.5, V2 = 0–1, and V3 = 0–2 m depth) on a vineyard area. Soil type was Inceptisol. The authors reported that soil ER values varied from 3–151 ohm-m for 0–0.5 m depth, 30–511 ohm-m for 0–1 m, and 9–750 ohm-m for 0–2 m depth. Lee and Yoon [29] investigated the theoretical relationship between elastic wave velocity and electrical resistivity. The authors measured the elastic wave velocity and electrical resistivity in several types of soils including sand, silty sand, silty clay, silt, and clay–sand mixture and the temperature compensated electrical resistivity probe was used for measuring. The authors said that the electrical resistivity showed at ranges of 1.23–2.17, 1.08–1.91, 1.01–1.40, 0.33–0.44, and 6.39–7.14 ohm-m in the order of soil types previously mentioned. Merritt et al. [30] developed a methodology for measurement and modeling of the moisture–electrical resistivity relationship of fine-grained unsaturated clay-based soils and electrical anisotropy. Soil resistivity measurements were conducted for four different soil types: Silty-clay, fine sand, clayey sandy silt, and siltstone. The results showed at approximately ranges of 10–100, 100–150, 50–800, 100–10000 ohm-m. The authors reported that the soil resistivity increases with decreasing moisture content. Kim et al. [31] evaluated the effects of soil properties and electrical conductivity on the water content reflectometer calibration for landfill cover soils. For this aim, the electrical conductivity measurements were performed for a set of 28 soils which have different soil textures by using high-frequency time domain reflectometer (TDR). The authors reported that the soils with a greater clay content or organic content have higher electrical conductivity than the soils with silts and sands. This means that the ER values of the clay soils should be low.

The traditional soil sampling is made by using the borehole method to determine the soil physical, chemical, or biological properties of soil layer in laboratory conditions. In this method, the laboratory calibration of the soil ER with soil moisture should be done. However, laboratory calibration may not give the correct relationship between soil moisture and electrical resistivity for real soil conditions [32]. On the other hand, in the precision farming domain, the autonomous and continuously measurement of the apparent soil ER has some advantages such as fast measurement and low cost for mapping both the horizontal and the vertical spatial variability in large farmlands. Moreover, the user setup configuration or calibration in this system is not required. Both Veris and ARP systems have been developed to achieve these advantages and measure the soil resistivity or conductivity as mobile for precision farming applications. The cost of the basic Veris system is 11.500 USD [33]. However, there is not any price about the ARP system in the literature. However, Andrenelli et al. [34] reported that the daily cost is 3000 Euro for ARP system usage. In our proposed system, the study was carried out within a project and the total budget was 8000 USD for all the system.

Both the ARP and the Veris system are semi-mounted type measuring platforms that are attached to the tractor or any other vehicle such as ATV. In this context, these systems need a traction system and at least one operator for their operation. Moreover, these systems are not lightweight and have low maneuverability. On the other hand, the benefits of the designed system are obvious: Easy to manufacture, compact measurement system with robot, lightweight, low manufacturing and operation cost, high maneuverability, and used autonomously.

This study was undertaken on silty-clay type soil by using the measurement system developed by us. In this study, the apparent soil ER values were measured between 30.757 and 70.732 ohm-m. The measurement results were shown similarity with the abovementioned literatures for silty-clay soils [25,27,30]. In addition to our results, for the obtained apparent soil ER values to be more significant, soil penetration resistance should be simultaneously measured and correlated with soil moisture content and bulk density [35–42]. No faults were detected in the electromechanical, data acquisition, and software parts of the system during field operation. The experimental results showed that our measurement system is suitable for map-based precision farming applications.

5. Conclusions

In this study, a new design for real-time apparent soil ER measuring system and its mapping capability has been presented for map-based precision farming applications. Although the laboratory analysis is usually the reliable method for determining most soil properties, real-time measurements to monitor the soil properties have advantages and benefits for precision farming applications. The DC apparent soil ER measurement method is one of the simplest geophysical techniques and still employed extensively because of its easy-to-use, no calibration required, and relatively easy interpretation in all engineering studies. However, a robot-based mobile Wenner measurement platform has not been found in the agriculture literature. The apparent soil ER map created by the developed software can be a useful source for precision farming applications across different fields. For researchers, data collection, analysis, and interpretation from farmlands have always been hard, time-consuming, and tedious studies in agricultural applications. The results of the study show that using this system is important for researches and professional applications of soil science.

Author Contributions: İ.Ü. was responsible for the project administration, conceptualization, data curation, formal analysis, methodology, software, and writing—original draft. Ö.K. was responsible for funding acquisition, investigation, resources, validation, and visualization. S.S. was responsible for the supervision, funding acquisition, and writing—review and editing. All authors have read and agreed to the published version of the manuscript.

Funding: This work is financially supported by The Scientific Research Projects Coordination Unit of Akdeniz University (project number: FBA-2017-1980).

Acknowledgments: We are very grateful to the Akdeniz University Technical Sciences Vocational School technicians for their cooperation and effort in supporting the experiment.

Conflicts of Interest: The authors declare no conflict of interest.

References

1. Peng, X.H.; Zhang, B.; Zhao, Q.G. A review on relationship between soil organic carbon pools and soil structure stability. *Acta Pedol. Sin.* **2004**, *41*, 618–623.
2. Tremsin, V.A. Real-Time Three-Dimensional Imaging of Soil Resistivity for Assessment of Moisture Distribution for Intelligent Irrigation. *Hydrology* **2017**, *4*, 54. [CrossRef]
3. Aizebeokhai, A.P. Assessment of soil salinity using electrical resistivity imaging and induced polarization methods. *Afr. J. Agric. Res.* **2014**, *9*, 3369–3378.
4. Kim, J.; Yoon, H.; Cho, S.; Kim, Y.; Lee, J. Four Electrode Resistivity Probe for Porosity Evaluation. *Geotech. Test. J.* **2011**, *34*, 668–675.
5. Fedotov, G.N.; Tretyakov, Y.D.; Pozdnayakov, A.I.; Zhukov, D.V. The Role of Organomineral Gel in the Origin of Soil Resistivity: Concept and Experiments. *Eurasian Soil Sci.* **2005**, *38*, 492–500.
6. Laloy, E.; Javaux, M.; Vanclooster, M.; Roisin, C.; Bielders, C. Electrical Resistivity in a Loamy Soil: Identification of the Appropriate Pedo-Electrical Model. *Vadose Zone J.* **2011**, *10*, 1023–1033. [CrossRef]
7. Heil, K.; Schmidhalter, U. Comparison of the EM38 and EM38-MK2 electromagnetic induction-based sensors for spatial soil analysis at field scale. *Comput. Electron. Agric.* **2015**, *110*, 267–280. [CrossRef]
8. Romero-Ruiz, A.; Linde, N.; Keller, T.; Or, D. A review of geophysical methods for soil structure characterization. *Rev. Geophys.* **2018**, *56*, 672–697. [CrossRef]

9. Samouelian, A.; Cousin, I.; Bruand, A.T.A.; Richard, G. Electrical resistivity survey in soil science: A review. *Soil Tillage Res.* **2005**, *83*, 173–193. [CrossRef]

10. Mostafa, M.; Anwar, M.B.; Radwan, A. Application of electrical resistivity measurement as quality control test for calcareous soil. *HBRC J.* **2018**, *14*, 379–384. [CrossRef]

11. Miccoli, I.; Edler, F.; Pfnür, H.; Tegenkamp, C. The 100th anniversary of the four-point probe technique: The role of probe geometries in isotropic and anisotropic systems. *J. Phys. Condens. Matter* **2015**, *27*, 1–29. [CrossRef] [PubMed]

12. Wenner, F. A Method of Measuring Resistivity. National Bureau of Standards. *Sci. Bull.* **1915**, *12*, 478–496.

13. McCorkle, W.H. Determination of Soil Moisture by the Method of Multiple Electrodes. *Bull. Tex. Agric. Exp. Stn.* **1931**, *B873*, 425–434.

14. Bertermann, D.; Schwarz, H. Laboratory device to analyse the impact of soil properties on electrical and thermal conductivity. *Int. Agrophys.* **2017**, *31*, 157–166. [CrossRef]

15. Rhoades, J.D.; Ingvalson, R.D. Determining salinity in field soils with soil resistance measurements. *Soil Sci. Soc. Am. Proc.* **1971**, *35*, 54–60. [CrossRef]

16. Afip, I.A.; Taib, S.N.L.; Jusoff, K.; Afip, L.A. Measurement of Peat Soil Shear Strength Using Wenner Four-Point Probes and Vane Shear Strength Methods. *Int. J. Geophys.* **2019**, *2019*, 1–12. [CrossRef]

17. Faleiro, E.; Asensio, G.; Denche, G.; Garcia, D.; Moreno, J. Wenner Soundings for Apparent Resistivity Measurements at Small Depths Using a Set of Unequal Bare Electrodes: Selected Case Studies. *Energies* **2019**, *12*, 695. [CrossRef]

18. Pan, L.; Adamchuk, V.I.; Prasher, S.; Gebbers, R.; Taylor, R.S.; Dabas, M. Vertical soil profiling using a galvanic contact resistivity scanning approach. *Sensors* **2014**, *14*, 13243–13255. [CrossRef]

19. Visconti, F.; de Paz, J.M. Electrical Conductivity Measurements in Agriculture: The Assessment of Soil Salinity. In *New Trends and Developments in Metrology*, 2nd ed.; Cocco, L., Ed.; IntechOpen: London, UK, 2016; Volume 1, pp. 99–126.

20. Heil, K.; Schmidhalter, U. Theory and Guidelines for the Application of the Geophysical Sensor EM38. *Sensors* **2019**, *19*, 4293. [CrossRef]

21. Heil, K.; Schmidhalter, U. The Application of EM38: Determination of Soil Parameters, Selection of Soil Sampling Points and Use in Agriculture and Archaeology. *Sensors* **2017**, *17*, 2540. [CrossRef]

22. Sudduth, K.A.; Kitchen, N.R.; Bollero, G.A.; Bullock, D.G.; Wiebold, W.J. Comparison of electromagnetic induction and direct sensing of soil electrical conductivity. *Agron. J.* **2003**, *95*, 472–482. [CrossRef]

23. Lueck, E.; Ruehlmann, J. Resistivity Mapping with GEOPHILUS ELECTRICUS—Information about Lateral and Vertical Soil Heterogeneity. *Geoderma* **2013**, *199*, 2–11. [CrossRef]

24. Ünal, İ.; Topakci, M. Design of a Remote-controlled and GPS-guided Autonomous Robot for Precision Farming. *Int. J. Adv. Robot. Syst.* **2015**, *12*, 1–10. [CrossRef]

25. Hunt, R.E. *Geotechnical Engineering Investigation Handbook*, 2nd ed.; Taylor & Francis: Boca Raton, FL, USA, 2005.

26. Giao, P.H.; Chung, S.G.; Kim, D.Y.; Tanaka, H. Electric imaging and laboratory resistivity testing for geotechnical investigation of Pusan clay deposits. *J. Appl. Geophys.* **2003**, *52*, 157–175. [CrossRef]

27. Juandi, M.; Syahril, S. Empirical relationship between soil permeability and resistivity, and its application for determining the groundwater gross recharge in Marpoyan Damai, Pekanbaru, Indonesia. *Water Pract. Technol.* **2017**, *12*, 660–666. [CrossRef]

28. Rossi, R.; Pollice, A.; Diago, M.P.; Oliveira, M.; Millan, B.; Bitella, G.; Tardaguila, J. Using an automatic resistivity profiler soil sensor on-the-go in precision viticulture. *Sensors* **2013**, *13*, 1121–1136. [CrossRef]

29. Lee, J.S.; Yoon, H.K. Theoretical relationship between elastic wave velocity and electrical resistivity. *J. Appl. Geophys.* **2015**, *116*, 51–61. [CrossRef]

30. Merritt, A.J.; Chambers, J.E.; Wilkinson, P.B.; West, L.J.; Murphy, W.; Gunn, D.; Uhlemann, S. Measurement and modelling of moisture—Electrical resistivity relationship of fine-grained unsaturated soils and electrical anisotropy. *J. Appl. Geophys.* **2016**, *124*, 155–165. [CrossRef]

31. Kim, K.; Sim, J.; Kim, T.H. Evaluations of the effects of soil properties and electrical conductivity on the water content reflectometer calibration for landfill cover soils. *Soil Water Res.* **2017**, *12*, 10–17. [CrossRef]

32. Parate, H.; Kumar, M.; Descloitres, M.; Barbiero, L.; Ruiz, L.; Braun, J.; Sekhar, M.; Kumar, C. Comparison of electoral resistivity by geophysical method and neutron probe logging for soil moisture monitoring in a forested watershed. *Curr. Sci.* **2011**, *100*, 1405–1412.

33. New Soil Maps Spark Change. Available online: https://www.farmprogress.com/new-soil-maps-spark-change (accessed on 18 November 2019).

34. Andrenelli, M.C.; Magini, S.; Pellegrini, S.; Perria, R.; Vignozzi, N.; Costantini, E.A.C. The use of the ARP© system to reduce the costs of soil survey for precision viticulture. *J. Appl. Geophys.* **2013**, *99*, 24–34. [CrossRef]

35. Costantini, A. Relationships Between Cone Penetration Resistance, Bulk Density and Moisture Content in Uncultivated, Repacked and Cultivated Hardsetting and NonHardsetting Soils From the Coastal Lowlands of South-Eastern Queensland. *N. Z. J. For. Sci.* **1996**, *26*, 395–412.

36. Vaz, C.M.P.; Hopmans, J.W. Simultaneous Measurement of Soil Penetration Resistance and Water Content with a Combined Penetrometer–TDR Moisture Probe. *Soil Sci. Soc. Am. J.* **2001**, *65*, 4–12. [CrossRef]

37. Prikner, P.; Lachnit, F.; Dvorak, F. A new soil core sampler for determination bulk density in soil profile. *Plant Soil Environ.* **2004**, *50*, 250–256. [CrossRef]

38. Weber, R.; Zalewski, D.; Hryńczuk, B. Influence of the soil penetration resistance, bulk density and moisture on some components of winter wheat yield. *Int. Agrophys.* **2004**, *18*, 91–96.

39. Vaz, C.M.P.; Manieri, J.M.; de Maria, I.C.; van Genuchten, M.T. Scaling the dependency of soil penetration resistance on water content and bulk density of different soils. *Soil Sci. Soc. Am. J.* **2013**, *77*, 1488–1495. [CrossRef]

40. Siqueira, G.M.; Dafonte, J.D.; Lema, J.B.; Armesto, M.V.; Silva, E.F. Using Soil Apparent Electrical Conductivity to Optimize Sampling of Soil Penetration Resistance and to Improve the Estimations of Spatial Patterns of Soil Compaction. *Sci. World J.* **2014**, *2014*, 1–12. [CrossRef]

41. Hosseini, M.; Movahedi, N.; Reza, S.A.; Dehghani, A.A.; Zeraatpisheh, M. Modeling of soil mechanical resistance using intelligent methods. *J. Soil Sci. Plant. Nutr.* **2018**, *18*, 939–951. [CrossRef]

42. Uusitalo, J.; Ala-Ilomaki, J.; Lindeman, H.; Toivio, J.; Siren, M. Modelling soil moisture-soil strength relationship of fine-grained upland forest soils. *Silv. Fenn.* **2019**, *53*, 1–16. [CrossRef]

Article

An ISE-based On-Site Soil Nitrate Nitrogen Detection System

Yanhua Li [1], Qingliang Yang [1], Ming Chen [1], Maohua Wang [1,2] and Miao Zhang [1,2,*]

[1] Key Laboratory on Modern Precision Agriculture System Integration Research of Ministry of Education, China Agricultural University, Beijing 100083, China; 15001228421@163.com (Y.L.); qingliangyang@cau.edu.cn (Q.Y.); mercury@cau.edu.cn (M.C.); wangmh@cau.edu.cn (M.W.)

[2] Key Lab of Agricultural Information Acquisition Technology of Ministry of Agriculture and Rural Affairs, China Agricultural University, Beijing 100083, China

* Correspondence: zhangmiao@cau.edu.cn

Received: 20 August 2019; Accepted: 23 October 2019; Published: 28 October 2019

Abstract: Soil nitrate–nitrogen (NO_3^--N) is one of the primary factors used to control nitrogen topdressing application during the crop growth period. The ion-selective electrode (ISE) is a promising method for rapid lower-cost in-field detection. Due to the simplification of sample preparation, the accuracy and stability of ISE-based in-field detection is doubted. In this paper, a self-designed prototype system for on-site soil NO_3^--N detection was developed. The procedure of spinning centrifugation was used to avoid interference from soil slurry suspension. A modified Nernstian prediction model was quantitatively characterized with outputs from both the ISE and the soil moisture sensor. The measurement accuracy of the sensor fusion model was comparable with the laboratory ISE detections with standard sample pretreatment. Compared with the standard spectrometric method, the average absolute error (AE) and root-mean-square error (RMSE) were found to be less than 4.7 and 6.1 mg/L, respectively. The on-site soil testing efficiency was 4–5 min/sample, which reduced the operation time by 60% compared with manual sample preparation. The on-site soil NO_3^--N status was dynamically monitored for 42 consecutive days. The declining peak of NO_3^--N was observed. In all, the designed ISE-based detection system demonstrated a promising capability for the dynamic on-site monitoring of soil macronutrients.

Keywords: on-site detection; ion-selective electrode (ISE); soil nitrate nitrogen (NO_3^--N); soil moisture; sensor fusion

1. Introduction

The ion-selective electrode (ISE) transfers the ionic activity (or concentration) of the target ion dissolved in testing solutions into electromotive force (EMF). Theoretically, the measured EMF is related to the logarithm of the ionic activity according to the Nernst equation. Because of the importance of fertilizer in agricultural production, ISEs have been used in soil nitrate–nitrogen (NO_3^--N) analysis for more than half a century [1]. A prototype ISE based on an in-field nitrate monitoring system was first developed in 1994 and has been successively improved by Canadian researchers [2–4]. Soil samples were collected at a depth of 0–15 cm with an autosampler. GPS information was recorded at the same time. Programmable processes of soil bulk crushing and plant residue removing were designed. NO_3^--N extraction was obtained by mixing the collected soil with de-ionized distilled water (DDW). The influence of soil texture was considered in sensor calibration. The fifth generation of the modified system demonstrated a satisfactory correlation with the standard method. An R^2 of 0.92 was found in testing of 13 sets of samples. The problem of random ISE signal disturbance caused by soil slurry was claimed.

In 2001, a portable ISE detection kit was developed for direct in-field measurement of soil chemical properties, including pH, mineral Na^+, mineral K^+, and NO_3^--N [5]. More than 500 soil samples were collected. However, the NO_3^--N testing results demonstrated obvious variations from the standard spectrometric method. At the same time, researchers from the University of Missouri compared extractants for ISE-based soil macronutrient detection. Kelowna solution was chosen for the extraction of soil available K^+, PO_4^{3-}, and NO_3^--N. Extracted soil solution was manually obtained using the recommended soil testing protocol. Feasibility was evaluated with 37 samples. ISE based laboratory soil NO_3^--N detection demonstrated good accuracy with standard deviations ranging from 8.04 to 19.7 mg/L [6,7]. Multiple studies were conducted on ISEs foron-the-go soil macronutrient monitoring by Adamchuk et al. For the purpose of achieving on-the-go soil testing, the "Direct Soil Measurement" (DSM) system was designed and then validated, updated, and commercially transformed in 2005. The ISEs of NO_3^-, K^+, and pH were integrated to form the sensing unit. De-ionized(DI) water was applied for the cleaning of the ISE sensing array. Sensing results were directly collected without pretreatment operations of stirring and filtration. Compared with laboratory detection, the DSM results of NO_3^-, K^+, pH were reported with coefficients of determination (R^2) of 0.41–0.51, 0.61–0.62, and around 0.9, respectively [8,9]. Insufficient sample extraction was considered to be a possible reason for the unsatisfactory accuracy level. Sethuramasamyraja et al. improved the soil pretreatment process of the system by integrating a mechanical agitation operation into the sample extractant process. The "Integrated Agitated Soil Measurement" (ASM) results of the soil pH were comparable to laboratory testing with an R^2 value of 0.99. However, the predicted NO_3^- value still demonstrated great deviation from standard spectrometric results with an R^2 value of 0.48 [10]. On the basis of the ASM system, the latest "On-the-Spot Analyzer" (OSA) system was developed for the simultaneously measurement of soil properties at a predefined soil depth. ISEs were brought into direct contact with the conditioned soil slurry, after the testing stand was moved to the experimental field and the topsoil was removed. Once sensors readings were retrieved, the analyzer was removed to another testing spot. Forty-five sets of surface topsoil samples with NO_3^--N concentrations ranging from 0 to 30 mg/kg were measured on the spot. The correlation coefficient R^2 was increased to 0.87 [11]. The improved detection accuracy with the OSA system demonstrated promising potential for the achievement of automated measurements.

As far as we are concerned, most of the in-field soil testing discussed above involves reduced soil pretreatment operations due to the system's simplicity and efficiency. The testing error, produced by "soil particle suspension disturbance", reached a magnitude of 26.6 mg/kg with an average relative error of 50% according to our preliminary laboratory validation of ISE-based NO_3^--N detection with 15 soil samples [12]. Besides, soil slurry would contaminate the membrane of ISE. The response slope of NO_3^- ISE was determined to be 44.4 and 25.4 mV/decade after continuous testing for 4 and 12 h, respectively [13]. Thus, it was necessary to obtain a transparent soil extract to enhance the accuracy and lifetime of the ISE. Pan et al. [14] tried to separate the clear soil NO_3^--N extractant from sample slurry through the short-time process of spinning centrifugation. Seven soil samples were used for the optimization of the centrifugation operation. Clear soil extractant was obtained by spinning for 30 s at the centrifugation speed of 1000 rpm. Compared with the direct soil slurry detection, the NO_3^--N detection relative error decreased from 64% to 5%. Yanhua et al. [15] attempted to evaluate the effects of uncalibrated soil moisture on NO_3^--N with six samples at the laboratory. The moisture of the tested samples was pre-manipulated to 2%–25%. The ISE based NO_3^- ISE results were uniformly smaller than the standard spectrometric results when the influence of soil moisture was neglected. A soil moisture percentage of 25% produced a maximum absolute error of 30 mg/kg. An error of no less than 5.0 mg/kg occurred even when the soil moisture was 5%.

For the purpose of improving the accuracy of on-site soil NO_3^--N detection, a self-designed prototype system was designed by making use of the sensor fusion method. Both the NO_3^- ISE and soil moisture sensor were employed as the sensing unit. The specific objectives were, first, to integrate necessary soil pretreatment steps, e.g., sample weighting and extractant spinning centrifugation into an on-site testing bench. Second, we investigated a modified Nernst model for the prediction of soil NO_3^--N with the real-time data provided by the ISE and the moisture sensor. Finally, we evaluated the feasibility of the system.

2. Materials and Methods

2.1. Reagents and Apparatus

A soil moisture sensor (ECH2O-5TE, Decagon, WA, USA) produced volumetric moisture readings that were used to determine the soil's net weight. The sensor was claimed to have a detection precision of $\pm3\%$ m^3/m^3. Reagents used were all Analytical grade. The testing solution was prepared with Deionized Water (Di-water). Standard soil chemical properties were provided by the soil testing center of the China Agricultural University with commercial analytical instruments. Detection was carried out according to the guidance of soil testing and fertilizer recommendations [16]. Soil moisture was oven dried at the temperature of 65 °C for 8 h (SG-GDJ50, SIOM, Shanghai, China). Soil NO_3^--N was detected with a UV-VIS spectrometer (UV2450, SHIMAZU, Kyoto, Japan) at 210 nm. H_2SO_4 (70%) was applied to the soil extractant for acidification. The Total-N (TN) soil concentration was determined with Kjeldahl determination (KJELTEC 8400, FOSS, Hillerød, Denmark). Soil available phosphate (AP) was detected based on Molybdenum Blue Colorimetry at 660 nm (UV2450, SHIMAZU, Kyoto, Japan). The Organic Carbon (OC) concentration was measured based on dry combustion at 550 °C for 24 h (SG-SJ1700, SIOM, Shanghai, China). Flame photometry (420, Cole-Parmer, IL, USA) was used to measure the Available potassium (AK) content of the soil. Commercial nitrate ISE (No.9707BNWP, Thermo Scientific Orion, MA, USA) with a detection limit of 1.4 mg/L was also employed in this study.

The analytical grade chemicals used for the calibrations of ISE and the detection of standard soil macronutrients were purchased from Sinopharm Chemical Reagent Beijing Co. Ltd.

2.2. Sensor Fusion Model

The detected NO_3^--N content would be greatly underestimated if soil moisture interference was not involved in the compensation of the sample net weight. In this study, volumetric soil moisture information was obtained during the on-site soil sampling. The volumetric moisture was converted into the gravimetric moisture for the correction of the sample's net (dry) weight. The detailed procedure was discussed in a previously published paper [15]. A sensor fusion model was designed for the NO_3^--N prediction, as illustrated in Equations (1)–(3). Compared to the conventional Nernst model, the ratio of extractant to soil weight of the sensor fusion model achieved real-time correction instead of using a constant value, as used in most of the previous studies.

$$\omega = \frac{\rho_w \times (\theta - \theta_0)}{\rho_s} = \frac{1}{\rho_s} \times (\theta - \theta_0) \tag{1}$$

$$N = \omega + \frac{\omega m + m}{M} \tag{2}$$

$$C_i = 1000N \cdot A_r 10^{\frac{E-E_0}{S}} \tag{3}$$

where ρ_s represents the pre-determined bulk density of dry soil (1.19 g/cm^3); ρ_w represents the density of deionized water (1.0 g/mL); θ_0 represents the pre-determined volumetric moisture ratio (−1.51%); θ represents the soil volumetric moisture (%); ω represents the soil mass moisture (%); M represents the weight of the raw soil sample (g); m represents the volume of soil extractant (mL); N represents the ratio of extractant to the net weight of soil (mL/g); Ar represents the relative atomic mass, which, for nitrogen, is 14; C_i represents the concentration of nitrate in the tested sample (m/V, mg/L); E represents the EMF value produced by ISE (mV); E_0 represents the intercept potential of the Nernstian model of the tested ISE (mV); and S represents the response slope of the Nernstian model of the tested ISE (mV/decade), where decade means 10 times the change in the target concentration.

2.3. System Design

The on-site soil NO_3^--N detection bench consisted of five major units, including the extractant preparing unit (A), extractant clarification unit (B), electrode holder unit (C), leveling unit (D), and electronic control circuit unit (E), as illustrated in Figure 1a,b. Centrifuge (B9) was employed to achieve separation of the clarified extractant from the soil slurry. The centrifuge process was conducted at a speed of 1000 rpm for 3 min. The manually collected soil sample was weighed with electronic scales with a precision of 0.1 g (A10). Stepper motors of A1 and B1 were employed to achieve vertical movements of two mechanical arms for extractant injection and transportation. The proximity sensors of A5 and B4 were used to define the working scale of the vertical slide table (A3/B3). The precision of vertical movement was measured to be 0.05 cm. Rotary table B7 was driven by step motor B7. Centrifuge B9 had 12 container positions, so B7 would rotate by 30 each time with a control precision of 0.5°. Transportation of DDW and the sample extract was achieved by peristaltic pumps A4/B5 through tubes of A6/A7. The stirring operation was performed with Blender A8. ISE testing was conducted by hanging the sensor on C2. To keep the balance of A10 and B9, the bench employed leveling meter D2, positioner D3, and screw adjuster E1.

The detection bench was manipulated in a programmable way by the self-designed electronic control circuit unit, as shown in Figure 1c. The STM 32 Microchip Controller Unit (MCU) was applied as the main processor. The underlying hardware of step motors 1–3 and peristaltic pumps 1–3 were motivated with the drive unit according to the pre-designated flowchart. A proximal sensing signal was sent to the MCU when the mechanical arms were close to the vertical limitation of 10 cm. A Bluetooth connection was formed among the control circuit, ISE datalogger, and Android terminal devices, e.g., smartphones. Sensor readings and user commands were communicated. A schematic diagram of the circuit is illustrated in Figure 1c.

The rural smartphone popularity was reported to be 32% in China [17]. Considering the interface resource, flexible communication mode, convenient data storage, and upload capability, application software running on Android terminal devices was also developed in this study. The interface of the smartphone App is shown in Figure 1d. Predetermined soil sample profile information, including soil texture, bulk density, sample weight, DDW volume, and electroconductivity, should be input, saved, and downloaded to the control circuit. The parameters of the sample pretreatment operation, e.g., stirring time, rinsing method, and motor speed, are chosen according to the testing mode. Testing setups were employed with the calibration solution number, testing duration, sample number, file save option, and real-time display. A Location-Based Service (LBS) was embedded to provide the sample's geographic position. The Bluetooth setup was operated on the App.

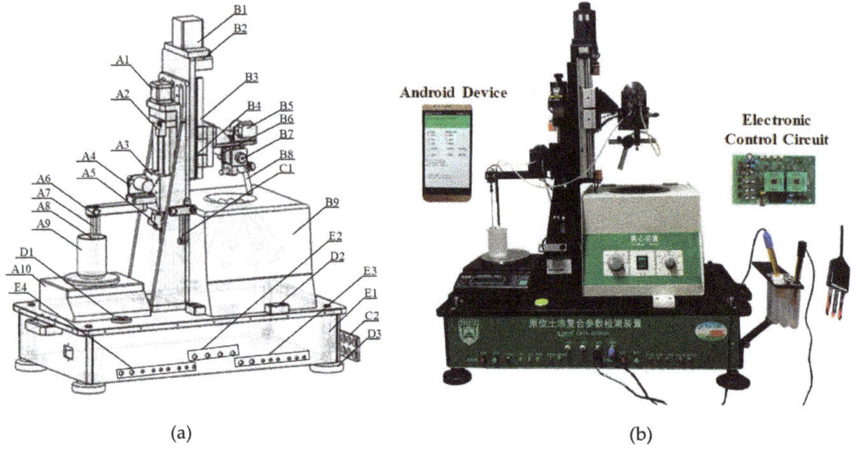

Figure 1. *Cont.*

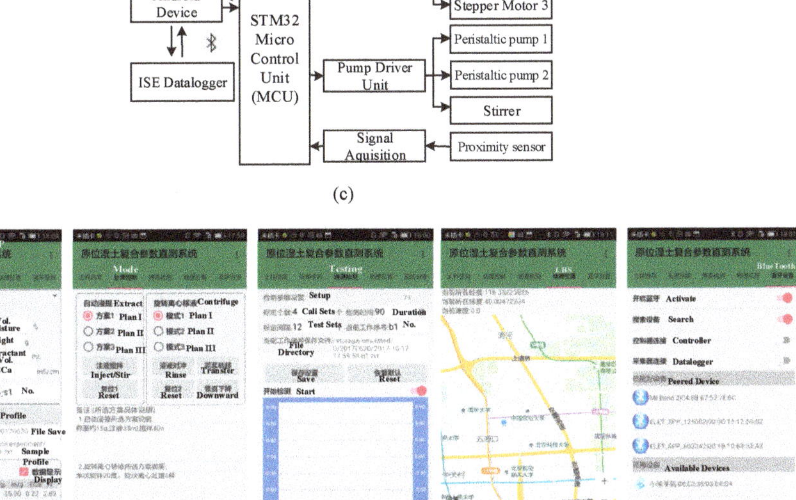

(c)

(d)

Figure 1. Diagram of the on-site detection bench: (**a**) System Design A1, Stepper motor 1 A2. Proximity sensor 1 A3, Vertical slide table 1 A4, Peristaltic pump A5, Proximity sensor 2 A6, Injecting tube A7, Outlet tube A8, Blender A9, Soil sample container A10, Electronic weight scale B1, Stepper motor 2 B2, Proximity sensor 3 B3, Vertical slide table 2 B4, Proximity sensor 4 B5, Peristaltic pump2 B6, Rotary table B7, Stepping motor 3 B8, Pipe hanger B9,Centrifuge C1, Electrode hanger 1 C2, Electrode hanger 2 D1, Horizontal Lever meter D2, Positioner D3, Leveling screw E1, Circuit controller E2, ISE connector E3, Control switches and indicator lights E4, Control switches and indicator lights; (**b**) Physical picture of the hardware; (**c**) Diagram of the Electric Control Circuit Design; (**d**) Android App for Smartphones.

2.4. Field Test Design

Fresh soil samples were manually collected at a depth of 0–25 cm from a demonstration summer corn planting farm (70 L × 24 W m^2) from April 30 to Aug 31, 2016 (40°8'37" N, 116°11'31" E). Soil sampling information is shown in Figure 2. The cornfield was divided into 12 fertility zones with a varied N application rate from 0 to 3 N, where 1 N equals the application of 375 kg/ha of compound fertilizer (Total content ≥ 40%, N:P$_2$O$_5$:K$_2$O, 28%:6%:6%, Shidanli Co. Ltd., Shandong, China) and 75 kg/ha of urea; $\frac{1}{2}$ N represents half of the 1 N rate; 0 N means no fertility; and 3 N means triple the rate. A total of 11 groups of soil samples were collected. Raw soil samples, detected in the field by the self-designed bench without moisture compensation, were recorded as ISE$_{raw}$. ISE results, provided by the self-designed detection bench by the sensor fusion model, were recorded as ISE$_{OS}$. Laboratory ISE soil testing results were labeled ISE$_{LT}$, in which soil samples was treated with conventional soil pretreatments. Soil samples measured by the standard UV-VIS spectrometer were provided by the soil testing center of China's Agricultural University. The nitrate–nitrogen content was recorded to be Stand$_{Spec}$.

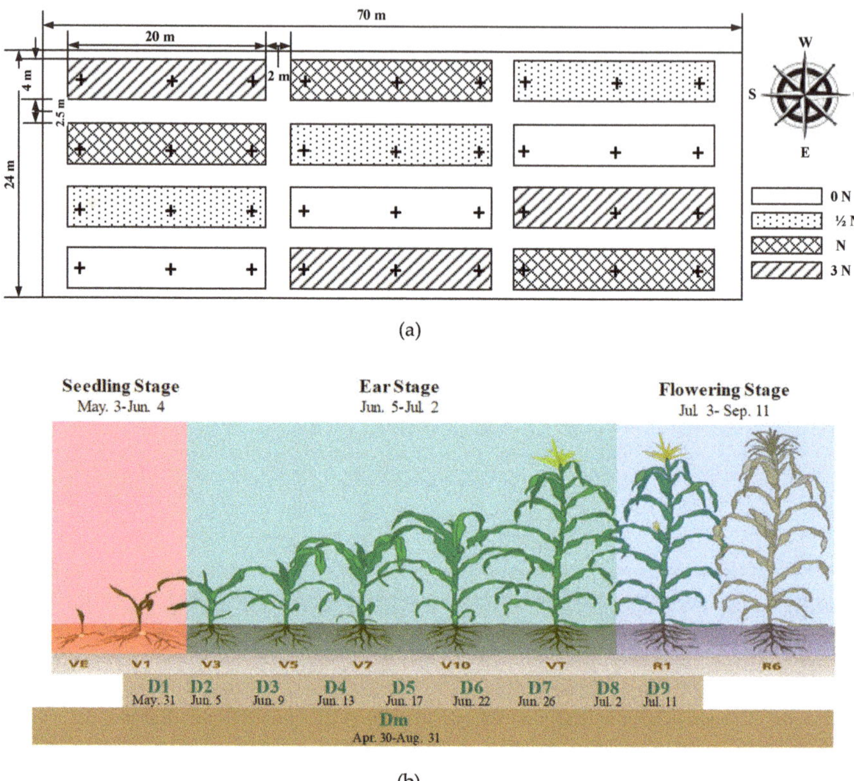

Figure 2. Soil Sampling Information: (**a**) Sampling space position inside the field (**b**) Sampling time.

Forty-two sets of raw samples, labeled as D_m, with broader time variance, were randomly sampled in the field from April 30 to August 31. The D_m testing group was used to evaluate the performance of the designed sensor fusion model. Differences among $Stand_{Spec}$, ISE_{raw}, ISE_{OS}, and ISE_{LT} were compared. The evaluation results are illustrated as Figure 3.

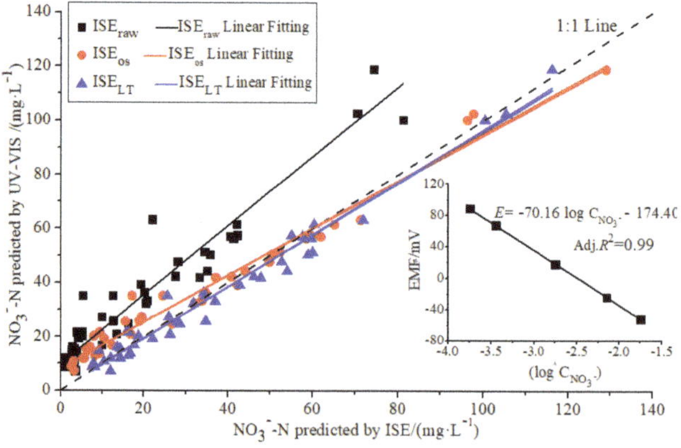

Figure 3. Comparison of soil NO_3^--N predicted with ISE_{raw}, ISE_{OS}, and ISE.

As demonstrated in Figure 2a, three sampling positions were marked with the plus cross icon in each of the 12 zones. One representative soil sample per zone was obtained by thoroughly mixing these three cores. A total number of 108 sets of fresh soil samples were collected for 42 days, which covered the summer corn growth stages from trifoliate to silking. The first 12 samples were collected on May 30, which were labeled as group D1. Then, the 7 continuous groups of samples, marked D2–D8, were obtained from June 5 until July 2, commonly at intervals of 3 days. The last group of soil samples (D9) was collected on July 11. Soil samples were applied to validate the feasibility of the on-site NO_3^--N testing system.

The soil properties provided by the standard testing center are summarized in Table 1.

Table 1. Soil sample information.

	No.	Mass Moisture	Nitrate Nitrogen ($mg·L^{-1}$)	Total -N [1] ($g·kg^{-1}$)	Available-P [1] ($mg·L^{-1}$)	Organic Matter [1] ($g·kg^{-1}$)	Available-K [1] ($mg·L^{-1}$)
D_m	42	2.5%–30.2%	11.2–87.7	0.3–10.5	9.8–32.5	3.2–9.0	8.3–121.3
D1	12	12.5%–16.3%	33.1–159.8	0.3–9.9	2.4–43.3	1.3–11.2	10.3–98.8
D2	12	13.3%–16.9%	31.6–345.0	-	-	-	-
D3	12	13.2%–17.6%	27.5–272.0	-	-	-	-
D4	12	11.4%–15.4%	16.2–189.7	-	-	-	-
D5	12	10.6%–13.7%	19.3–260.5	-	-	-	-
D6	12	9.2%–15.1%	19.3–256.9	-	-	-	-
D7	12	23.8%–26.4%	12.9–72.3	-	-	-	-
D8	12	14.3%–17.3%	9.5–32.6	-	-	-	-
D9	12	14.8%–18.1%	5.2–16.7	-	-	-	-

[1] Soil Total-N, Available-P, Organic Matter, and Available-K were tested in two groups of soil samples. D_m was 42 soil samples evaluated using the sensor fusion model. D1 was 12 samples used for the evaluation of the on-site bench. Detection was not conducted in D2–D9, because these soil properties were considered to be stable during the same corn growth season.

3. Results and Discussion

3.1. Validation of the Sensor Fusion Model

The sensor fusion compensation model, described in Equations (1)–(3), was evaluated with 42 soil samples, as demonstrated in Figure 3. The soil testing results of ISE_{raw} were, on average, 46.8% smaller than $Stand_{Spec}$. The maximum deviation was calculated as 44.8 mg/L. ISE_{OS} and ISE_{LT} demonstrated a good correlation with the standard spectrometric results. Absolute error values of 0.2–17.2 and 0–9.8 mg/L were obtained, respectively. The measurement accuracy of ISE_{OS} was increased by more than 50% compared with that of ISE_{raw}. The soil moisture compensation model eliminated the testing error.

3.2. Evaluation of the On-Site Soil NO_3^--N Detection

Soil NO_3^--N detection results were compared among three different methods—standard spectrometric results, laboratory ISE testing, and on-site ISE based monitoring—as shown in Table 2. The testing efficiency was also evaluated. The time duration and the labor force consumed for dealing with a dozen soil samples were compared among UV-VIS, ISE_{OS}, and ISE_{LT}. The results are summarized in Table 3.

Table 2. Statistical analysis of the linear regression fitting results.

	Detection Range $(mg·L^{-1})$	Linear Fitting Model	*Adj. R^2*	*F-Value*	*P-Value*	*Sig.*	AE $(mg·L^{-1})$	MRE (%)	RMSE $(mg·L^{-1})$
ISE_{OS}	5.0–156.3	y = 1.02x − 0.57	0.98	6055.8	0.0	*	0.1–19.9	13.9	6.1
ISE_{LT}	5.9–150.5	Y = 0.98x − 0.71	0.98	5488.9	0.0	*	0.0–18.4	13.7	5.5

* represents that the linear fitting model is significant.

Table 3. Comparison of the testing duration and labor force among $Stand_{Spec}$, ISE_{OS}, and ISE_{LT}.

Measurement [1]		$Stand_{Spec}$		ISE_{LT}		ISE_{OS}	
Testing Duration (min) [2]	OPERATIONS	Quantitative Weighing	12	Quantitative Weighing	12	Sample Weighing	2
		Extractant adding	12	Extractant adding	12	Extractant Injection	16
		Shaking	20	Shaking	20		
		Stabilization	20	Stabilization	20	Centrifuge Filtration	3
		Filtration	4				
		Titration	24	Filtration	4		
		Detection	15	Detection	24	Detection	24
	Total		107		92		45
Labor force Intensity		Intensive physical work. Participation in the overall process		Intensive physical work. Participation in the overall process		Light physical work. Participation in sample pickup and weighting.	

[1] Soil samples detected by $Stand_{Spec}$ and ISE_{LT} should be pretreated according to the soil testing recommendations. The shaking time required is 20 min. The optimal stabilization time is 20 min.; Soil samples detected by ISE_{OS} did not undergo quantitative weighting. Fresh soil samples were first weighed after moisture measurement. A peristaltic pump was used for extractant injection. The extractant injection rate was 36 s/sample. The stirring process was used for 40 s/sample. The centrifuge filtration rate was 40 s/12 samples. A stable ISE reading was obtained when the variation of EMF less was than ±1 mV. The ISE detection rate was 4–5 min/sample. [2] Time used for processing 12 soil samples.

As illustrated in Table 2, the linear regression fitting results of ISE_{OS}, ISE_{LT}, and UV-VIS were $y_{UV-VIS} = 1.02ISE_{OS} − 0.57$, $y_{UV-VIS} = 0.98ISE_{LT} − 0.71$. Both linear fitting curves were close to the 1:1 line. The ISE detection accuracy demonstrated a slight variation with the change in soil $NO_3^−$-N content. The accuracy was derived as ±30%, ±16% and 5% (Full Scale, FS) at the $NO_3^−$-N content ranges of 0–30, 31–90, and 91–200 mg/L, respectively. The maximum error (with the possibility of ±90%) was less than 10 mg/L. The intersection was close to 1. Adj. R^2 values were both 0.98. The ISE results demonstrated close consistency with UV-VIS. The absolute error values among ISE_{OS}, ISE_{LT}, and UV-VIS were calculated to be 0.1–19.9 and 0.0–18.4 mg/L with average values of 4.7 and 4.0 $mg·L^{-1}$, respectively. The RMSEs were found to be 6.1 and 5.5 mg/L. No significant difference was found between the results of ISE_{OS} and ISE_{LT}.

The ISE_{OS} demonstrated obvious advantages in terms of the testing efficiency and labor force intensity, as shown in Table 3. Compared with the conventional soil pretreatment protocols conducted before UV-VIS and ISE_{LT}, the self-designed on-site detection bench was decreased by 45 mins. The total time consumption was reduced to 40% of the duration of the conventional spectrometry method.

Integrated with the multi-sensor, centrifuge filtration, and programmable fluidic control, the self-designed on-site soil $NO_3^−$-N detection bench produced a reliable result with an efficient operation, which demonstrated a promising perspective for the infield monitoring applications.

3.3. $NO_3^−$-N Variation Monitoring

Based on the workbench, the on-site $NO_3^−$-N variation was monitored from the trifoliate stage to the silking stage of summer corn. Samples collected from three 1N zones were selected to demonstrate the $NO_3^−$-N content change with corn growth, as shown in Figure 4. The $NO_3^−$-N content was at a level of around 70–100 mg/L at the beginning of D1. NO_3^--N demonstrated great variation in characteristics with time and at different sample sites. However, an obvious NO_3^--N decrease occurred uniformly at an amplitude of 80 mg/L across all three testing sites from D6 to D7. According to the definition of corn growth, D6 was the V_T period and D7 was in the R_1 period, as shown in Figure 2b. The monitoring results perfectly fit the nitrogen growth law of corn. After that growth stage, no clear nitrogen absorption was verified. The NO_3^--N content stayed at the level of 13.2–17.0 mg/L.

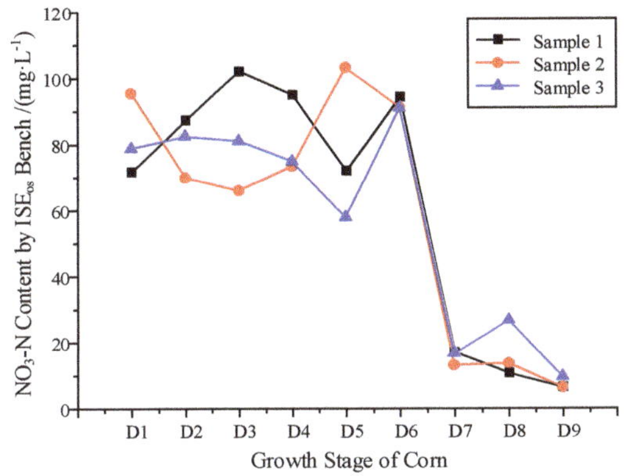

Figure 4. Monitoring of NO_3^--N Variation by the On-site detection Bench.

4. Conclusions

In this paper, a self-designed prototype system for on-site soil NO_3^--N detection based on ISE was designed and tested. Sensor fusion of ISE and a moisture sensor effectively eliminated 50% of the testing error. The performance of the on-site soil NO_3^--N system demonstrated good consistency with the UV-VIS testing and laboratory ISE testing methods. Compared with the UV-VIS method, the average absolute error was determined to be 4.7 mg·L^{-1}. The RMSE was found to be 6.1 mg/L. In addition, the detection duration decreased to 40% of that of the spectrometric method.

Author Contributions: Conceptualization, Y.L. and M.Z.; Methodology, Y.L. and M.Z.; Software, Y.L. and Q.Y.; Validation, Y.L., Q.Y. and M.C.; Formal Analysis, Y.L. and M.Z.; Investigation, Y.L. and M.Z.; Resources, M.W. and M.Z.; Data Curation, Y.L. and M.Z.; Writing—Original Draft Preparation, Y.L. and M.Z.; Writing—Review & Editing, Y.L. and M.Z.; Visualization, Y.L. and Q.Y.; Supervision, M.Z.; Project Administration, M.W. and M.Z.; Funding Acquisition, M.W. and M.Z.

Funding: This research was financially supported by the National Key Research and Development Program (Grant No. 2016YFD0700300-2016YFD0700304 & 2016YFD0800900-2016YFD0800907) and Key Laboratory of Technology Integration and Application in Agricultural Internet of Things, Ministry of Agriculture, P. R. China (2016KL03).

Conflicts of Interest: The authors declare no conflict of interest.

References

1. Myers, R.J.K.; Paul, E.A. Nitrate ion electrode method for soil nitrate-nitrogen determination. *Can. J. Soil Sci.* **1968**, *48*, 369–371. [CrossRef]
2. Thottan, J.; Adsett, J.F.; Sibley, K.J.; Macleod, C.M. Laboratory evaluation of the ion selective electrode for use in an automated soil nitrate monitoring system. *Commun. Soil Sci. Plant Anal.* **1994**, *25*, 3025–3034. [CrossRef]
3. Sibley, K.J.; Adsett, J.F.; Struik, P.C. An on-the-go-soil sampler for an automated soil nitrate mapping system. *Trans. ASABE* **2008**, *51*, 1895–1904. [CrossRef]
4. Sibley, K.J.; Astatkie, T.; Brewster, G.; Struik, P.C.; Adsett, J.F.; Pruski, K. Field-scale validation of an automated soil nitrate extraction and measurement system. *Precis. Agric.* **2009**, *10*, 162–174. [CrossRef]

5. Davenport, J.R.; Jabro, J.D. Assessment of hand held ion selective electrode technology for direct measurement of soil chemical properties. *Commun. Soil Sci. Plant Anal.* **2001**, *32*, 3077–3085. [CrossRef]

6. Kim, H.J.; Hummel, J.W.; Birrell, S.J. Evaluation of nitrate and potassium ion-selective membranes for soil macronutrient sensing. *Trans. ASABE* **2006**, *49*, 597–606. [CrossRef]

7. Kim, H.J.; Hummel, J.W.; Sudduth, K.A.; Motavalli, P.P. Simultaneous Analysis of Soil Macronutrients Using Ion Selective Electrodes. *Soil Sci. Soc. Am. J.* **2006**, *71*, 1867–1877. [CrossRef]

8. Adamchuk, V.I.; Lund, E.D.; Dobermann, A.; Morgan, M.T. On-the-go mapping of soil properties using ion-selective electrodes. In *Precision Agriculture*; John, V.S., Armin, W., Eds.; Wageningen Academic Publishers: Wagenin, The Netherlands, 2003; pp. 27–33.

9. Adamchuk, V.I.; Lund, E.D.; Sethuramasamyraja, B.; Morgan, M.T. Direct measurement of soil chemical properties on-the-go using ion-selective electrodes. *Comput. Electron. Agric.* **2005**, *48*, 272–294. [CrossRef]

10. Sethuramasamyraja, B.; Adamchuk, V.I.; Dobermann, A.; Marx, D.B.; Jones, D.D.; Meyer, G.E. Agitated soil measurement method for integrated on-the-go mapping of soil pH, potassium and nitrate contents. *Comput. Electron. Agric.* **2008**, *60*, 212–225. [CrossRef]

11. Adamchuk, V.; Dhawale, N.; Renelaforest, F. Development of an on-the-spot analyzer for measuring soil chemical properties. In Proceedings of the International Conference on Precision Agriculture, Sacramento, CA, USA, 20–23 July 2014.

12. Zhang, L.; Zhang, M.; Ren, H.; Pu, P.; Kong, P.; Zhao, H. Comparative Investigation on Soil Nitrate-nitrogen and Available Potassium Measurement Capability by Using Solid-State and PVC ISE. *Comput. Electron. Agric.* **2015**, *112*, 83–91. [CrossRef]

13. Zhang, L. Study on Establishing the Electrochemical Methods for Fast Determination of Soil Available Macronutrients Based on Ion-Selective Electrodes. Ph.D. Thesis, China Agricultural University, Beijing, China, 2015. (In Chinese with English Abstract).

14. Kong, P.; Zhang, M.; Ren, H.; Li, Y.; Pu, P. Rapid Pretreatment Method for Soil Nitrate Nitrogen Detection based on Ion-selective Electrode. *Trans. Chin. Soc. Agric. Mach.* **2015**, *46*, 102–107, (In Chinese with English Abstract).

15. Li, Y.; Zhang, M.; Pan, L.; Zheng, J. ISE-base Sensor Fusion Method for Wet Soil Nitrate-nitrogen Detection. *Trans. Chin. Soc. Agric. Mach.* **2016**, *47*, 285–290, (In Chinese with English Abstract).

16. Bai, Y.L.; Yang, L.P. Soil testing and fertilizer recommendation in Chinese agriculture. *Soil Fertil. Sci. China* **2006**, *2*, 3–7, (In Chinese with English Abstract).

17. Cyberspace Administration of China. The 44th China Statistical Report on Internet Development. Available online: http://www.cac.gov.cn/2019-08/30/c_1124938750.htm (accessed on 30 August 2019).

Article

Simplifying Sample Preparation for Soil Fertility Analysis by X-ray Fluorescence Spectrometry [†]

Tiago Rodrigues Tavares [1], **Lidiane Cristina Nunes** [2], **Elton Eduardo Novais Alves** [3], **Eduardo de Almeida** [4], **Leonardo Felipe Maldaner** [1], **Francisco José Krug** [2], **Hudson Wallace Pereira de Carvalho** [4] and **José Paulo Molin** [1,*]

[1] Laboratory of Precision Agriculture (LAP), Department of Biosystems Engineering, "Luiz de Queiroz" College of Agriculture (ESALQ), University of São Paulo (USP), Piracicaba, São Paulo 13418900, Brazil; tiagosrt@usp.br (T.R.T.); leonardofm@usp.br (L.F.M.)

[2] Laboratory of Analytical Chemistry (LQA), Center for Nuclear Energy in Agriculture (CENA), University of São Paulo (USP), Piracicaba, São Paulo 13416000, Brazil; lcnunes@cena.usp.br (L.C.N.); fjkrug@cena.usp.br (F.J.K.)

[3] Laboratory of 14 Carbon (LC14), Center for Nuclear Energy in Agriculture (CENA), University of São Paulo (USP), Piracicaba, São Paulo 13416000, Brazil; elton.alves@usp.br

[4] Laboratory of Nuclear Instrumentation (LIN), Center for Nuclear Energy in Agriculture (CENA), University of São Paulo (USP), Piracicaba, São Paulo 13416000, Brazil; edualm@cena.usp.br (E.d.A.); hudson@cena.usp.br (H.W.P.d.C.)

[*] Correspondence: jpmolin@usp.br; Tel.: +55-19-3447-8502

[†] This paper is an extended version of our paper published in "Sample preparation for assessing the potential use of XRF on soil fertility analysis". In Proceedings of the 5th World Workshop on Proximal Soil Sensing, Columbia, MO, USA, 28–31 May 2019.

Received: 26 August 2019; Accepted: 15 November 2019; Published: 20 November 2019

Abstract: Portable X-ray fluorescence (pXRF) sensors allow one to collect digital data in a practical and environmentally friendly way, as a complementary method to traditional laboratory analyses. This work aimed to assess the performance of a pXRF sensor to predict exchangeable nutrients in soil samples by using two contrasting strategies of sample preparation: pressed pellets and loose powder (<2 mm). Pellets were prepared using soil and a cellulose binder at 10% w w^{-1} followed by grinding for 20 min. Sample homogeneity was probed by X-ray fluorescence microanalysis. Exchangeable nutrients were assessed by pXRF furnished with a Rh X-ray tube and silicon drift detector. The calibration models were obtained using 58 soil samples and leave-one-out cross-validation. The predictive capabilities of the models were appropriate for both exchangeable K (ex-K) and Ca (ex-Ca) determinations with R^2 ≥ 0.76 and RPIQ > 2.5. Although XRF analysis of pressed pellets allowed a slight gain in performance over loose powder samples for the prediction of ex-K and ex-Ca, satisfactory performances were also obtained with loose powders, which require minimal sample preparation. The prediction models with local samples showed promising results and encourage more detailed investigations for the application of pXRF in tropical soils.

Keywords: precision agriculture; X-ray fluorescence; spectroscopy; soil nutrients; proximal soil sensing; soil testing

1. Introduction

Brazil is the fourth largest consumer of fertilizers in the world [1] due to the predominance of acidic and low fertility tropical soils. Thus, the diagnosis of soil fertility is crucial for the correct management of fertilizers and limestone in crops. Per year, it is estimated that about 1 million soil tests are carried out by Brazilian fertility analysis laboratories. There is an expectation of increasing this number, due to the expansion of agricultural areas [2], as well as the adoption of soil mapping techniques for precision agriculture (PA) practices, which demand information in a high spatial and temporal density [3]. In addition, as commented by Demattê et al. [2], traditional soil analyses face other challenges related to the time required for performing the laboratory measurements (about 3 to 15 days), and also the hazardous reagents still used in some tests (e.g., dichromate and sulfuric acid).

The establishment of a robust method for the direct analysis of soils using sensing techniques—allowing farmers and laboratories to increase the number of analyses in a practical and clean way, without relying exclusively on traditional fertility soil tests—is a current need in Brazil and all over the world [3]. This task is a great multidisciplinary challenge for the researchers involved [4]. Hence, discussions have recently begun between academics [2] and companies for the development of hybrid laboratories. The term hybrid refers to labs, where analyses performed by sensor systems are used in combination with the traditional methods, allowing one to use the sensing techniques to predict some soil attributes. Hybrid laboratories are compatible with controlled-environment and on-field analyses (e.g., using a mobile soil-testing lab [5]). This is an interesting strategy, which should boost worldwide research in the coming years to seek the best set of sensors compatible with direct analysis of soils, as well as the best strategy for calibration of the predictive models.

X-ray fluorescence (XRF) is a spectroanalytical technique compatible with direct soil analysis, which can be applied with minimal or no sample preparation [6]. The recent technological advance of the optical and electronic components allowed the development and miniaturization of this technology, and it has become attractive for use in hybrid laboratories and in situ analyses. Some studies have already pointed out the potential of using XRF sensors in proximal soil sensing (PSS) approaches [7,8]. Despite that, XRF has been poorly explored for assessments of physical and chemical attributes of tropical soils, mainly under the context of PSS and PA.

To use XRF sensors as a practical analytical method in hybrid laboratories—in order to ensure a massive increase in the number of samples analyzed—it should be compatible with a simple soil sample preparation (e.g., just air-dried and sieved rapidly). Recent studies involving XRF sensors for practical analysis of soil attributes have used dried samples with particle sizes smaller than 2 mm [8–11]. It is a consensus that pellet preparation after grinding the soil allows one to explore the potential of the XRF technique in soil analysis [12]. The preparation of a pellet is recommended for analyses with the XRF technique because it improves the homogeneity of the material and also allows one to control the density, porosity and surface roughness characteristics, reducing the physical matrix effects [13]. Although it is known that the preparation of pellets guarantees better precision in the measurements performed with the XRF [14,15], recent studies have assumed that, when analyzing soil samples with particle sizes smaller than 2 mm, its heterogeneity and physical matrix effects can be neglected.

XRF analyses are more flexible with regards to sample preparation because—unlike other elemental analysis techniques, such as laser induced breakdown spectroscopy (LIBS)—they also allow one to evaluate loose powder [14]. However, we did not find any study comparing XRF performance for the prediction of fertility attributes on soil samples that were just dried and sieved (<2 mm) with samples prepared with the optimal sample preparation method. Therefore, the level of performance loss when neglecting the physical matrix effect and heterogeneity is unknown. For a robust development of the XRF technique as a practical tool for soil fertility analysis, one of the key points is to understand the tradeoff between analytical performance and sample preparation, in order to reduce or eliminate these procedures based on the analytical potential of the sensor for each sample condition. Such knowledge is important for the development of PSS applications using this tool.

To evaluate the possibility of simplifying the sample preparation procedure for XRF analyses, this work aimed to assess the performance of a portable XRF (pXRF) to predict exchangeable nutrients in soil samples prepared using two contrasting types of sample preparation: pellets and loose powder (≤2 mm). The effect of sample preparation in the spatial distribution of nutrients on the sample surface was evaluated using a benchtop microprobe X-ray fluorescence spectrometry (µ-XRF). Moreover, a procedure for preparation of soil pellets, involving planetary ball milling and the use of binding agents was also assessed.

2. Material and Methods

2.1. Soil Samples

A set of 58 soil samples were selected for the comparison of their exchangeable nutrients content with the X-ray fluorescence produced by the pellet and loose powder samples. These samples were collected from 0 to 0.2 m soil depth in an agricultural field located at the southeast region of Brazil, in the municipality of Piracicaba, state of São Paulo (at coordinates 22°41′57.24″ S and 47°38′33.33″ W, WGS84 datum). The soil is classified as Lixisol [16] with a clayey texture and high nutrient variability.

The soil samples were air-dried and sieved (<2 mm) and after that, three subsamples were separated: (i) 0.8 g was used for pelletizing, (ii) 10 g was analysed as loose powder, and (iii) about 30 g was used for the reference measurements.

2.2. Sample Preparation

For pelletizing, the samples (particles < 2 mm) were initially dried at 105 °C for 24 h and thereafter ground in a planetary ball mill (Retsch model PM 200 mill, Germany) (Figure 1A) by using two grinding tungsten carbide jars (50 mL; Retsch, Germany) with 10 tungsten carbide balls (10 mm diameter) (Figure 1B). Grinding was performed at 400 rpm for 5 min clockwise/5 min counter clockwise with a 10-s stop before changing the rotation direction.

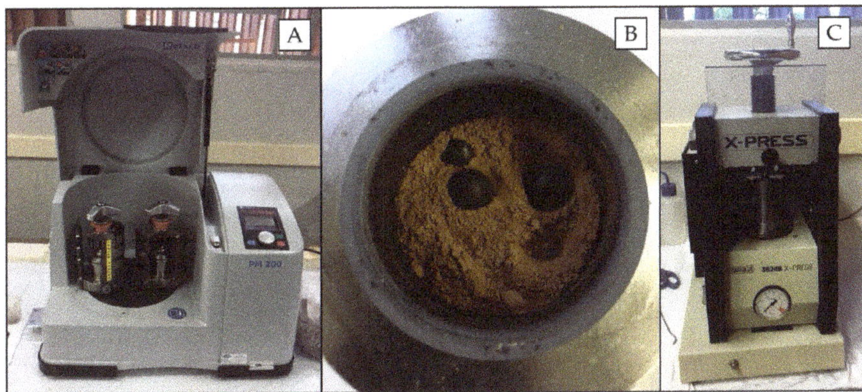

Figure 1. Planetary ball mill (**A**), loose soil inside the tungsten carbide jars with the tungsten carbide balls (**B**) and hydraulic press (**C**), which were used at work.

Preliminary experiments were carried out by just pressing the soil samples without a binder. It was observed that for the sandier sample (clay content of 175 g dm^{-3}) the resulting pellets were friable and easily crumbled (Figure 2A). Therefore, binder addition was decisive for improving the quality of the pellets. The binders tested were chosen based on their similarity to the analytical blank, i.e., lower analyte mass fractions of elements evaluated in the soil fertility (e.g., P, K, Ca, and Mg). In this case, binding agents, such as a microcrystalline cellulose powder (Sigma-Aldrich, Merck, Darmstadt, Germany), and cellulose (SPEX 3642, Metuchen, NJ, USA) were evaluated in the proportion of 10 and 15% w w^{-1}, with grinding/homogenization times of 10, 15, and 20 min.

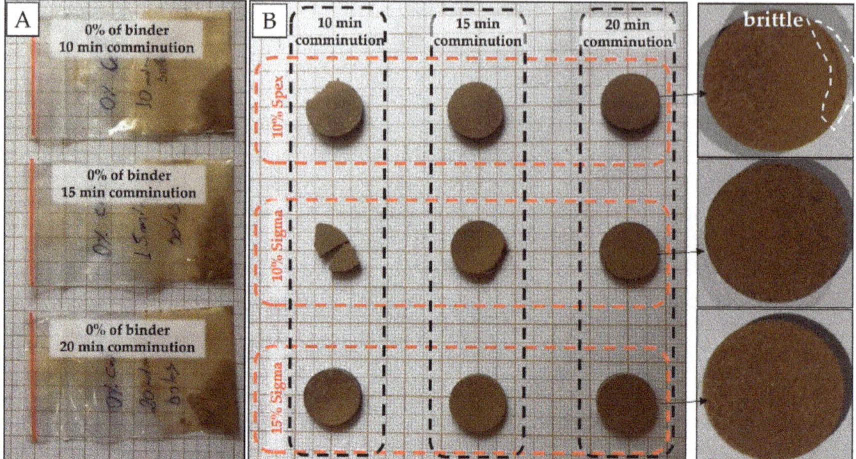

Figure 2. Soil samples without the addition of binder and with different grinding times (**A**); pellets resulting from tests with different grinding times, cellulose concentrations and brands (**B**).

The grinding and homogenized samples were pelletized in a hydraulic press (SPEX 3624B X-Press) (Figure 1C) by transferring 0.8 g of the powdered material to a stainless steel set and applying 8.0 t cm^{-2} for 3 min. Cylindrical pellets were approximately 15 mm diameter and 2 mm thick, with mass per unit area of 0.45 g cm^{-2}. The pellets were visually inspected, evaluating their homogeneity aspect and integrity. Furthermore, an XRF spectra (obtained with pXRF, as described in Section 2.5) of a pellet and a loose soil sample were also compared in order to assess possible contamination during the milling process. Further experiments were performed with 10% w w^{-1} cellulose binder (Sigma-Aldrich, Merck, Darmstadt, Germany) and 20 min of grinding in a planetary ball mill.

Sample presentation in the form of loose powder was also considered for the analysis. The air-dried samples were sieved in a sieve with apertures of 2 mm. Ten grams of test sample was transferred to an XRF polyethylene cup of 31 mm (n. 1530, Chemplex Industries Inc., Palm City, FL, USA) assembled with a 4-μm thick polypropylene film (n. 3520, SPEX, USA).

2.3. Soil Laboratory Analysis

Soil testing conducted by a commercial laboratory determined the exchangeable (ex-) contents of P, K, Ca and Mg via ion exchange resin extraction. Clay content was quantified by the Bouyoucos hydrometer method in dispersing solution. The pseudo total content (ptc) of P, K, Ca and Mg were also analyzed following the USEPA Method 3051A [17]. The latter methods involve the extraction of ions using HNO_3 and HCl. The multielement quantification was made by inductively-coupled plasma optical emission spectrometry (ICP OES). The term ptc is used, because it is not a total digestion method. Despite this, this method presents proportional recoveries to the most aggressive methods for the determination of elements in tropical soils [18], allowing one to understand the relationship between exchangeable and total content of the elements evaluated. In addition, it is a method that requires less time for digestion, less consumption of acids and lower risks of environmental contamination [19].

2.4. μ-XRF Chemical Images

μ-XRF is a type of energy dispersive X-ray fluorescence that employs a micrometric beam with a shape and size defined by a primary optic element; this can be done by a simple collimator, an optical capillary or a focusing mirror [20,21]. In this work, the μ-XRF technique was employed to characterize, on the sample surface, the influence of the sample preparation method on the spatial distribution of elements of interest.

The net intensities for K and Ca Kα emission lines were characterized with high spatial resolution on the surface of loose soil and pellet samples. A benchtop μ-XRF system (Orbis PC EDAX, United States) furnished with a Rh anode X-ray tube was used. The detection was carried out by a 30 mm^2 silicon drift detector (SDD). The μ-XRF tube current and voltage was operated at 15 kV and 200 μA, respectively; the beam size used was 30 μm and no primary filter was used; the live time was set to 2 s per spot; and the analysis was carried out under vacuum. In each sample, 800 points (matrix of 32 $\times$ 25 points) were evaluated in an area of about 2.32 mm^2 (1.60 $\times$ 1.45 mm).

Chemical images showing the variability of K and Ca, produced by Orbis Vision software, were linearly interpolated using Origin Lab 2016. The mean, maximum and minimum values, as well as the coefficient of variation (CV)—the ratio between the standard deviation and the mean expressed in percentage—were also calculated. P and Mg Kα emission lines were not identified in the samples, which did not allow the evaluation of these element lines. Similar μ-XRF analysis procedures are described by Rodrigues et al. [21].

2.5. pXRF Measurements and Its Performance Evaluation

The measurements were carried out using a portable X-ray fluorescence spectrometer (portable ED-XRF), Tracer III–SD model (Bruker AXS, Madison, USA), equipped with a 4 W Rh X-ray tube and 12 mm^2 of active area, and a X-Flash®Peltier-cooled SDD, with 2048 channels (Bruker AXS, Madison, USA). The tube operated at 23 μA and 15 kV, and emission intensities were measured for 90 s without vacuum. The voltage configuration was chosen based on the interest in low atomic number elements and the current, in order to keep deadtime below 15%, and avoiding spectral distortions and artifacts. Soil samples were measured in triplicate at different portions of its surface. To ensure the same attenuation conditions of the loose soil samples, the pellets were placed on a 4-μm-thick polypropylene thin-film.

All data were acquired using the software Bruker S1PXRF®(Bruker AXS, Madison, USA). The data were obtained through the deconvolution process using the Artax®(Bruker AXS, Madison, USA). The Kα emission characteristic lines of the elements P, K, Ca and Mg were evaluated. However, only K and Ca presented detectable emission lines, which were evaluated by their signal-to-noise ratio (SNR) and intensity, through the counts of photons per second (cps). The SNR was determined by dividing the characteristic X-ray net intensities by the background square root [22], and the cps were obtained by the ratio of total X-ray intensity to detector live time. The standard deviation (SD) behavior of the intensity of the K and Ca Kα emission lines within the replicates was also evaluated for both sample preparations.

The intensity of K and Ca Kα emission lines, obtained from pellets and loose soil samples, were compared with the exchangeable contents of these nutrients. Calibration models were built using simple linear regressions (a univariate model), with the independent variable being pXRF data and the dependent variable being the soil property measured via commercial laboratory procedures. The prediction models were validated using "leave-one-out" full cross-validation. The quality of the developed calibration was assessed with the coefficient of determination (R^2), the root mean square error (RMSE) and the ratio of performance to interquartile range (RPIQ), as recommended by Bellon-Maurel et al. [23]. Arbitrary groups were used for simplification of interpretation, as proposed by Nawar and Mouazen [24]: (1) excellent models (RPIQ > 2.5), (2) very good models (2.5 > RPIQ > 2.0), (3) good model (2.0 > RPIQ > 1.7), fair (1.7 > RPIQ > 1.4), and very poor model (RPIQ < 1.4). The descriptive statistics of soil fertility and the correlation between pseudo total and exchangeable contents were also determined.

3. Results

3.1. Soil Pelletizing Procedure

The pelletizing of sandy soil samples (e.g., about 175 g dm^{-3} of clay content) was only possible with the addition of binder. In the initial tests, which evaluated different grinding times, they did not form pellets without the addition of binder (Figure 2A). In general, pellets produced after 10 and 15 min of grinding were brittle, except for pellets containing 15% w w^{-1} of binder (Figure 2B), which, in turn, were less homogeneous with white spots on their surface. The best cohesion between particles was obtained with 20 min of grinding. For this milling time, the pellet with 15% w w^{-1} of binder was slightly less heterogeneous than the pellet with 10% w w^{-1}. In relation to the brand, microcrystalline cellulose powder (Sigma-Aldrich, Merck, Darmstadt, Germany) presented more cohesive pellets. The best results were observed for pressed pellets prepared from soil mixed with cellulose binder at 10% w w^{-1} and ground for 20 min.

Tungsten (W) contamination was observed in these ground soil samples. This contamination was caused by the tungsten carbide ball mill and it was evidenced by the W L-emission lines presented in the pellet spectrum, which were not observed in the loose soil spectrum (Figure 3). In the XRF spectra, W presents L and M-emission lines with energy lower than 15 keV, as highlighted in the red lines in the spectrum of Figure 3. In this range, the main W emission lines present energy of 1.77 (Mα), 1.83 (Mβ), 8.39 (Lα), 9.67 (Lβ_1) and 9.95 keV (Lβ_2). The effect of this contamination is best seen on L-emission lines, as they do not overlap with any other emission lines present. The W M-α line overlaps with Si (1.74 keV) and can also promote the enhancement of the Al Kα line (Al K edge = 1.55 keV) due to secondary radiation excitation (chemical matrix effect). Thus, this W interference must be considered and corrected in the case of Al quantification. For the Ca and K determinations performed in this work, contamination with W was not a limiting factor.

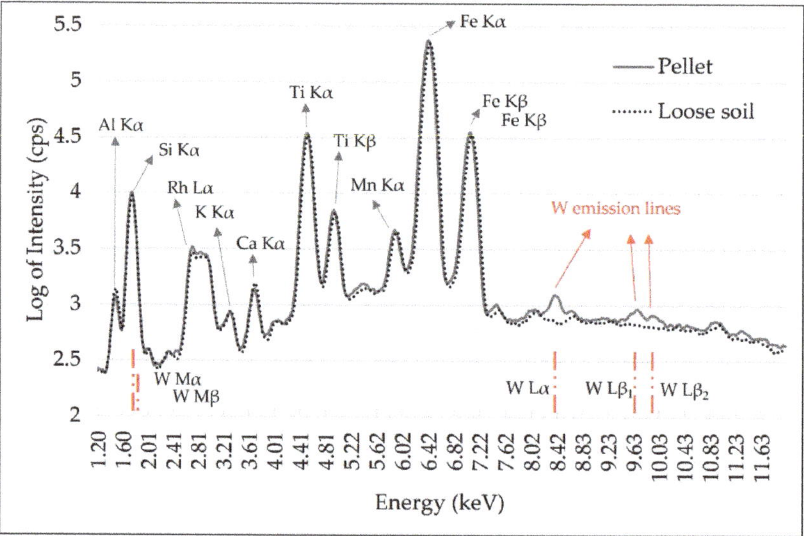

Figure 3. X-ray fluorescence spectra obtained with pXRF equipment using the pellet and loose soil sample. Tungsten (W) emission lines were identified with red lines. The emission spectra intensity is shown in the logarithm to reduce the differences in scales between the emission lines, allowing a better qualitative assessment.

3.2. µ-XRF Chemical Images and Sample Homogeneity

The spatial distribution patterns of Ca and K at the pellet and loose soil surface are shown in Figure 4. For both elements, the preparation of pellets resulted in more homogeneous surfaces than those observed for loose soil. The particle size reduction—required for pellet production—allows one to homogenize the distribution of the different elements in the sample, which occurs because it fragments the regions where these elements are agglomerated (nuggets). In the loose soil sample, the presence of nuggets can be observed for the Ca and K (Figure 4B,C, respectively), which do not appear in the pelletized samples. The homogenization promoted by pelletizing drastically reduced the CV of both Kα emission line intensities, oscillating from 100.17% to 13.03% for Ca and from 46.09% to 18.01% for K, respectively.

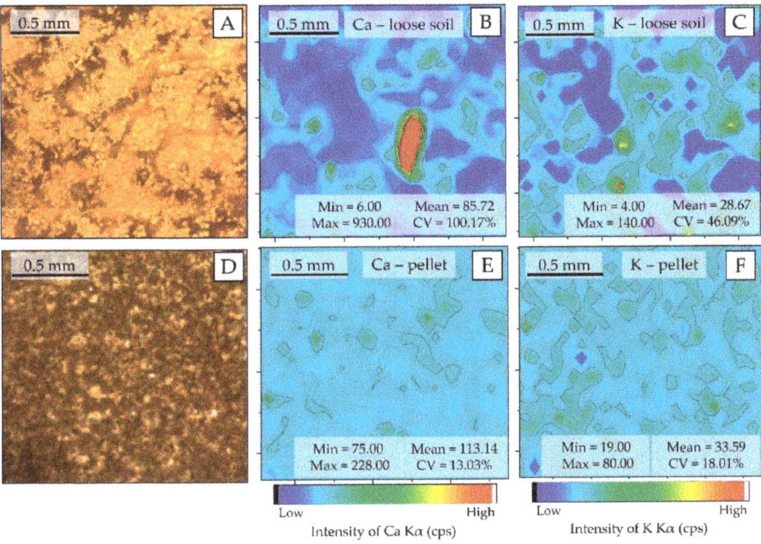

Figure 4. µ-XRF chemical images showing the spatial distribution of Ca and K in the loose soil (**B** and **C**, respectively); Ca and K in the pellet (**E** and **F**, respectively). Photo of the analyzed area of (**A**); the loose soil (**A**) and pellet (**D**).

3.3. Soil Exchangeable Nutrient Prediction Using a pXRF Spectrometer

Soil samples were characterized by clayey texture, low variability of clay content (between 345 and 511 g dm^{-3}) and high variability of exchangeable nutrients. According to the local fertility interpretation [25], the level of ex-P content oscillates between low to medium; and the level of ex-K, ex-Ca and ex-Mg content varies between medium to very high. These samples are also characterized by a significant correlation between available and pseudo total contents for all nutrients. A descriptive summary of these analyses is presented in Table 1.

The qualitative evaluation of the XRF spectra (Figure 3) allowed us to identify the K and Ca emission lines, but no fluorescence emission was detected for Mg and P. The XRF intensity and the SNR of K and Ca were slightly higher for loose soil than pellet samples, and both had a highly significant correlation (r > 0.9) between pellet and loose soil, indicating that the changes promoted by sample preparation were well standardized for all samples (Figure 5). Despite the small gain in fluorescence intensity (an average of 4.21 and 15.31 cps for K and Ca, respectively) and in SNR (an average of 0.63 and 2.49 for K and Ca, respectively), when evaluating the behavior of the emission line intensity SD for the replicates, we can observe that the loose soil samples presented greater variation in relation to the pellets, both for K and Ca (Figure 5E,F, respectively). The replicate SD is an indicator of the reading precision. In this work, the lower precision of the loose soil samples might be related to their lower homogeneity in relation to the pelletized ones. Despite this loss of precision among the different replicates obtained in loose soil samples, the triplicate scans smoothed this effect. After averaging the replicates, the distribution of the XRF intensity showed a similar distribution between both sample preparations (Figure 5A,B).

Table 1. Descriptive statistics of exchangeable and pseudo total nutrients and the correlation between the respective nutrients.

	Exchangeable Nutrients				Pseudo Total Contents			
	ex-P	ex-K	ex-Ca	ex-Mg	ptc P	ptc K	ptc Ca	ptc Mg
	mg dm^{-3}		mmol$_c$ dm^{-3}		mg kg^{-1}			
Min	7.00	1.70	27.00	11.00	405.31	154.04	492.42	411.59
Mean	20.20	5.14	49.12	26.28	489.93	318.10	750.51	607.38
Max	46.00	10.30	78.00	54.00	669.18	477.18	1225.91	789.65
SD	8.42	1.78	12.89	10.66	55.74	79.00	159.46	112.31
CV (%)	41.69	34.60	26.25	40.57	11.38	24.84	21.25	18.49
Correlation with pseudo total	0.79 *	0.67 *	0.83 *	0.52 *				

* Significant correlation at the probability level of 0.01.

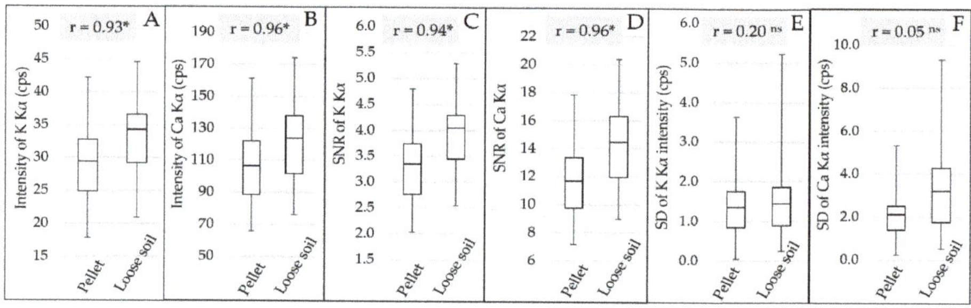

Figure 5. Box plot of the Kα emission line intensity of K and Ca (**A** and **B**, respectively), after averaging the replicates. Box plot of the signal-to-noise ratio (SNR) of the Kα emission lines of K and Ca (**C** and **D**, respectively), after averaging the replicates. Box plot of the standard deviation (SD) of the Kα emission line intensity of K and Ca (**E** and **F**, respectively) for the replicates. The Pearson correlation between the pellet and loose soil is also presented (correlations followed by * were significant at the probability level of 0.01; correlations followed by ^{ns} were not significant).

Regarding the regression analysis, there was a slight reduction of precision in the calibration of ex-K and ex-Ca in loose powder soil samples, marked by a slight increase in error and reduction in R^2 (Figure 6). Comparing the loose soil samples in relation to the pellet samples, the prediction error of ex-K increased from 0.65 to 0.78 mmol$_c$ dm^{-3}. Similarly, for the ex-Ca prediction, the error increased from 5.89 to 6.12 mmol$_c$ dm^{-3}. Moreover, concerning the R^2 values, it oscillated from 0.87 to 0.81 for ex-K, and from 0.78 to 0.76 for ex-Ca. Besides that, all prediction models, obtained from both pellet and loose soil, showed excellent performance in their validation with RPIQ values above 2.5.

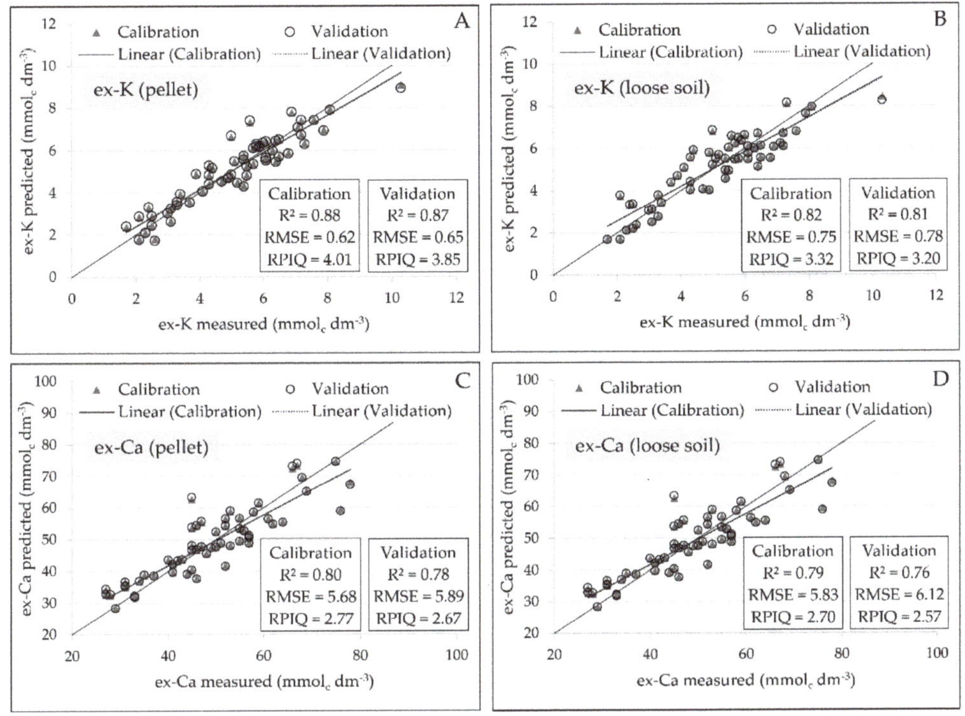

Figure 6. Scatter plots of measured versus predicted ex-K, for the pellets and loose soil (**A** and **B**, respectively), and of measured versus predicted ex-Ca, for the pellets and loose soil (**C** and **D**, respectively). Models were obtained using simple linear regressions with the Kα emission lines of each element (n = 58) and the validation was performed by "leave-one-out" full cross-validation.

4. Discussion

Although the XRF technique measures the total content of elements present in the soil, these sensors have been suggested as an auxiliary technique for evaluating fertility attributes [8–10]. In addition, interest in such equipment has recently increased due to its portability, enabling on-field studies [7] and in controlled environments such as hybrid laboratories [2,11]. Such applications are only compatible with minimal or no sample preparation.

Even if the XRF technique is flexible concerning sample preparation, there is a consensus that pellet preparation increases the data precision [15]. Pellet preparation consists of conforming and binding the samples into a specific shape. For pellet formation, the soil samples must be grinded to an extremely fine powder by using a grinder; furthermore, a proper binding agent can also be necessary [26]. In this work, a procedure for soil pellet preparation was evaluated, with and without binding agents. Sandy soils are more likely to not form pellets without the use of a binder [12]. In our work, it was perceived for samples with clay content around 175 g dm^{-3}. During testing for optimizing binder concentration and ball milling time, it was observed that increasing the grinding time, as well as the binder concentration, improves the cohesion between the particles of the pellet. In contrast, higher concentrations of binder (e.g., 15% w w^{-1}, in this work) make it difficult to homogenize it. It is known that the smaller the particle size to be pressurized is, the more resistant and cohesive the pellet will be [27]. However, pellet preparation in different soil sample sets can be optimized with different binder concentrations and milling time. To optimize these conditions, we suggest conducting preliminary tests, as described in this paper.

The reduction of the particle size, before pressing the material, promotes sample homogenization [26]. Both the element spatial patterns and the SD behavior on the replicates, clearly showed gain on homogeneity after pelletizing. The lower homogeneity of loose soil samples should be considered for determining the number of replicates during data acquisition with a pXRF device. Due to the high variability of element distribution in loose soil samples, a greater number of scans has to be acquired for more accurate representation of the sample surface, and then, different spectra can be averaged [12]. In this work, pXRF spectra were obtained in triplicate (at different positions) and were sufficient for a good characterization of both sample preparation samples.

One point to consider while milling samples is the possibility of contamination. Two types of contamination can occur, cross-contamination, due to inefficient cleaning when changing samples, and/or contamination with milling surfaces (e.g., agate and tungsten carbide), due to the abrasion between the sample and mill components. In addition to the careful execution of cleaning procedures, the milling surfaces should be considered in terms of hardness and elemental composition to avoid the risk of sample contamination [12]. In this work, the grinding in a ball mill made of tungsten carbide contaminated the samples with W. Soil samples usually have hard minerals, like silicates, and W should not be measured when tungsten carbide milling jars and balls are used for milling soil [28]. Specifically, in XRF analysis, W contamination can also be a problem for the determination of Si and Al, as described in the previous section. Iwansson and Landström [29] showed that this kind of contamination is higher in quartz-rich samples and increases with grinding time. When the element contamination (e.g., W) is the same one that is to be quantified, the values must be adjusted, discarded, or over-looked to avoid misinterpretation of the results [29].

Pelletizing also reduces surface roughness effects and increases the density of the material [26]. Theoretically, reducing sample roughness means decreasing the physical matrix effect, which would attenuate part of the fluorescence produced by the analytes, and increasing the density of samples and fluorescence intensity [14]. Thus, a higher fluorescence yield and an upsurge in the SNR were expected for the emission lines of the pelletized samples, due to the increase of their density and reduction of the physical matrix effect [13]. Nevertheless, this behavior was not observed in this work. In turn, the addition of binder (10% w w^{-1}), as well as the contamination by W, and the differences in homogeneity found (Figure 4B,E), appear to be the factors that influenced this behavior, slightly altering the chemical composition of the pellet samples and, consequently, their fluorescence production.

Soil samples are naturally heterogeneous and therefore comminution procedures are generally recommended for improving matrix homogenization, and should yield homogeneous pellets [26]. Ultimately, this can avoid heterogeneity effects, such as grain size effect, mineralogical effect, and segregation, factors that cause errors in the XRF analysis [30]. However, in this work, the prediction models of ex-Ca and ex-K using pellets showed just a small performance gain over those obtained in loose powder soil samples. Using loose powder soil, prediction models for ex-Ca and ex-K calibrated with the 58 local samples obtained excellent performances, with RPIQ values over 2.5. These results are promising and encourage more detailed investigations on the application of the XRF technique. They even lead us to new questions such as: (i) how long will this calibration remain robust over different cropping seasons? (ii) is it possible to further reduce sample preparation without losing analytical quality? (iii) what would be the analytical performance in samples with field conditions (e.g., with different humidity and particle size patterns)? (iv) how to determine fields where the XRF sensor will have potential as an auxiliary tool alongside traditional fertility analysis methods?

Different works on temperate soils have shown good performance in predicting fertility attributes such as pH [9], cation exchange capacity (CEC) [10], base saturation (V%) [31], soil texture [32] and total content of different elements [8,11]. In Brazilian tropical soils, satisfactory performances have already been obtained for predictions of organic carbon and organic matter [33] and textural attributes [34]. Although, so far, the prediction of available nutrients has not been explored. The possibility of XRF application on soil samples that have just been dried and sieved (<2 mm), with satisfactory predictions of ex-K and ex-Ca using local models, is a promising alternative to increasing the efficiency of analytical procedures. This may intensify the amount of analysis in tropical soil samples without the need for wet chemistry methods. In addition, this level of sample preparation is compatible with evaluations using vis-NIR diffuse reflectance, opening the potential to exploit joint XRF and vis-NIR sensors on these types of samples. Moreover, the vis-NIR technique has great potential for obtaining information about texture, and organic and mineralogical components, which can synergistically complement the XRF information for a more complete characterization of the attributes of soil fertility [11].

A simple calibration method was applied in this work, using only the emission line of the elements of interest for predictive modeling. XRF spectra are multi-informational, allowing the measurement of several soil properties from a single scan. This is possible because each spectrum stores a large amount of information along with its emission lines and different types of scattering (e.g., Compton and Thomson scattering), not strictly related to the elementary constituents of the samples. An example is the prediction of organic carbon and organic matter in soil samples using the information contained in the scattering region of the spectra, as explored by Morona et al. [33]. In this sense, predictive modeling based on multivariate statistics and machine learning methods are an alternative to better exploit the hidden information present in XRF spectra [35], and it can enable robust determinations of fertility attributes that have an indirect relationship with inorganic soil constituents such as pH, CEC, V% and texture.

The evaluation of Mg and P using the XRF technique is challenging as they are light elements that produce fluorescence emission at low energy levels (between 1.2 and 2.2 keV) that are absorbed by atmospheric gases (N_2, O_2, and Ar) before reaching the detector. The maximum pseudo total content (Table 1) of P (670 mg kg^{-1}) and Mg (790 mg kg^{-1}) in our samples was not enough to produce X-ray fluorescence intensities detectable by the equipment. Therefore, even if there is a correlation between their pseudo total and available contents, direct calibrations for ex-P and ex-Mg (made with their own emission lines) have not been possible using the XRF spectra so far. However, this does not preclude an attempt of indirect calibrations, using other information present in the spectrum. Furthermore, the determination of these elements can be improved by using a vacuum system and changing the X-ray tube conditions to lower voltage (<10 keV) and increasing the current (10 to 15% of deadtime), which reduces air attenuation over Mg and P emission lines, and increases the fluorescence yield of these chemical elements. Some portable XRF equipment already allows the use of this condition and future work should be done to evaluate the use of a vacuum to improve the detection of light elements in soil samples to predict fertility attributes.

In addition, this study was conducted under a clayey lixisol, which is a representative and common type of soil in Brazilian tropical areas [36]. Therefore, this pioneering evaluation provided useful information to help XRF users—who aim to use this technique as a tool for practical soil analysis—to understand that the expected effects of sample preparation related to heterogeneity and physical matrix effects can be neglected. However, it is fundamental to also validate other types of soils with contrasting textural classes, which can proportionate distinct levels of physical matrix effects.

5. Conclusions

The addition of a binder was decisive for improving the quality of the pellets. The best results for soil preparation in the form of pellets were obtained with samples prepared with cellulose binder at 10% w w^{-1} and ground for 20 min.

Pressed pellets allowed a slight gain in performance over loose powder samples for the prediction of ex-K and ex-Ca. In spite of that, predictions in loose powder soil for ex-Ca and ex-K, calibrated with 58 local samples, obtained excellent performances in their validation, showing that it is possible to reduce the optimal sample preparation of XRF analyses for predicting soil nutrients. However, loose samples are less homogenous than pellets, and scanning loose soil samples in replicates is important for smoothing this effect.

The prediction models of ex-K and ex-Ca calibrated with local samples presented promising results. More detailed investigations are necessary to foster the application of the XRF technique in agricultural soil samples for determination of soil fertility attributes. Finally, XRF can serve as a complementary method to traditional laboratory analyses.

Author Contributions: Conceptualization, T.R.T., L.C.N. and J.P.M.; methodology, T.R.T., L.C.N., E.E.N.A., E.d.A., and H.W.P.d.C.; validation, L.C.N., E.E.N.A. and J.P.M.; formal analysis, T.R.T., L.C.N. and E.E.N.A.; investigation, T.R.T.; resources, T.R.T.; data curation, T.R.T. and L.F.M.; writing—original draft preparation, T.R.T.; writing—review and editing, L.C.N., E.E.N.A., E.d.A., H.W.P.d.C. and J.P.M.; visualization, T.R.T. and L.F.M.; supervision, H.W.P.d.C., F.J.K. and J.P.M.; project administration, T.R.T. and J.P.M.; funding acquisition, T.R.T., J.P.M. and H.W.P.d.C.

Funding: The T.R.T. and E.E.N.A. were funded by São Paulo Research Foundation (FAPESP), grant number 2017/21969-0 and 2018/08877-2, respectively; and also partial funded by the Brazilian Federal Agencies: Coordination for the Improvement of Higher Education Personnel (CAPES) – Finance Code 001, and the National Council for Scientific and Technological Development. (CNPq). XRF facilities were funded by FAPESP, grant 2015-19121-8, and "Financiadora de Estudos e Projetos" (FINEP) project "Core Facility de suportes às pesquisas em Nutrologia e Segurança Alimentar na USP", grant 01.12.0535.0.

Acknowledgments: We would like to thank the technicians Fátima Patreze and Liz Mary Bueno de Moraes, from Analytical Chemistry Laboratory, for the support with the sample preparation; and, also the technician Marina Colzato, from the Laboratory of Environmental Analysis, for the support with soil analysis.

Conflicts of Interest: The authors declare no conflict of interest. The funders had no role in the design of the study; in the collection, analyses, or interpretation of data; in the writing of the manuscript, or in the decision to publish the results.

References

1. FAO. *World Fertilizer Trends and Outlook to 2020*; Food and Agriculture Organization of the United Nations (FAO): Rome, Italy, 2017.

2. Demattê, J.A.M.; Dotto, A.C.; Bedin, L.G.; Sayão, V.M.; Souza, A.B. Soil analytical quality control by traditional and spectroscopy techniques: Constructing the future of a hybrid laboratory for low environmental impact. *Geoderma* **2019**, *337*, 111–121. [CrossRef]

3. Viscarra Rossel, R.A.; Bouma, J. Soil sensing: A new paradigm for agriculture. *Agric. Syst.* **2016**, *148*, 71–74. [CrossRef]

4. Molin, J.P.; Tavares, T.R. Sensor systems for mapping soil fertility attributes: Challenges, advances and perspectives in Brazilian tropical soils. *Eng. Agric.* **2019**, *39*, 126–147. [CrossRef]

5. Pandey, S.; Bhatta, N.P.; Paudel, P.; Pariyar, R.; Maskey, K.H.; Khadka, J.; Panday, D. Improving fertilizer recommendations for Nepalese farmers with the help of soil-testing mobile van. *J. Crop. Improv.* **2018**, *32*, 19–32. [CrossRef]

6. Gredilla, A.; de Vallejuelo, S.F.O.; Elejoste, N.; de Diego, A.; Madariaga, J.M. Non-destructive Spectroscopy combined with chemometrics as a tool for Green Chemical Analysis of environmental samples: A review. *TrAC Trend Anal. Chem.* **2016**, *76*, 30–39. [CrossRef]

7. Weindorf, D.C.; Chakraborty, S. Portable X-ray Fluorescence Spectrometry Analysis of Soils. In *Methods of Soil Analysis*; Hirmas, D., Madison, W.I., Eds.; Soil Science Society of America: Wisconsin, WI, USA, 2016; pp. 1–8. [CrossRef]

8. Nawar, S.; Delbecque, N.; Declercq, Y.; Smedt, P.; Finke, P.; Verdoodt, A.; Meirvenne, M.V.; Mouazen, A.M. Can spectral analyses improve measurement of key soil fertility parameters with X-ray fluorescence spectrometry? *Geoderma* **2019**, *350*, 29–39. [CrossRef]

9. Sharma, A.; Weindorf, D.C.; Man, T.; Aldabaa, A.A.A.; Chakraborty, S. Characterizing soils via portable X-ray fluorescence spectrometer: 3, Soil reaction (pH). *Geoderma* **2014**, *232*, 141–147. [CrossRef]

10. Sharma, A.; Weindorf, D.C.; Wang, D.; Chakraborty, S. Characterizing soils via portable X-ray fluorescence spectrometer: 4. Cation exchange capacity (CEC). *Geoderma* **2015**, *239*, 130–134. [CrossRef]

11. O'Rourke, S.M.; Stockmann, U.; Holden, N.M.; McBratney, A.B.; Minasny, B. An assessment of model averaging to improve predictive power of portable vis-NIR and XRF for the determination of agronomic soil properties. *Geoderma* **2016**, *279*, 31–44. [CrossRef]

12. Jantzi, S.C.; Motto-Ros, V.; Trichard, F.; Markushin, Y.; Melikechi, N.; Giacomo, A. Sample treatment and preparation for laser-induced breakdown spectroscopy. *Spectrochim. Acta Part B* **2016**, *115*, 52–63. [CrossRef]

13. Takahashi, G. Sample preparation for X-ray fluorescence analysis III. Pellets and loose powder methods. *Rigaku J.* **2015**, *31*, 26–30.

14. Shibata, Y.; Suyama, J.; Kitano, M.; Nakamura, T. X-ray fluorescence analysis of Cr, As, Se, Cd, Hg, and Pb in soil using pressed powder pellet and loose powder methods. *X-Ray Spectrom.* **2009**, *38*, 410–416. [CrossRef]

15. Krug, F.J.; Rocha, F.R.P. *Métodos de Preparo de Amostras Para Análise Elementar*; EditSBQ: São Paulo, Brazil, 2016; 572p. (In Portuguese)

16. IUSS Working Group WRB. World reference base for soil resources 2014. In *World Soil Resources Reports No. 106*; Schad, P., van Huyssteen, C., Micheli, E., Eds.; FAO: Rome, Italy, 2014; 189p, ISBN 978-92-5-108369-7.

17. Element, C.A.S. Method 3051A microwave assisted acid digestion of sediments, sludges, soils, and oils. *Z. Für Anal. Chem.* **2007**, *111*, 362–366.

18. Nogueirol, R.C.; De Melo, W.J.; Bertoncini, E.I.; Alleoni, L.R.F. Concentrations of Cu, Fe, Mn, and Zn in tropical soils amended with sewage sludge and composted sewage sludge. *Environ. Monit. Assess.* **2013**, *185*, 2929–2938. [CrossRef] [PubMed]

19. Silva, Y.J.A.B.; Nascimento, C.W.A.; Biondi, C.M. Comparison of USEPA digestion methods to heavy metals in soil samples. *Environ. Monit. Assess.* **2014**, *186*, 47–53. [CrossRef] [PubMed]

20. Tian, S.; Lu, L.; Xie, R.; Zhang, M.; Jernstedt, J.A.; Hou, D.; Ramsier, C.; Brown, P.H. Supplemental macronutrients and microbial fermentation products improve the uptake and transport of foliar applied zinc in sunflower (*Helianthus Annuus* L.) plants. Studies utilizing micro X-ray florescence. *Front. Plant. Sci.* **2015**, *5*, 808. [CrossRef]

21. Rodrigues, E.S.; Gomes, M.H.F.; Duran, N.M.; Cassanji, J.G.B.; da Cruz, T.N.M.; Sant'Anna Neto, A.; Savassa, S.M.; de Almeida, E.; Carvalho, H.W.P. Laboratory Microprobe X-Ray Fluorescence in Plant Science: Emerging Applications and Case Studies. *Front. Plant. Sci.* **2018**, *9*, 1588. [CrossRef]

22. Ernst, T.; Berman, T.; Buscaglia, J.; Eckert-Lumsdon, T.; Hanlon, C.; Olsson, K.; Palenik, C.; Ryland, S.; Trejos, T.; Valadez, M.; et al. Signal-to-noise ratios in forensic glass analysis by micro X-ray fluorescence spectrometry. *X-Ray Spectrom.* **2014**, *43*, 13–21. [CrossRef]

23. Bellon-Maurel, V.; Fernandez-Ahumada, E.; Palagos, B.; Roger, J.M.; McBratney, A. Critical review of chemometric indicators commonly used for assessing the quality of the prediction of soil attributes by NIR spectroscopy. *TrAC Trend. Anal. Chem.* **2010**, *29*, 1073–1081. [CrossRef]

24. Nawar, S.; Mouazen, A.M. Predictive performance of mobile vis-near infrared spectroscopy for key soil properties at different geographical scales by using spiking and data mining techniques. *Catena* **2017**, *151*, 118–129. [CrossRef]

25. Van Raij, B.; Andrade, J.C.; Cantarela, H.; Quaggio, J.A. *Análise química para avaliação de solos tropicais*; IAC: Campinas, Brazil, 2001; 285p. (In Portuguese)

26. Gondal, M.A.; Hussain, T.; Yamani, Z.H.; Baig, M.A. The role of various binding materials for trace elemental analysis of powder samples using laser-induced breakdown spectroscopy. *Talanta* **2007**, *72*, 642–649. [CrossRef] [PubMed]

27. Carvalho, G.G.A.; Santos, D., Jr.; Gomes, M.S.; Nunes, L.C.; Guerra, M.B.B.; Krug, F.J. Influence of particle size distribution on the analysis of pellets of plant materials by laser-induced breakdown spectroscopy. *Spectrochim. Acta Part B* **2015**, *105*, 130–135. [CrossRef]

28. Jantzi, S.C.; Almirall, J.R. Elemental analysis of soils using laser ablation inductively coupled plasma mass spectrometry (LA-ICP-MS) and laser-induced breakdown spectroscopy (LIBS) with multivariate discrimination: Tape mounting as an alternative to pellets for small forensic transfer specimens. *Appl. Spectrosc.* **2014**, *68*, 963–974. [CrossRef] [PubMed]

29. Iwansson, K.; Landström, O. Contamination of rock samples by laboratory grinding mills. *J. Radioanal. Nucl. Chem.* **2000**, *244*, 609. [CrossRef]

30. Morikawa, A. Sample preparation for X-ray fluorescence analysis II. Pulverizing methods of powder samples. *Rigaku J.* **2014**, *30*, 23–27.

31. Rawal, A.; Chakraborty, S.; Li, B.; Lewis, K.; Godoy, M.; Paulette, L.; Weindorf, D.C. Determination of base saturation percentage in agricultural soils via portable X-ray fluorescence spectrometer. *Geoderma* **2019**, *338*, 375–382. [CrossRef]

32. Zhu, Y.; Weindorf, D.C.; Zhang, W. Characterizing soils using a portable X-ray fluorescence spectrometer: 1. Soil texture. *Geoderma* **2011**, *167*, 167–177. [CrossRef]

33. Morona, F.; dos Santos, F.R.; Brinatti, A.M.; Melquiades, F.L. Quick analysis of organic matter in soil by energy-dispersive X-ray fluorescence and multivariate analysis. *Appl. Radiat. Isotopes* **2017**, *130*, 13–20. [CrossRef]

34. Silva, S.; Poggere, G.; Menezes, M.; Carvalho, G.; Guilherme, L.; Curi, N. Proximal sensing and digital terrain models applied to digital soil mapping and modeling of Brazilian Latosols (Oxisols). *Remote Sens.* **2016**, *8*, 614. [CrossRef]
35. Panchuk, V.; Yaroshenko, I.; Legin, A.; Semenov, V.; Kirsanov, D. Application of chemometric methods to XRF-data—A tutorial review. *Anal. Chim. Acta* **2018**, *1040*, 19–32. [CrossRef]
36. EMBRAPA Solos. *Brazilian Soil Classification System*, 5th ed.; EMBRAPA: Brasília, Brazil, 2018.

Article

Comparison of Calibration Approaches in Laser-Induced Breakdown Spectroscopy for Proximal Soil Sensing in Precision Agriculture

Daniel Riebe [1], Alexander Erler [1], Pia Brinkmann [1], Toralf Beitz [1], Hans-Gerd Löhmannsröben [1,*] and Robin Gebbers [2]

[1] Physical Chemistry, University of Potsdam, Karl-Liebknecht-Str. 24-25, 14476 Potsdam, Germany; riebe@uni-potsdam.de (D.R.); aerler@uni-potsdam.de (A.E.); pbrinkma@uni-potsdam.de (P.B.); beitz@uni-potsdam.de (T.B.)

[2] Leibniz Institute for Agricultural Engineering and Bioeconomy (ATB), Max-Eyth-Allee 100, 14469 Potsdam, Germany; rgebbers@atb-potsdam.de

* Correspondence: loeh@chem.uni-potsdam.de

Received: 28 October 2019; Accepted: 25 November 2019; Published: 28 November 2019

Abstract: The lack of soil data, which are relevant, reliable, affordable, immediately available, and sufficiently detailed, is still a significant challenge in precision agriculture. A promising technology for the spatial assessment of the distribution of chemical elements within fields, without sample preparation is laser-induced breakdown spectroscopy (LIBS). Its advantages are contrasted by a strong matrix dependence of the LIBS signal which necessitates careful data evaluation. In this work, different calibration approaches for soil LIBS data are presented. The data were obtained from 139 soil samples collected on two neighboring agricultural fields in a quaternary landscape of northeast Germany with very variable soils. Reference analysis was carried out by inductively coupled plasma optical emission spectroscopy after wet digestion. The major nutrients Ca and Mg and the minor nutrient Fe were investigated. Three calibration strategies were compared. The first method was based on univariate calibration by standard addition using just one soil sample and applying the derived calibration model to the LIBS data of both fields. The second univariate model derived the calibration from the reference analytics of all samples from one field. The prediction is validated by LIBS data of the second field. The third method is a multivariate calibration approach based on partial least squares regression (PLSR). The LIBS spectra of the first field are used for training. Validation was carried out by 20-fold cross-validation using the LIBS data of the first field and independently on the second field data. The second univariate method yielded better calibration and prediction results compared to the first method, since matrix effects were better accounted for. PLSR did not strongly improve the prediction in comparison to the second univariate method.

Keywords: laser-induced breakdown spectroscopy; LIBS; proximal soil sensing; soil nutrients; elemental composition

---

1. Introduction

Precision agriculture (PA) requires reliable, affordable, immediately available soil data with sufficient spatial and temporal resolution [1]. Soil maps for PA are typically derived from soil sampling with subsequent laboratory analysis or from mapping with automated mobile proximal soil sensors. Soil sampling and laboratory analysis is time consuming and becomes prohibitively expensive if conducted on a fine grid [2]. As an alternative, mobile proximal soil sensors can measure several hundred points (different locations) per hectare [3–7]. However, most of the current soil sensors neither cover the whole range of soil fertility parameters nor do they directly measure them. The automated mobile proximal soil sensing systems frequently used in practical PA as well as research include geoelectrical, potentiometric pH, gamma-ray and spectral-optical sensors. Among these, pH electrodes most directly access a soil fertility parameter, namely active acidity [8]. Gamma-ray sensors detect radiation from K decay and from other isotopes. This can be correlated with K in clay minerals and with plant available K^+ [9,10]. Spectral-optical sensors include spectrometers and multi-wavelength sensors in the visible and near-infrared region. Correlations with several soil fertility parameters were observed, particularly with organic matter [11]. In research, visible and near-infrared spectrometers were used to map several fertility parameters at the same time [12,13]. For practical applications, cheaper and more robust dual or multi-wavelength sensors for organic matter and soil moisture were commercialized (e.g., by Veris Technologies and Precision Planting). However, the relationships between optical soil properties and soil fertility parameters are variable due to large overlaps of absorption bands. Thus, optical online sensors require careful calibration for each field [11,14]. Geoelectric sensors traditionally formed the backbone of PA soil analysis efforts and they are still widely used due to their robust nature making them both dependable and suitable for field applications [15–17]. However, apparent soil electrical conductivity (ECa) is affected by many soil parameters, including water content, texture, salinity, bulk density and temperature [18]. Therefore, reference sampling in each field is required [19]. In a comparative study, Piiki et al. [20] recently addressed the issue of obtaining as much direct information with as little calibration as possible. In that study, X-ray fluorescence (XRF) spectroscopy produced the most reliable predictions of soil parameters due to its direct detection of elemental compositions. XRF has gained interest in recent years due to the availability of handheld sensors [21–23]. The drawbacks of XRF include the harmful X-rays, long measurement times and the restriction to heavier elements.

Laser-induced breakdown spectroscopy (LIBS) is a promising alternative to XRF for determining element mass fractions in soils. In this method, an intense pulse of laser radiation is focused onto the soil, where it ablates material from the surface and creates a microplasma. Subsequently, excited atoms and ions in the plasma emit specific radiation which can be analyzed to elucidate the elemental composition of the sample [24–28]. Ablation and plasma excitation are both highly complex phenomena. Since the interaction of the laser radiation with the sample surface is influenced by its composition and structure, a matrix-dependence of the signal response is observed. Matrix effects result from the light-to-sample coupling, collisional interactions within the plasma, and plasma temperature, among others, all of which influence the ratio of neutral and ionized species and self-absorption. These matrix effects and spectral interferences were already investigated in different types of soil [29,30]. In LIBS soil analyses, the whole spectrum of elements can be accessible. Depending on the calibration effort, elements can be determined qualitatively or quantitatively. This allows the direct analysis of macro and micro nutrients since no or only minimal sample preparation is necessary. In order to achieve the power density required for plasma generation, the laser light pulse, of typically nanosecond duration, is usually focused to a spot of about 10–500 µm diameter. As a consequence, the soil micro-heterogeneity has to be considered in order to obtain representative results. This can be achieved by averaging multiple spectra.

Additional advantages of LIBS include the measurement speed, safety, as well as the portability of the technique. These attributes make the method particularly interesting for on-site soil mapping. While large-scale soil mapping applications of LIBS have not been reported yet and early LIBS investigations of soils were focused on pollutants [31], the detection of nutrients in soils was already demonstrated in some publications. Diaz et al. [32] were able to determine detection limits of P, Fe, Mg, Ca and Na by univariate calibration in fertilizer/soil mixtures. Yongcheng et al. [33] could improve the prediction of the Mg mass fraction by using a multivariate regression model that incorporates the lines of other metals present in the soil. Nicolodelli et al. [34] investigated the feasibility of measuring C in soils with a low resolution spectrometer. Rühlmann et al. [35] compare univariate and multivariate data evaluation approaches for the quantification of the nutrient Ca in reference soils. A review of recent work on the application of LIBS for investigating agricultural materials can be found in [36].

In this work, LIBS spectra of soils were measured in the laboratory as a fundamental study and to provide a basis for the future application of LIBS directly on agricultural fields. The aim of the study was the evaluation of different quantification approaches for LIBS data which consider the matrix effect. Another goal was to explore how a calibration obtained for one field can be transferred to another one. The accuracy of standard-free LIBS approaches [31,37] for the analysis of the very complex matrix soil is still unsatisfactory. Therefore, the focus of this work was to compare different univariate or multivariate calibration methods for the determination of nutrient mass fractions. The target parameters of the investigation were the major nutrients Ca and Mg as well as the minor nutrient Fe. One aim of this work was to examine whether the quantification of nutrients on the size scale of a field can be carried out by univariate calibration or whether a multivariate method, namely partial least squares regression (PLSR) has to be applied. Different calibration strategies, such as the generation of calibration standards by standard addition to a single reference soil sample and the use of multiple reference soil samples, were compared. The heterogeneity of the soil samples was characterized by principal component analysis (PCA). The plasma was generated using UV radiation (355 nm), which was recommended for soil investigations [30] in contrast to the widely applied NIR radiation (1064 nm).

2. Materials and Methods

Soil samples and reference analysis. A total of 139 samples from two agricultural fields near the village of Wilmersdorf in Northeast Germany (53°06′ N, 13°54′ E) were investigated. The regional soilscape was formed by the last glaciation about 10,000 years ago and the following postglacial processes. The parent material of the soil consists of calcareous glacial till covered by sandy deposits. Soil texture varies between sand, loamy sand, and sandy loam in the topsoil. The main soil types are alfisols. The samples were collected in 2011 to obtain reference data for proximal soil sensing as published by Schirrmann, Gebbers and Kramer [14]. Reference analysis for elements was carried out by inductively coupled plasma optical emission spectroscopy (ICP-OES) after aqua regia extraction by a certified laboratory. As target parameters, Ca and Mg were selected as examples for major nutrients and Fe as an example for minor nutrients.

ECa mapping: ECa [mS/m] was mapped with a Veris 3100 system (Veris Technologies, Salina, KS, USA) in 2011. Only data from the shallow measurement were used. Data were interpolated by block kriging (a) on a regular grid for visualization and (b) on the sampling locations for correlation analysis.

LIBS setup and measurement parameters. The plasma was created using a Nd:YAG laser (Quanta-Ray, Spectra-Physics, Santa Clara, CA, USA, $\lambda = 355$ nm, $E = 90$ mJ). Emissions were collected by a concave mirror, coupled into an optical fiber and guided to an echelle spectrometer (Aryelle Butterfly, LTB, Berlin, Germany) equipped with an ICCD camera (iStar, AndorTechnology, Belfast, UK). The spectrometer has two separate wavelength ranges (UV range: 190–330 nm, VIS range: 275–750 nm) and a resolution of 20–30 pm. A total of 200 single shot spectra were recorded per sample in the UV as well as in the VIS range. The sample holder was rotated and linearly translated during measurements forming a spiral-like trace of ablation events. Optimization of LIBS spectra led to the following measurement parameters: a detection delay of 2 µs, a measurement window of 10 µs as well as a constant amplification factor of the iCCD camera.

Sample treatment for LIBS. The soil samples were mixed with starch (19 wt% final mass fraction), ground in an agate ball mill and subsequently pressed into pellets. One pellet was created for each field sample point. For standard addition, the sample point of the first field with the lowest intensity of the respective element peak was chosen for each investigated element. To that end, the soils were mixed with the respective amounts of target elements added as salts ($CaCO_3$, $MgCl_2$ and FeS, the amount of starch was reduced in order to keep soil fraction constant) and also formed into pellets. The salts were obtained from Sigma Aldrich (St. Louis, MO, USA).

Data pretreatment. Pretreatment of the LIBS spectra consisted of outlier (e.g., spectra where the laser did not fire) removal using the following procedure. The total intensity of each spectrum for a given sample was calculated. The median of these values was derived and only the spectra in the range 0.75 × median ≤ total intensity ≤ 1.25 × median were averaged to yield a single spectrum for each soil sample. For PLSR calibration (third method, see below) spectra were mean centered. Logarithms of the known element mass fractions were used as y variables when the data distribution was strongly skewed. For univariate calibration, the peaks of investigated elements were integrated for each averaged spectrum individually. For multivariate analysis, the entire spectra of either the UV or the VIS region were used.

Calibration. *Standard addition univariate calibration.* The first calibration method was based on standard addition using just one soil sample and applying the derived univariate calibration model to the LIBS data of both fields. Using the standard addition method potentially has several advantages. First, the technique can be used to determine the mass fraction of one or several nutrients in one soil sample by extrapolating the mass fraction of the base sample from the samples with added known quantities of an element. Second, as applied in this work, reference analysis of the pure soil sample by the traditional digestion method and in combination with samples where increasing amounts of the nutrients were added can provide LIBS calibration curves. One additional benefit is that standard addition yields response curves where the matrix effect of the local soil type is accounted for. The nutrients are advantageously added as salts. However, care must be taken that the salts are not hygroscopic. The third advantage is that the large range of nutrient mass fractions covered for the determination of calibration curves facilitates assigning lines to elements and finding the best lines to use for the quantification of each element. Potentially, different lines could be chosen for different mass fraction ranges. The most intense lines might be useful at low element mass fractions, while at high element mass fractions, these intense lines can be self-absorbed. Therefore, choosing weaker lines can become beneficial.

Reference univariate calibration. The second method derived a univariate calibration model from the reference analytics of all samples from one field. The prediction is validated using the second field. This approach accounts for the matrix effects even better, as the variation of the matrix across a field is also reflected. Furthermore, the error in the laboratory-based reference analytics of one soil sample has a large influence on the LIBS-based results. Therefore, a larger data pool with known reference values and a broader variety of matrices should be regarded. In our case, the data of the first field is used for the calibration and the data of the second field is used for the validation. The concept of this procedure is that reference data gained in one year could be used to build a calibration for a specific field. In subsequent years, it will not be necessary to take new samples and the calibration can therefore also be applied to fields in the near surrounding and to fields with a similar soil type. Furthermore, a much larger number of data points can be evaluated, as LIBS can be employed directly on the field, allowing a closer spatial mapping of the fields.

Reference multivariate calibration. The third approach was a multivariate calibration based on PLSR. PLSR was done using the kernel algorithm and a maximum of 7 components. The LIBS spectra of the first field were used for training. Validation was carried out by 20-fold cross-validation using the data of the first field and independently by testing the model trained on the data of the first field with the data of the second field.

Software: Origin (OriginLab, Northampton, MA, USA) was used for PCA. Unscrambler X (Camo Analytics, Oslo, Norway) was used to perform PLSR.

3. Results and Discussion

3.1. LIBS Spectra

LIBS spectra of soils are rich in lines, mostly due to the presence of Fe and other transition metals (Figure 1). Additionally, lines of most minor and major mineral nutrients are found, the lines of metal nutrients are particularly intense.

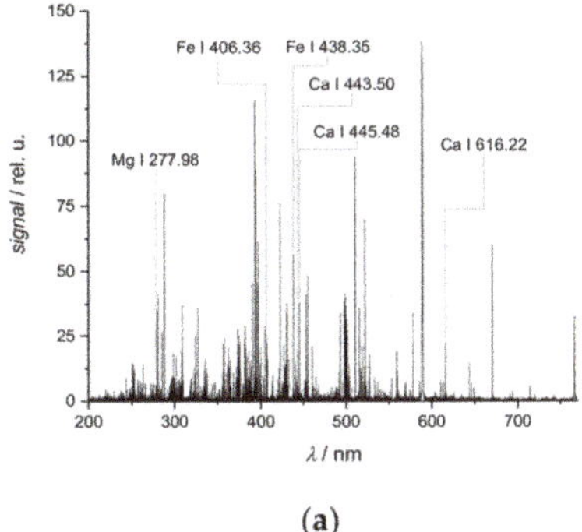 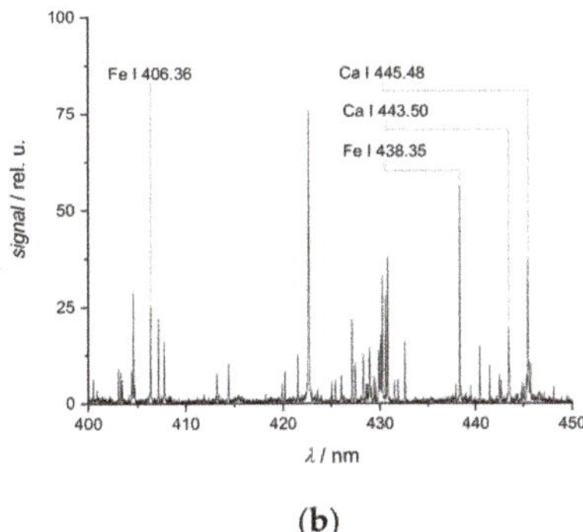

(a)   (b)

Figure 1. Typical soil laser-induced breakdown spectroscopy (LIBS) spectrum (composite of UV and VIS part of the LIBS spectrum), (**a**) full wavelength range, (**b**) detail view of wavelength range 400–450 nm.

But the plethora of lines complicates their assignment to elements, as peak overlapping is frequently observed. The line identification is not straightforward as the total elemental composition of soils is generally not known. Otherwise, the limitation of the measurement campaign on one or two fields is advantageous for peak assignment as the soil heterogeneity is not as great as on a larger scale. Most peaks are generally found in spectra of all samples (different points on the field) but their intensities still differ substantially on the scale of one field. This observation is due to variations of the element mass fractions across the field, which is also the origin of the strong matrix effects encountered and necessitates the determination of element distributions in precision agriculture.

3.2. Standard Addition Calibration

Ca mass fractions vary greatly (from 500 ppm$_w$ to 5 wt%) in the investigated soils. A corresponding calibration curve has to reflect a mass fraction range of nearly three orders of magnitude and is best presented in a double-logarithmic diagram. The fitting function $I = a\,w^c$, where I is line intensity, w the mass fraction and a and c are the optimized constants, essentially equivalent to taking the logarithm of the data prior to linear fitting, was used. The calibration curves for three different Ca lines (at 443.496 nm, 445.478 nm, 616.217 nm) are shown in Figure 2. These lines provided best results in terms of repeatability (precision), linearity ($R^2 > 0.993$) and usability over the entire range of relevant mass fractions.

The Fe mass fraction in the investigated soil (from 0.5 wt% to 6 wt%) is on a higher mass fraction level but covers a narrower range than the Ca mass fraction. Thus, a linear calibration plot was selected (Figure 3a). The LIBS spectrum contains a very high number of Fe lines. Due to the high mass fraction, many of them show strong effects of self-absorption. This requires a careful selection of lines. The lines at 406.359 nm and 438.354 nm (see Figure 3a) are best suited for calibration and yield $R^2 > 0.988$ for $\lambda = 406.359$ nm and $R^2 > 0.992$ for $\lambda = 438.354$ nm. The repeatability is not as good as for Ca as the error bars show, but quantification is still possible.

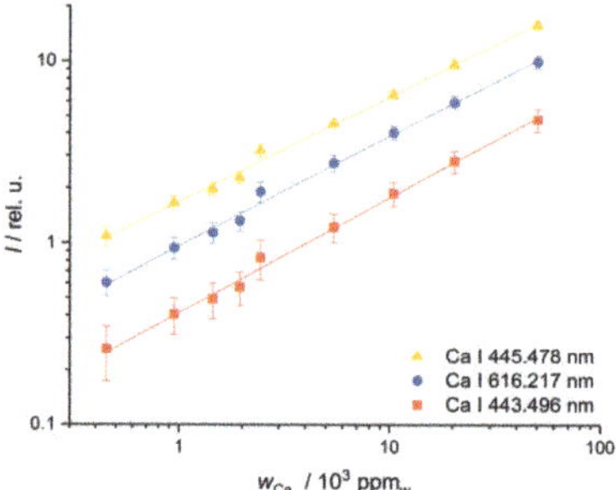

Figure 2. Calibration curves of different Ca lines in soil, integrated line intensity over calculated Ca mass fraction: R^2 (443.496 nm) = 0.994, R^2 (445.478 nm) = 0.996, R^2 (616.217 nm) = 0.993 (individual wavelengths are offset for visibility).

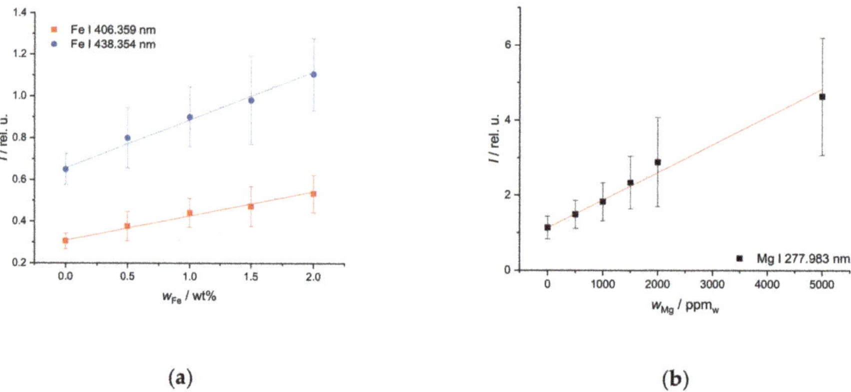

(a)   (b)

Figure 3. Response curve of (**a**) FeS and (**b**) $MgCl_2$ in soil, integrated line intensity over calculated (**a**) Fe and (**b**) Mg mass fractions, (**a**) R^2 (406.359 nm) = 0.988, R^2 (438.354 nm) = 0.992, (**b**) R^2 (277.983 nm) = 0.989.

Mg appears in the soils at low to intermediate mass fractions (from 700 ppm_w to 4000 ppm_w) and covers a narrow mass fraction range. Therefore, a linear calibration curve was selected as well. Useful Mg lines are only found in the UV region. The repeatability is worse at higher mass fractions as evidenced by higher error bars. The line at 277.983 nm shown in Figure 3b yields the best result in terms of linearity over a range of mass fractions (R^2 = 0.989).

For all chosen elements, lines could be found in the available spectral range that are free from overlap, less prone to self-absorption effects and provide good dynamic ranges. The detection limits usually are in the range of 100 ppm which is adequate for the soils in question. Lower detection limits would only be needed for micronutrients.

3.3. Application of Standard Addition Calibration to the First Field

In order to obtain the mass fraction distribution of the nutrients on a field, the standard addition calibration was applied to LIBS spectra of all samples (points on the field) for predicting the element mass fractions. A comparison of the values thus obtained from the LIBS measurements and the reference values obtained by ICP-OES yields insights into the performance of the calibration method. This is shown for the nutrient Ca in Figure 4a, where the Ca mass fractions measured by LIBS are the averages of the results for the three lines in the LIBS spectrum. The line in Figure 4a presents the identity of values obtained by LIBS and ICP-OES. The Ca mass fractions obtained by LIBS scatter around this line and show a very good agreement over the range of mass fractions investigated. Thus, a univariate calibration, built using only one reference sample, can be used for predicting the Ca mass fraction for the entire field. Absolute deviations for the different sample points on the field are shown in Figure 4b with the highest errors in the range of 1%. The deviations seem accumulated around certain points of the field as sample numbers indicate proximity. Because samples were taken along predefined lines across the field, this could be an indication of different soil types and thus matrix effects at those positions.

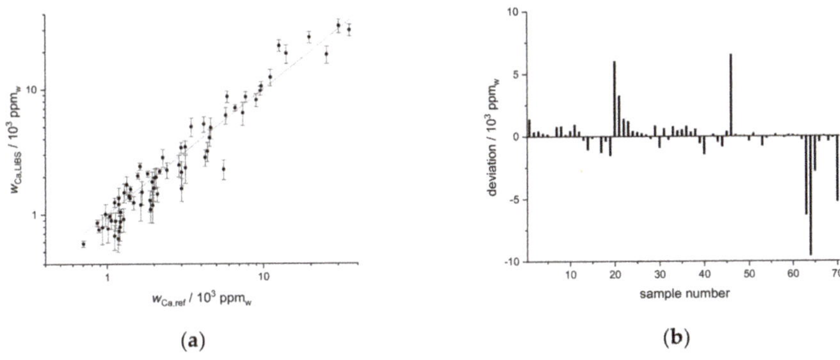

Figure 4. (a) Mass fraction of Ca predicted by LIBS univariate calibration based on standard addition ($w_{Ca,LIBS}$) compared to reference values obtained by ICP-OES upon aqua regia extraction ($w_{Ca,ref}$) for the first field, $R^2 = 0.91$, (b) deviation of predicted from reference value sorted by sample number.

Results were not as promising for Fe (Figure 5a). Although a rough agreement is present in values around 1 wt%, lower and especially higher Fe mass fractions are overestimated by as much as twice the reference value. Reasons could be the relatively narrow range of mass fractions and especially the high overall Fe mass fractions, leading to e.g., self-absorption. The results of the Mg prediction are shown in Figure 5b. Agreement of the results is much better, although in the low mass fraction region, a larger spread of the data points is observed. Overall, this kind of calibration is satisfactory for the major nutrients Ca and Mg but the predictive power is limited for the minor nutrient Fe.

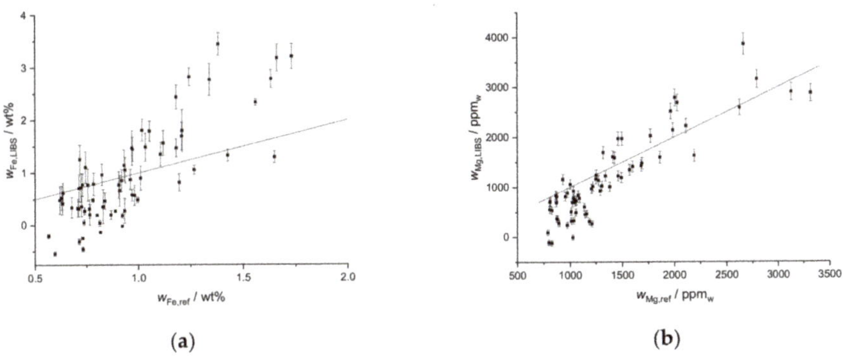

Figure 5. Mass fractions of (**a**) Fe and (**b**) Mg predicted by LIBS univariate calibration based on standard addition compared to reference values obtained by ICP-OES upon aqua regia extraction for the first field.

3.4. Application of Standard Addition Calibration to Second Field

As a further validation, the same calibration curves were also applied to the second field data, which does not contain the soil sample used for obtaining the standard addition calibration. Therefore, the results of this second field are a measure of the predictive power for local agriculturally-used areas in the surrounding of the calibration field. It is likely that the soil types on these fields will have a similar characteristic.

The prediction results of the three elements for both fields are similar (Figures 6 and 7). The best agreements are again achieved for Ca and Mg, especially the correlation for Ca is excellent. The prediction for Ca seems to be robust and can be also used for different fields of similar soil types. The deviation of the Mg prediction (Figure 7) is a bit stronger than for the first field. While it could still be usable in practice to get a rough estimation of the Mg mass fraction, a better method is desirable. The agreement of the values of the second field in the Fe plot (Figure 6b) is actually better than for the first field from which the soil for the calibration was taken, although the range of mass fractions is larger. However, the deviation is still present, and the calibration can only be used as a rough estimation.

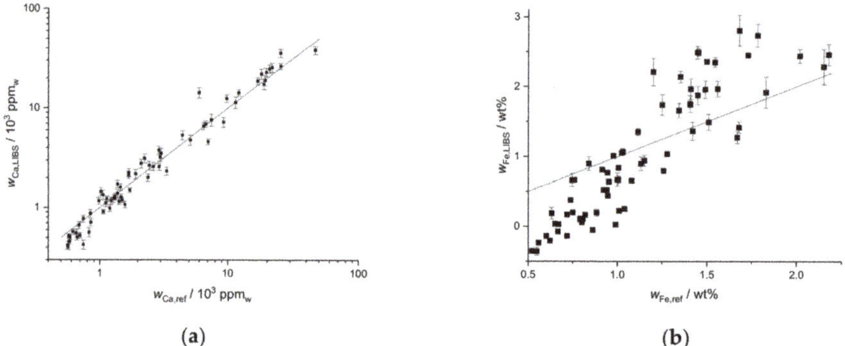

(a)   **(b)**

Figure 6. Mass fraction of (**a**) Ca and (**b**) Fe predicted by LIBS univariate calibration based on standard addition compared to reference values obtained by ICP-OES upon aqua regia extraction for the second field, R^2 (Ca) = 0.60.

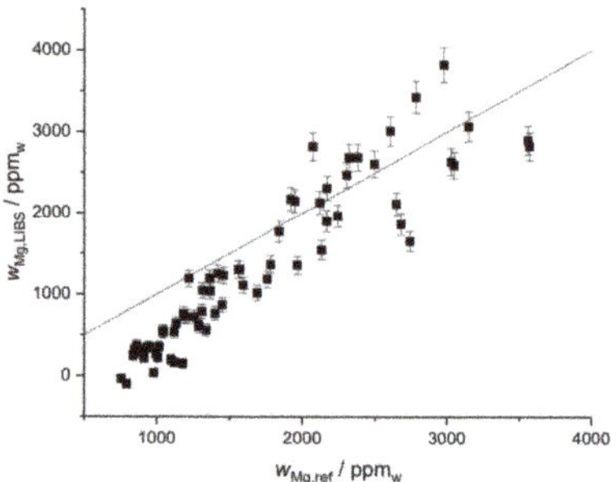

Figure 7. Mass fraction of Mg predicted by LIBS univariate calibration based on standard addition compared to reference values obtained by ICP-OES upon aqua regia extraction for the second field.

Although the calibration method established from one sample works fine for the element Ca, even on different fields, the applicability to other nutrients is limited. The most important reason for this finding is the matrix effect, which is not sufficiently taken into account by this method. On a future field campaign, the discussed peaks in the LIBS spectra can, however, be used to find points of extreme element mass fractions, which are then crucial sampling points from which a better calibration can be built.

3.5. Reference Univariate Calibration

The results of the reference univariate calibration using all samples from the first field and its application to the second field are shown below. The calibration data of the LIBS response for three different wavelengths as a function of the Ca mass fraction obtained by the reference analysis are shown in Figure 8a. For all wavelengths, a good regression (e.g., R^2 (445.478 nm) = 0.94) between data of the LIBS and reference analytical method is obtained, the coefficients of determination for the lines at 443.496 nm and 445.478 nm are better than those for the line at 616.217 nm. The calibration curves are applied to the LIBS data of the second field (Figure 8b), the Ca mass reported as measured by LIBS are the averages of the results for the three lines. This procedure results in predicted values that are in very good agreement with the reference data (R^2 = 0.93).

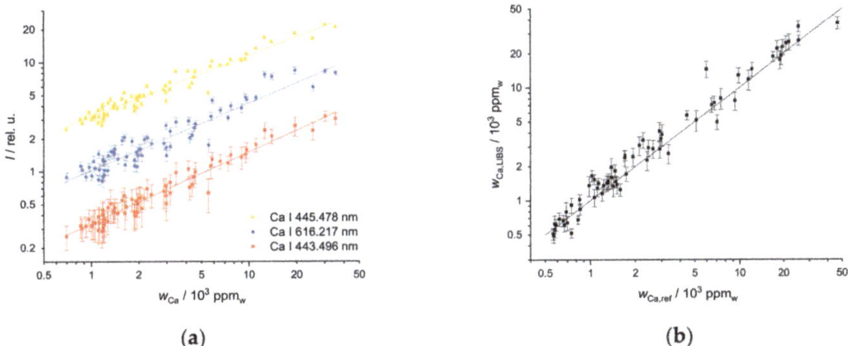

(a)   (b)

Figure 8. (**a**) Univariate calibration of Ca mass fraction calculated for data of the first field, R^2 (443.496 nm) = 0.93, R^2 (445.478 nm) = 0.94, R^2 (616.217 nm) = 0.86 (individual wavelengths are offset for visibility), (**b**) Calibration applied to the second field, R^2 = 0.93.

The same procedure was applied to Fe (Figure 9). The calibration plots (Figure 9a) took two signals at the wavelengths at 406.359 nm and 438.354 nm into account but yielded a poorer prediction than that found for Ca. However, the larger amount of data used for the calibration compared to the standard addition method results in a better prediction for the second field (Figure 9b). Here, the matrix effects are better accounted for. Although the result is acceptable, a univariate calibration might not be ideal for Fe.

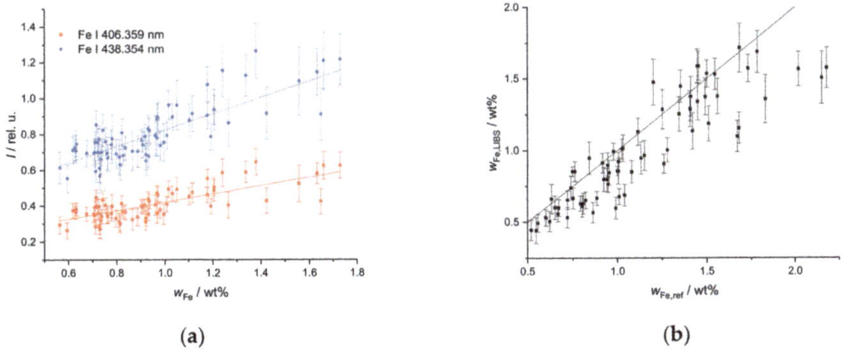

(a)   (b)

Figure 9. (**a**) Univariate calibration of Fe mass fraction calculated for data of the first field, R^2 (406.359 nm) = 0.53, R^2 (438.354 nm) = 0.61; (**b**) Calibration applied to the second field, R^2 = 0.51, without two most obvious outliers (same samples as for Ca) R^2 = 0.68.

While the calibration plot for Mg is characterized by a linear relationship but relatively large error bars (Figure 10a), the validation plot (Figure 10b) obtained on the second field actually has a better coefficient of determination than the calibration fit itself. Although this can be explained by a few high error intervals of the intensities used for the calibration, it is still a promising result.

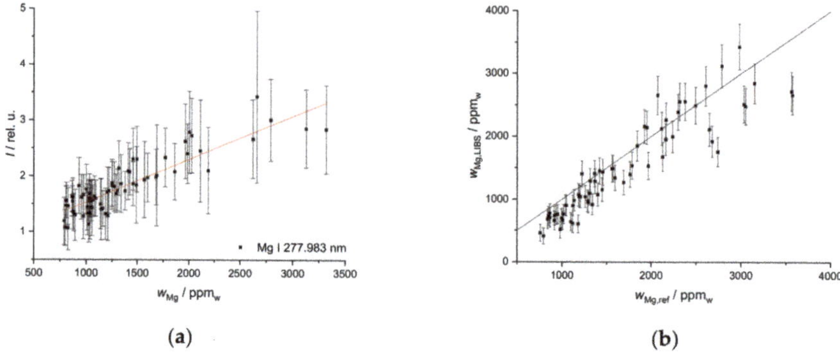

(a)   (b)

Figure 10. (**a**) Univariate calibration of Mg mass fraction calculated for data of the first field, the line used is Mg I 277.983 nm, $R^2 = 0.71$, (**b**) Calibration applied to the second field, $R^2 = 0.51$, without two most obvious outliers (same samples as for Ca) $R^2 = 0.76$.

Overall, it can be concluded that using actual field samples for building the calibration improves the performance for the univariate calibration. Such univariate methods can thus be used for predicting values. In the future, online LIBS measurements on the field can be used to determine which samples to use for laboratory analyses. If these samples represent a good range of the values of interest, a strong calibration can be built, which can in turn be used to predict values for all points of the online measurements.

3.6. Multivariate Analysis

One possibility for characterizing the heterogeneity of soil compositions of the individual fields, and of the first field in relation to the second field, is to apply PCA. For that purpose, the complete LIBS spectra were introduced into PCA. In the score plot (Figure 11) the first two components, which explain 93% of the variance, are presented. Grouping the data points by fields implies that the heterogeneity of the second field (red points) is lower than that of the first (black points). This also explains why the univariate calibrations were largely successful and justifies using the first field as training data and the second field as validation data in multivariate analyses.

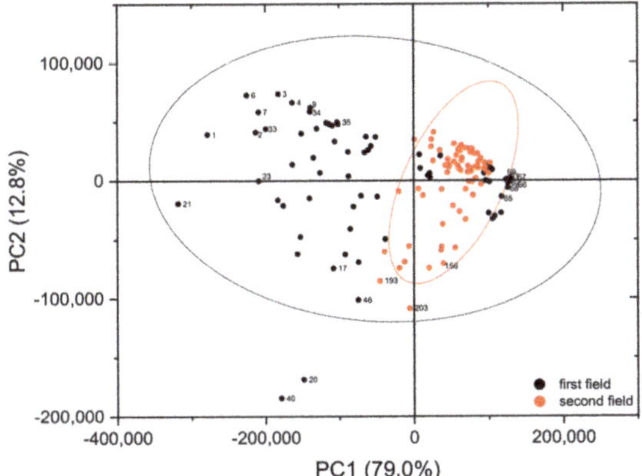

Figure 11. PCA of all LIBS spectra of the first (black points) and the second field (red points).

As a multivariate regression method, PLSR was performed on our data. Validation of the results of the regression model was taken into account in two different ways. First, 20-fold cross-validation was applied on the data of the first field. This is demonstrated for Ca data of the first field in Figure 12a. Correlation plots of Ca mass fractions predicted by LIBS and by ICP-OES are shown. The red squares are the result of the PLSR calibration, and the blue points represent 20-fold cross-validation. For PLSR, the logarithms of Ca values were taken, and 5 components were used.

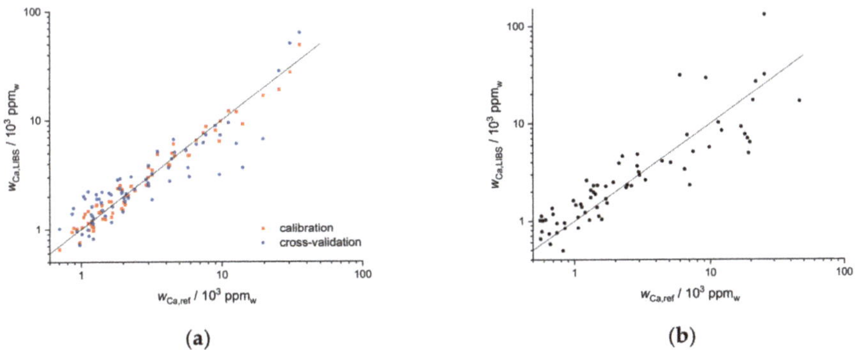

(a)   (b)

Figure 12. (**a**) Partial least squares regression (PLSR) of Ca data for the first field (UV range), red squares represent calibration data ($R^2 = 0.96$), blue points represent 20-fold cross-validation ($R^2 = 0.83$), (**b**) application of PLSR of Ca data to the second field ($R^2 = 0.58$).

The PLSR cross-validation data ($R^2 = 0.83$) demonstrate the potential of multivariate regression for predicting mass fractions on the same field. A second, additional validation is obtained by applying the calibration obtained of the first field to the second field, demonstrating how universal a calibration could become. This is shown in Figure 12b for Ca and gives an impression of the prediction quality of PLSR calibrations for additional fields with similar soil types ($R^2 = 0.58$).

PLSR was also applied to the elements Fe and Mg (Figure 13). For both elements, the logarithms of the mass fractions were used for PLSR. Seven components were used for PLSR of the Fe data, and 6 components were used for PLSR of the Mg data. The cross-validation plots of both elements scatter a bit stronger than the Ca plot (R^2 (Fe) = 0.70, R^2 (Mg) = 0.79). The application of the calibration data of the first to the second field results in a similar scattering of the predicted values. With increasing element mass fractions, the deviation of predicted values from reference values becomes systematically larger. LIBS underestimates high Fe and Mg mass fractions.

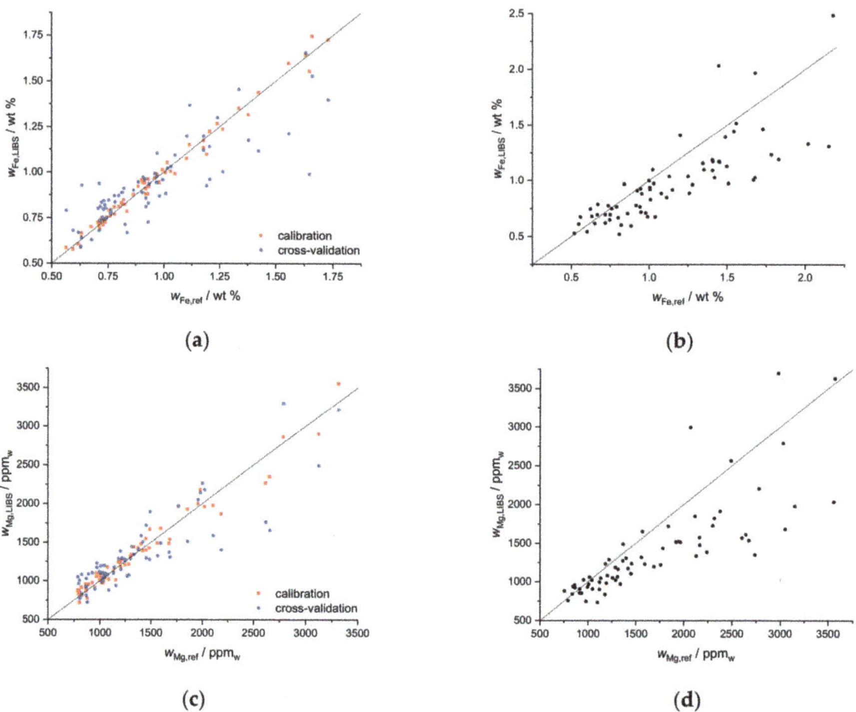

Figure 13. (**a**) PLSR of Fe data for the first field (UV range), red points represent calibration data ($R^2 = 0.986$), blue diamonds represent 20-fold cross-validation ($R^2 = 0.70$), (**b**) application of PLSR of Fe data to the second field $R^2 = 0.37$, (**c**) PLSR of Mg data for the first field (UV range), red points represent calibration data ($R^2 = 0.96$), blue diamonds represent 20-fold cross-validation $R^2 = 0.79$, (**d**) application of PLSR of Fe data to the second field $R^2 = 0.38$.

3.7. Field Maps

One main aim in precision farming is the mapping of nutrient distributions on agricultural fields. In connection with fertilizer recommendations based on measured data as well as models, these maps help the farmers identify nutrient-depleted areas. The change between nutrient-rich and depleted areas can be seen on the three maps of Figure 14. Thus, the eastern areas are richer in all three elements while the western areas are depleted.

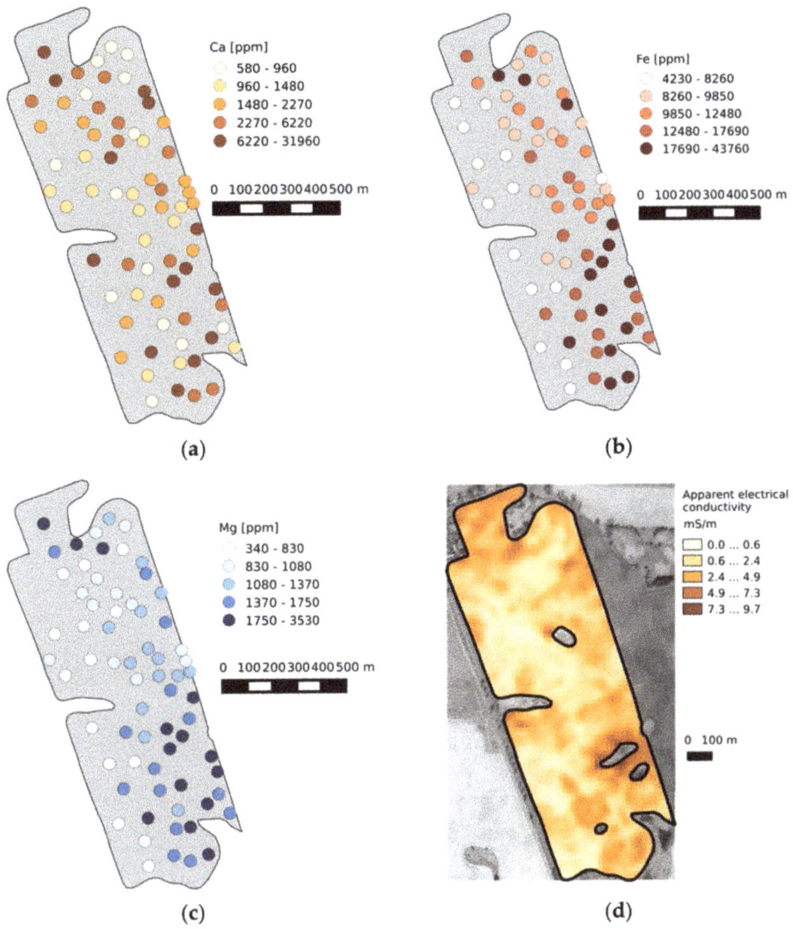

(a) (b) (c) (d)

Figure 14. Maps showing the element distribution of (**a**) Ca, (**b**) Fe, (**c**) Mg and (**d**) ECa on the first field (north is up).

This spatial trend was further explained by the evaluation of the ECa map (Figure 14d), which was recorded directly on the field. The ECa values were positively correlated with all elements as determined by the reference method and by LIBS.

4. Summary and Conclusions

LIBS is a promising method for efficient soil data collection at high spatial resolution, as required for precision agriculture. This is due to the capability of LIBS to rapidly measure mass fractions of many elements simultaneously without or only very little sample preparation. These advantages are contrasted by a strong matrix dependence of the LIBS signal which requires careful calibration and data evaluation methods.

In this work, different approaches for the evaluation of LIBS data were investigated. The first univariate method was based on standard addition to just one soil sample, the second was a univariate method based on a larger number of samples characterized by reference analytics. This second method accounted for matrix effects and thus yielded better predictions. The third method was a multivariate method (PLSR) which yielded better calibration curves, but did not substantially improve the prediction compared to the reference univariate method. Alternative and maybe better suited multivariate methods can potentially provide better results.

The work begun here should be continued by expanding the LIBS measurements to a larger database, containing more and especially different soil types. It could also be interesting to obtain soil samples from the same field in different years and verify if the same calibrations can be used. The classification of soil samples based on different soil types and the application of tailored calibration methods for these soil types can improve the accuracy and repeatability of measuring results in the future. Additionally, the amount of sample preparation performed in this work has to be reduced in order to transfer the method to the field.

Author Contributions: Conceptualization: D.R., T.B., H.-G.L. and R.G.; Data curation: D.R. and P.B.; Formal analysis: D.R., Funding acquisition: D.R., T.B., H.-G.L. and R.G.; Investigation: D.R., A.E., P.B. and T.B.; Methodology: D.R. and T.B.; Project administration: D.R., T.B., H.-G.L. and R.G.; Resources: H.-G.L. and R.G.; Supervision: H.-G.L.; Visualization: D.R.; Writing—Original draft: D.R. and T.B. Writing—Review & editing: H.-G.L. and R.G.

Funding: The authors gratefully acknowledge the financial support for this research received from the German Ministry of Education and Research (BMBF) in the framework of the BonaRes project Integrated System for Site-Specific Soil Fertility Management (I4S, grant no. 031A564H).

Acknowledgments: The authors would also like to thank Sascha Hons for support during the LIBS measurements. We acknowledge the support of the Deutsche Forschungsgemeinschaft and Open Access Publishing Fund of University of Potsdam

References

1. Stafford, J.V. Remote, non-contact and in-situ measurement of soil moisture content: A review. *J. Agric. Eng. Res.* **1988**, *41*, 151–172. [CrossRef]

2. Viscarra Rossel, R.A.; McBratney, A.B. Soil chemical analytical accuracy and costs: Implications from precision agriculture. *Aust. J. Exp. Agric.* **1998**, *38*, 765. [CrossRef]

3. Hummel, J.W.; Gaultney, L.D.; Sudduth, K.A. Soil property sensing for site-specific crop management. *Comput. Electron. Agric.* **1996**, *14*, 121–136. [CrossRef]

4. Viscarra Rossel, R.A.; McBratney, A.B.; Minasny, B. (Eds.) *Proximal Soil Sensing*; Springer: Dordrecht, The Netherlands, 2010.

5. Viscarra Rossel, R.A.; Adamchuk, V.I.; Sudduth, K.A.; McKenzie, N.J.; Lobsey, C. Proximal Soil Sensing: An Effective Approach for Soil Measurements in Space and Time. In *Advances in Agronomy*; Academic Press: Cambridge, MA, USA, 2011; pp. 243–291.

6. Viscarra Rossel, R.A.; Bouma, J. Soil sensing: A new paradigm for agriculture. *Agric. Syst.* **2016**, *148*, 71–74. [CrossRef]

7. Gebbers, R. Proximal soil surveying and monitoring techniques. In *Precision Agriculture for Sustainability*; Stafford, J., Ed.; Burleigh Dodds Science Publishing: Cambridge, UK, 2018; pp. 1–50.

8. Adamchuk, V.I.; Morgan, M.T.; Ess, D.R. An automated sampling system for measuring soil pH. *Trans. ASAE* **1999**, *42*, 885–891. [CrossRef]

9. Hyvönen, E.; Turunen, P.; Vanhanen, E.; Arkimaa, H.; Sutinen, R. Airborne Gamma-ray Surveys in Finland. In *Aerogeophysics in Finland 1972–2004: Methods, System Characteristics and Applications*; Special Paper, 39; Airo, M.-L., Ed.; Geological Survey of Finland: Espoo, Finland, 2005; pp. 119–134.

10. Heggemann, T.; Welp, G.; Amelung, W.; Angst, G.; Franz, S.O.; Koszinski, S.; Schmidt, K.; Pätzold, S. Proximal gamma-ray spectrometry for site-independent in situ prediction of soil texture on ten heterogeneous fields in Germany using support vector machines. *Soil Tillage Res.* **2017**, *168*, 99–109. [CrossRef]

11. Stenberg, B.; Viscarra Rossel, R.A.; Mouazen, A.M.; Wetterlind, J. Visible and Near Infrared Spectroscopy in Soil Science. In *Advances in Agronomy*; Academic Press: Cambridge, MA, USA, 2010; pp. 163–215.

12. Shibusawa, S.; Sato, H.; Sasao, A.; Hirako, S.; Otomo, A. A Revised Soil Spectrophotometer. *IFAC Proc. Vol.* **2000**, *33*, 231–236. [CrossRef]

13. Schirrmann, M.; Gebbers, R.; Kramer, E.; Seidel, J. Soil pH mapping with an on-the-go sensor. *Sensors* **2011**, *11*, 573–598. [CrossRef]

14. Schirrmann, M.; Gebbers, R.; Kramer, E. Performance of Automated Near-Infrared Reflectance Spectrometry for Continuous in Situ Mapping of Soil Fertility at Field Scale. *Vadose Zo. J.* **2013**, *12*, 1. [CrossRef]

15. Corwin, D.L. Past, present, and future trends of soil electrical conductivity measurement using geophysical methods. In *Handbook of Agricultural Geophysics*; Allred, B.J., Daniels, J.J., Ehsani, M.R., Eds.; CRC Press: Boca Raton, FL, USA, 2008; pp. 17–44.

16. Gebbers, R.; Lück, E.; Dabas, M.; Domsch, H. Comparison of instruments for geoelectrical soil mapping at the field scale. *Near Surf. Geophys.* **2009**, *7*, 179–190. [CrossRef]

17. Heil, K.; Schmidhalter, U. The application of EM38: Determination of soil parameters, selection of soil sampling points and use in agriculture and archaeology. *Sensors* **2017**, *17*, 2540. [CrossRef] [PubMed]

18. Corwin, D.L.; Lesch, S.M. Application of Soil Electrical Conductivity to Precision Agriculture. *Agron. J.* **2003**, *95*, 455. [CrossRef]

19. Adamchuk, V.I.; Viscarra Rossel, R.A.; Marx, D.B.; Samal, A.K. Using targeted sampling to process multivariate soil sensing data. *Geoderma* **2011**, *163*, 63–73. [CrossRef]

20. Piikki, K.; Söderström, M.; Eriksson, J.; John, J.M.; Muthee, P.I.; Wetterlind, J.; Lund, E. Performance evaluation of proximal sensors for soil assessment in smallholder farms in Embu County, Kenya. *Sensors* **2016**, *16*, 1950. [CrossRef] [PubMed]

21. Zhu, Y.; Weindorf, D.C.; Zhang, W. Characterizing soils using a portable X-ray fluorescence spectrometer: 1. Soil texture. *Geoderma* **2011**, *167–168*, 167–177. [CrossRef]

22. Gebbers, R.; Schirrmann, M. Potential of using portable x-ray fluorescence spectroscopy for rapid soil analysis. In *Precision Agriculture '15*; Stafford, J.V., Ed.; Wageningen Academic Publishers: Wageningen, The Netherlands, 2015; pp. 27–34.

23. Nawar, S.; Delbecque, N.; Declercq, Y.; De Smedt, P.; Finke, P.; Verdoodt, A.; Van Meirvenne, M.; Mouazen, A.M. Can spectral analyses improve measurement of key soil fertility parameters with X-ray fluorescence spectrometry? *Geoderma* **2019**, *350*, 29–39. [CrossRef]

24. Fortes, F.J.; Moros, J.; Lucena, P.; Cabalín, L.M.; Laserna, J.J. Laser-Induced Breakdown Spectroscopy. *Anal. Chem.* **2013**, *85*, 640–669. [CrossRef]

25. Cremers, D.A.; Chinni, R.C. Laser-Induced Breakdown Spectroscopy—Capabilities and Limitations. *Appl. Spectrosc. Rev.* **2009**, *44*, 457–506. [CrossRef]

26. Hahn, D.W.; Omenetto, N. Laser-Induced Breakdown Spectroscopy (LIBS), Part I: Review of Basic Diagnostics and Plasma—Particle Interactions: Still-Challenging Issues within the Analytical Plasma Community. *Appl. Spectrosc.* **2010**, *64*, 335A–336A. [CrossRef]

27. Hahn, D.W.; Omenetto, N. Laser-Induced Breakdown Spectroscopy (LIBS), Part II: Review of Instrumental and Methodological Approaches to Material Analysis and Applications to Different Fields. *Appl. Spectrosc.* **2012**, *66*, 347–419. [CrossRef]

28. Zorov, N.B.; Popov, A.M.; Zaytsev, S.M.; Labutin, T.A. Qualitative and quantitative analysis of environmental samples by laser-induced breakdown spectrometry. *Russ. Chem. Rev.* **2015**, *84*, 1021–1050. [CrossRef]

29. Popov, A.M.; Zaytsev, S.M.; Seliverstova, I.V.; Zakuskin, A.S.; Labutin, T.A. Matrix effects on laser-induced plasma parameters for soils and ores. *Spectrochim. Acta Part B At. Spectrosc.* **2018**, *148*, 205–210. [CrossRef]

30. Zaytsev, S.M.; Krylov, I.N.; Popov, A.M.; Zorov, N.B.; Labutin, T.A. Accuracy enhancement of a multivariate calibration for lead determination in soils by laser induced breakdown spectroscopy. *Spectrochim. Acta Part B At. Spectrosc.* **2018**, *140*, 65–72. [CrossRef]

31. Ciucci, A.; Corsi, M.; Palleschi, V.; Rastelli, S.; Salvetti, A.; Tognoni, E. New Procedure for Quantitative Elemental Analysis by Laser-Induced Plasma Spectroscopy. *Appl. Spectrosc.* **1999**, *53*, 960–964. [CrossRef]

32. Díaz, D.; Hahn, D.W.; Molina, A. Evaluation of Laser-Induced Breakdown Spectroscopy (LIBS) as a Measurement Technique for Evaluation of Total Elemental Concentration in Soils. *Appl. Spectrosc.* **2012**, *66*, 99–106. [CrossRef]

33. Yongcheng, J.; Wen, S.; Baohua, Z.; Dong, L. Quantitative Analysis of Magnesium in Soil by Laser-Induced Breakdown Spectroscopy Coupled with Nonlinear Multivariate Calibration. *J. Appl. Spectrosc.* **2017**, *84*, 731–737. [CrossRef]

34. Nicolodelli, G.; Marangoni, B.S.; Cabral, J.S.; Villas-Boas, P.R.; Senesi, G.S.; dos Santos, C.H.; Romano, R.A.; Segnini, A.; Lucas, Y.; Montes, C.R.; et al. Quantification of total carbon in soil using laser-induced breakdown spectroscopy: A method to correct interference lines. *Appl. Opt.* **2014**, *53*, 2170. [CrossRef]

35. Rühlmann, M.; Büchele, D.; Ostermann, M.; Bald, I.; Schmid, T. Challenges in the quantification of nutrients in soils using laser-induced breakdown spectroscopy–A case study with calcium. *Spectrochim. Acta Part B At. Spectrosc.* **2018**, *146*, 115–121. [CrossRef]

36. Nicolodelli, G.; Cabral, J.; Menegatti, C.R.; Marangoni, B.; Senesi, G.S. Recent advances and future trends in LIBS applications to agricultural materials and their food derivatives: An overview of developments in the last decade (2010–2019). Part I. Soils and fertilizers. *TrAC Trends Anal. Chem.* **2019**, *115*, 70–82. [CrossRef]

37. Herrera, K.K.; Tognoni, E.; Omenetto, N.; Smith, B.W.; Winefordner, J.D. Semi-quantitative analysis of metal alloys, brass and soil samples by calibration-free laser-induced breakdown spectroscopy: Recent results and considerations. *J. Anal. At. Spectrom.* **2009**, *24*, 413–425. [CrossRef]

Article

Coupling Waveguide-Based Micro-Sensors and Spectral Multivariate Analysis to Improve Spray Deposit Characterization in Agriculture

Anis Taleb Bendiab [1,2], Maxime Ryckewaert [1], Daphné Heran [1], Raphaël Escalier [2], Raphaël K. Kribich [3], Caroline Vigreux [2,*] and Ryad Bendoula [1,*]

[1]   ITAP, Irstea, Montpellier SupAgro, Université de Montpellier, 361 rue Jean-François Breton, 34000 Montpellier, France
[2]   ICGM, UMR 5253, cc1503, Université de Montpellier, Place Eugène Bataillon, CEDEX 5, 34095 Montpellier, France
[3]   IES, UMR 5214, cc05005, Université de Montpellier, Bâtiment 5, 860 rue Saint-Priest, 34090 Montpellier, France
*   Correspondence: caroline.vigreux@umontpellier.fr (C.V.); ryad.bendoula@irstea.fr (R.B.)

Received: 7 August 2019; Accepted: 24 September 2019; Published: 26 September 2019

Abstract: The leaf coverage surface is a key measurement of the spraying process to maximize spray efficiency. To determine leaf coverage surface, the development of optical micro-sensors that, coupled with a multivariate spectral analysis, will be able to measure the volume of the droplets deposited on their surface is proposed. Rib optical waveguides based on Ge-Se-Te chalcogenide films were manufactured and their light transmission was studied as a response to the deposition of demineralized water droplets on their surface. The measurements were performed using a dedicated spectrophotometric bench to record the transmission spectra at the output of the waveguides, before (reference) and after drop deposition, in the wavelength range between 1200 and 2000 nm. The presence of a hollow at 1450 nm in the relative transmission spectra has been recorded. This corresponds to the first overtone of the O–H stretching vibration in water. This result tends to show that the optical intensity decrease observed after droplet deposition is partly due to absorption by water of the light energy carried by the guided mode evanescent field. The probe based on Ge-Se-Te rib optical waveguides is thus sensitive throughout the whole range of volumes studied, i.e., from 0.1 to 2.5 μL. Principal Component Analysis and Partial Least Square as multivariate techniques then allowed the analysis of the statistics of the measurements and the predictive character of the transmission spectra. It confirmed the sensitivity of the measurement system to the water absorption, and the predictive model allowed the prediction of droplet volumes on an independent set of measurements, with a correlation of 66.5% and a precision of 0.39 μL.

Keywords: optical micro-sensors; crop protection; precision agriculture; infrared spectroscopy; principal component analysis (PCA); partial least squares (PLS); droplet characterization

1. Introduction

Despite the various campaigns conducted to initiate the reduction of pesticides in agriculture, their use has further increased in recent years [1]. The main challenge is to reduce the use of products while improving plant protection. This challenge motivates numerous research projects that aim to find alternative solutions [2] or to develop decision support tools to reduce doses of applied pesticides [3,4]. Meeting this challenge could significantly reduce the environmental risks associated with agricultural activities. Improving spray efficiency can be a solution implemented to optimize plant protection use in the fields. However, the efficiency of the treatments depends largely on the spray drift which must be minimized, and on the foliar coverage, which must be as large as possible.

Different authors have addressed the spray drift. Examples include the work of Grella et al. [5] who have set up a 20 m-long device, with Petri dishes spaced every 0.5 m. The analysis of the different Petri dishes after spraying with water, in which a yellow marker (E-302 Tartrazine yellow dye tracer) was added, allowed them to compare different spray devices and to highlight those that generate the least important drift. To minimize spray drift, knowledge of droplet size is of paramount importance. Indeed, studies have shown, for example, that the use of sprayers distributing too fine drops leads to a greater drift of the spray and therefore reduces its effectiveness [6]. Techniques to measure the size of the spray droplets can therefore be extremely useful. Different methods already exist. Balsari et al. [7] used, for example, a laser diffraction system (Malvern SprayTec instrument) coupled with a SprayTec software to obtain droplet size data from different sprayers. We can also cite the PDPA (Phase Doppler Particle Analyze) method for example [8]. Other optical methods include the use of a non-contact optical sensor to detect and measure droplets in flight [9] or interferometric techniques (IPI: Interferometric Particle Imaging) [10].

The characterization of leaf coverage surface has also interested researchers. The techniques used were mostly based on water-sensitive papers (WSP) [11], more recently coupled with imaging [12–14]. In Reference [14], the authors propose a comparative study of three methods for using data collected on sensitive paper: visual assessment of coverage rate, visual droplet counting and measurement by an imaging system. The correlation obtained was very good only for coverage rates below 20%. In addition to techniques using water-sensitive papers, other methods imply the use of fluorescent tracers. We can mention the work of Llop et al. [15] in which the use of a fluorescent tracer (Helios SC 500; 0.1% *v/v*) traced the deposition on the leaves of the plants (in ng per cm^2) as well as the coverage rate of the leaves by the sprayed product. In Reference [16], the authors tried to correlate the surface coverage rate and droplet size distribution. To do this, they simultaneously sprayed sensitive papers and Petri dishes containing silicone oil with water to which a red Ponceau marker (with a concentration of 2%) had been added.

As existing qualitative analysis methods adopting the use of sensitive paper or tracers are often not "smart" enough to achieve rapid collection with a single click and automatically adapt to the environment (e.g., changing lighting), researchers have recently proposed an intelligent node based on a vision sensor to automatically collect images of droplet deposition [17]. Other authors have considered the use of a commercial leaf wetness sensor to monitor spraying and have compared the results obtained with the WSP technique [18].

The final objective of our work is to design low-cost micro-sensors that can be used directly in the field to quickly measure leaf coverage surface. It should be noted that our measurement of the coverage surface is an indirect measurement, since we intend to determine the covered surface via a measurement of the volume of droplets deposited on the sensor. Some authors have already proposed such a solution, consisting of an electronic detection system based on resistance [19,20]. However, there is not yet a good enough resolution to detect fine drops. In our case, we propose an integrated optics solution. In a previous work, we fabricated straight waveguides based on Ge-Se-Te chalcogenide films and we tested their sensitivity to water droplet deposition at the wavelength of 1.55 µm. As expected, water absorbed part of the evanescent wave of the guided light, leading to a decrease in the transmitted intensity. Moreover, we showed that the greater the droplet volume and the greater the number of drops, the greater the decrease in intensity [21]. Nevertheless, we observed a non-linearity of the results, with a lower sensitivity for large volumes. Moreover, we limited our study to volumes comprised between 2 and 10 µL, with a step of 2 µL. In this article, we propose to go a little further, even if we are still at the beginning of the development of sensors. The goal was to develop a probe able to measure smaller drops. The studied volumes ranged from 0.1 to 2.5 µL, and a step of 0.1 µL was chosen to estimate the probe precision. Furthermore, we propose, with the same straight waveguides as in our previous work [21], to no longer work at a single wavelength but to use spectroscopy, which, coupled with multivariate analysis, should allow us to increase sensitivity and to consider accessing a measurement of the composition of the sprayed liquid. Among the different methods of multivariate analysis, principal component analysis (PCA) and partial least squares (PLS) techniques, that are commonly used in chemometrics [22], were chosen.

2. Materials and Methods

2.1. Waveguide Manufacture and Characterization

(a) Manufacture

According to previous studies [21,23–25], rib-type waveguides were fabricated. The waveguiding structures were made from a 6 µm-thick $Ge_{25}Se_{65}Te_{10}$ cladding layer (which refractive index n is 2.55 at the wavelength $\lambda = 1.55$ µm) deposited onto silicon substrates and a 6 µm-thick $Ge_{25}Se_{55}Te_{20}$ core layer (n = 2.65 at $\lambda = 1.55$ µm) on top, the core layer being then lithographed, and unprotected parts etched down to 4 µm. Ge-Se-Te films were deposited by thermal co-evaporation, using a MEB 500 device from PLASSYS, Marolles-en-Hurepoix, France. Two current-induced heated sources were used to evaporate Tellurium and Selenium and an electron beam evaporator was used for Germanium. The three sources were placed in a specific configuration to obtain homogeneous films on an area of 4 cm in diameter. Silicon substrates of 3 cm × 4 cm were used to fabricate the components and microscope slides were used for further film characterization. Before deposition, the co-evaporation chamber was evacuated down to 10^{-5} Pa to avoid contamination. During deposition, the substrate holder was rotated for a best homogenization and heated at 70 °C thanks to a hot water circulation. The elemental deposition rates were controlled thanks to three independent thin film deposition controllers (from TELEMARK, Battle Ground, WA, USA and INFICON, Bad Ragaz, Switzerland). The deposition process was initiated when the three elemental deposition rates were stable and was stopped when a total thickness of 6 µm was reached for each film.

Once the cladding layer and the core layer were deposited onto silicon substrates, the geometry of the core layer was modified by photolithography and ion beam etching, successively. For the photolithographic step, the AZ4533 positive resin and the AZ726 developer from MicroChemicals (Ulm, Germany) were used and samples were insolated during 7 s under a 400 W-UV lamp, through a mask patterned with opaque bands of different widths (ranging from 4 to 42 µm). For the etching step, performed using a set-up provided by PLASSYS (Marolles-en-Hurepoix, France), a mixture of argon and few percent of oxygen was used. The working pressure was fixed to 4.5×10^{-2} Pa, the partial pressure P_{O2} being first set to 7.7×10^{-3} Pa. A voltage of 400 V was applied, and an etching duration of 50 min was chosen to achieve an etching depth of 4 µm [21].

(b) Characterization

After deposition, each film was analyzed in terms of thickness, composition, and refractive index thanks to a DEKTAK mechanical profilometer (BRUKER, Billerica, MA, USA), a S4500 Energy-Dispersive X-ray spectrometer (HITASHI, Tokyo, Japan) and by exploiting NIR spectra recorded in the range 500–2500 nm using a Cary 5000 spectrophotometer (AGILENT TECHNOLOGIES, Santa Clara, CA, USA), respectively. After etching, the etching depth was checked by profilometry and the component was observed from above and on the edge by using an Inspect S50 scanning electron microscope (FEI, Hillsboro, OR, USA).

2.2. Optical Setup

The fabricated straight waveguides provided either a multi-mode or a single-mode propagation behaviour, depending on their width and on the wavelength. The bench used for their characterization was equipped with a supercontinuum laser source (SM-100 NIR; LEUKOS, Limoges, France) combined with a pulse generator (4422; SEFRAM, Saint-Etienne, France) (Figure 1). A wavelength high pass filter (FELH1150nm; THORLABS, Newton, NJ, USA) was installed to fit to the Ge-Se-Te waveguides transmission range. To minimize the coupling losses, we used a 3.5 mm focal length lens (87–127 AR; Edmund Optics, Barrington, NJ, USA) to focus the signal from the supercontinuum fiber at the waveguide entrance with a 0.4 numerical aperture. The waveguide sample was held on a 3 axes translation stage platform (M-562; Newport) and fixed by a vacuum mount (HWV001, THORLABS, Newton, NJ, USA). Moreover, we used a convex lens (AL1815-C; THORLABS, Newton, NJ, USA) placed at 11 mm from the output of the waveguide sample to collimate the output beam. This beam hit a beam splitter (BSF10-C, THORLABS, Newton, NJ, USA), 10% of the beam was reflected onto a camera (WDY SMR S320-VS; NIT, Verrières le Buisson, France) to control the injection in the right RIB waveguide. The remaining 90% of the beam was transmitted to a lens (AL1225 M-C; THORLABS, Newton, NJ, USA) to inject light into the spectrometer (UV-VIS-NIR TF; ARCoptix, Neuchatel, Switzerland). This spectrometer performed a Fourier-Transform scanning in the NIR spectrum range (FT-NIR). The measurement spectral range was between 1200 and 2000 nm.

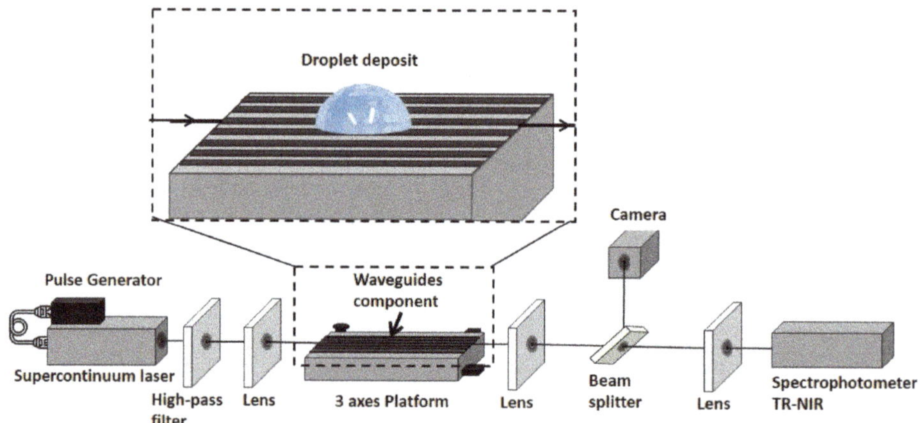

Figure 1. Spectrophotometric bench used to characterize droplet deposits onto the rib waveguide surface. Zoom on the component to visualize the droplet deposited on the top to perform the tests.

2.3. Droplet Deposition and Spectra Acquisition

The influence of the droplet deposition on the waveguide transmission was studied by recording the output spectrum before and after the droplet deposition onto the surface of a rib waveguide. It should be noted that a multi-mode (and therefore wide) guide ensuring the highest possible light intensity collection was selected for the experiments. The temperature of the room was regulated at 24 ± 1 °C. Demineralized water was used in all the experiments and different volumes of droplets were tested. A micropipette was used to dispense droplets from 0.1 to 2.5 μL in volume, in steps of 0.1 μL, on the surface of the rib waveguides. Between two droplet depositions, the sample was dried with dry air and the initial transmission of the waveguide was ensured not having been altered. All the experiments were performed three times on three different days. Each series, corresponding to each day, contains 25 spectra for deposit volumes from 0.1 to 2.5 μL. In total, 75 spectra were thus recorded. The series are called S1, S2 and S3.

For each spectrum, the transmittance T (λ) was calculated thanks to Equation (1) where I_0 is the spectrum recorded before the droplet deposit, I the spectrum after the droplet deposit and I_n the black spectrum (recorded when the shutter of the spectrometer is closed and no light can enter it):

$$T(\lambda) = \frac{I(\lambda) - I_n(\lambda)}{I_0(\lambda) - I_n(\lambda)}, \tag{1}$$

2.4. Multivariate Data Analysis

(a) Principal Component Analysis (PCA)

The objective of the multivariate data analysis, in spectroscopy, is to extract information from the different spectra by projecting results in a reduced basis containing much less dimensions than a spectrum acquisition points number (number of wavelengths) [22]. The hypothesis which is done is that this information will be obtained by studying the differences between the spectra. Ideally, spectra are expected to be identical for equal droplet volumes, and on the contrary, the spectra corresponding to the smallest volumes are expected to be a little different from the spectra obtained for the largest volumes. But several parameters can interfere with the measurement, such as the position of the droplet on the waveguide, the temperature or the light injection efficiency; indeed, this latter can vary due to misalignment between series of similar experiments or to laser sources fluctuations between reference (I_0) and measured (I) spectra; this makes it difficult to exploit the raw spectra. One of the objectives of the PCA will first be to identify the outliers, corresponding to aberrant or atypical spectra to clean the dataset. Another objective will consist in highlighting the differences or similarities between the spectra.

PCA consists in trying to define new variables, which will cause as little information loss as possible. Several formalisms can be used, such as matrix writing [26]. Thus, we can write the matrix X of the initial data (a matrix of dimensions $N \times P$, N corresponding to observations–light transmission and P to variables–wavelengths), as the sum of A terms (A being the dimensionality of the new basis, lower than N and P), which explains most of the data and a residual matrix E corresponding to "noise" (part of the data that are not explained by the analysis). Each of the A terms is itself the product of two vectors t_i and $p_i{}'$, so that the matrix X of the initial data can be written as in Figure 2:

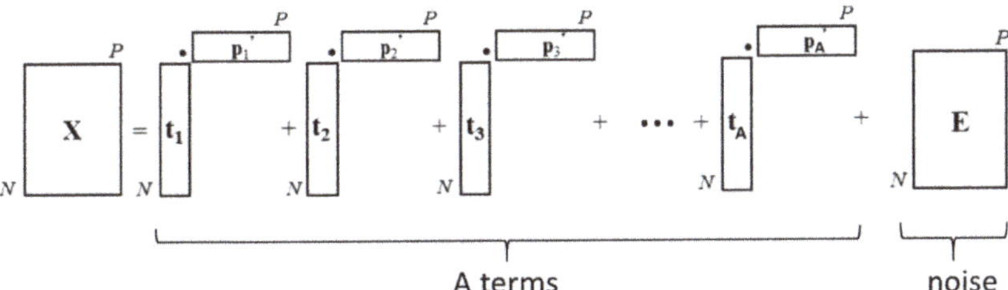

Figure 2. Writing of the matrix X of the initial data according to the PCA model.

The t_i are called score vectors, whereas the p_i are called loading vectors [26]. The p_i are the coordinates of a principal component (PC) in the original space (the wavelengths space, i.e., the PC is a spectrum) while the t_i are the coordinates of measured spectra in the PC basis. The p_1 are computed to maximize the variance of measured spectra versus PC1, and so on for the rest of the variance versus the other PCi. Thus, the PCi are ordered in decreasing percentage of explained variance versus the different recorded spectra, PC1 corresponding to the highest percentage. Usually, the score plot of the first two score vectors t_1 and t_2, which thus capture the essential information from the spectra, will highlight similarities or differences between the initial spectra, and will show outliers.

(b) Regression Model

In chemometrics, partial least squares (PLS) regression is the most common regression method for multivariate data [27]. In spectroscopy, PLS is used in most cases for relating spectral data to measured variables [28]. In this study, PLS regression was used to model droplet volume using $T(\lambda)$. A general PLS model was built using S1 and S2 calibration sets (50 samples) to predict the samples of the independent test set S3 (25 samples).

Before preprocessing, the data acquired in transmittance were transformed in absorbance $[-\log T(\lambda)]$. The remained spectra were smoothed by using a Savitzky–Golay algorithm [29] with a second order polynomial and 15 points. Then, standard normal variate (SNV) [30] was applied to remove the baseline. The number of latent variables was determined by comparing performances by leave-one-out cross-validation [31].

The performances of the cross-validations (using S1 and S2 sets), and prediction (using the independent set S3) were assessed through the number of latent variables used in the models, the coefficient of determination R^2 and the standard error of cross-validation (SECV) and standard error of prediction (SEP) corrected from the bias.

All the computations were performed with Matlab software (Matlab R2015b, Mathworks).

3. Results & Discussion

3.1. Characterization of the Component

Table 1 gives the characteristics of the two deposited layers. The thicknesses errors are less than 2.5%, the compositions are little different (higher atomic percentages in Te but lower atomic percentages in Se) with those expected and the refractive indices n deduced from the transmission spectra are little lower than the targets but the index difference Δn remains around 0.1, as wanted.

Table 1. Ge-Se-Te core and cladding layer compositions and optogeometrical properties.

Film	Thickness (µm) ± 0.01 µm	Composition ± 2 at.%	n at 1.55 µm ± 0.01	$\Delta n = n_{core} - n_{cladd}$
Cladding layer	5.85	$Ge_{25}Se_{61}Te_{14}$	2.49	0.11
Core layer	5.91	$Ge_{26}Se_{53}Te_{21}$	2.60	

Figure 3 shows the typical section of a rib waveguide and allows to confirm the etching depth of 4.5 µm measured by profilometry. This value is close to that expected after an etching of 50 min. A thin layer of resin can be seen, since the images were recorded before the step of removing the excess resin that had not been attacked during etching. Note that the 35 µm-wide rib waveguide presented in Figure 3 is the one used for the experiments.

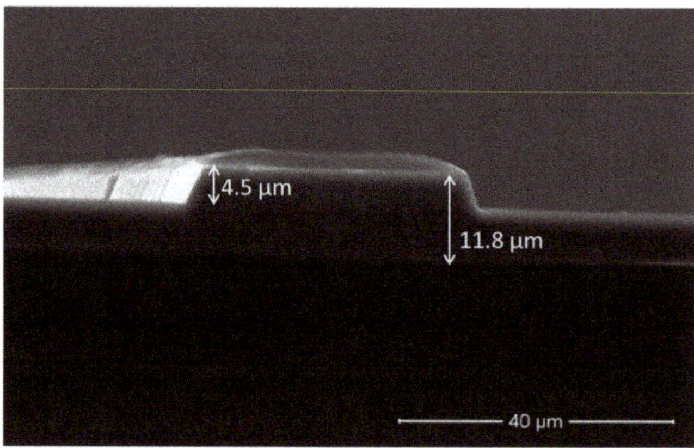

Figure 3. SEM picture of a typical rib waveguide.

3.2. Raw Spectra

A total of 75 relative transmission spectra (corresponding to the three series S1, S2, and S3) were recorded and are superimposed in Figure 4. The first remark is that the raw spectra are not noisy, even for low transmissions, which means that the electronic noise is negligible. The second remark is that there is a certain homogeneity in the spectra with the presence of a hollow at about 1450 nm, corresponding to an overtone of the absorption band related to the vibrational energy of the oxygen-hydrogen (O–H) bond in water. This means that our waveguide is sensitive throughout the whole range of volumes studied, i.e., from 0.1 to 2.5 µL, whereas resistance-based sensors did not reach the below 5 µL region [19,20]. However, one can notice that spectra are not gathered depending on the volume: This lack of repeatability is probably due to random fluctuations of the optical alignment. Moreover, four spectra are a little different, with a smaller amplitude in the transmittance, and correspond to the rejected outliers (dashed spectra in Figure 4). To finish, we can see that it is impossible from these raw spectra to extract a correlation between transmittance and volumes. This observation is also true if we focus on the water absorption wavelength (1450 nm), as shown in the insert in Figure 4. The non-correlation between the transmittance of the raw spectra and the droplet volumes justifies the use of a multivariate analysis to solve this problem.

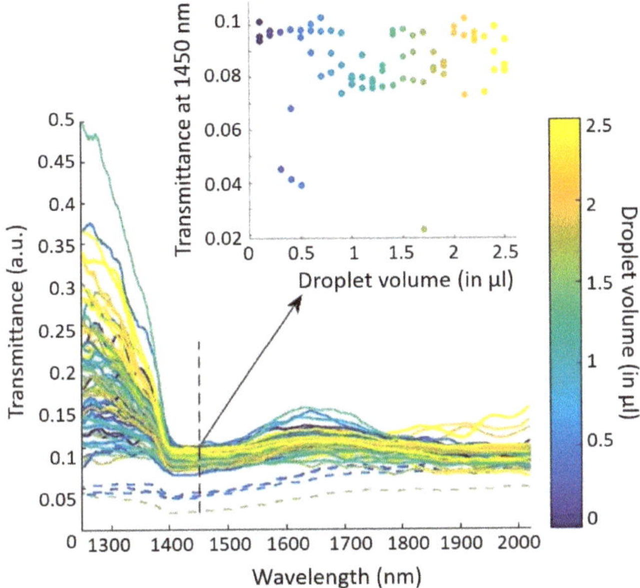

Figure 4. Transmittance spectra obtained from measurements. Outliers are highlighted in dashed lines. The insert shows the transmittance at 1450 nm for each volume.

3.3. Principal Component Analysis

A principal component analysis (PCA) was performed on the whole dataset. It allowed identifying two principal components PC1 and PC2, explaining 95.81% of the variance between the different recorded spectra (85.72% for PC1 and 10.09% for PC2). We can therefore consider that the main information is contained in these two principal components, the rest can be considered as noise.

To go further in the analysis, we plotted the scores on each principal component for each sample, corresponding to the different volumes and measurement days (3 times 25 different volumes). Note that scores represent new coordinates of the different samples on the generated components (Figure 5a for PC1 and Figure 6a for PC2). We also plotted the loadings versus wavelength for each principal component (Figure 5b for PC1 and Figure 6b for PC2). To end, we realized a score plot on PC1 and PC2 generated from the whole dataset (Figure 7).

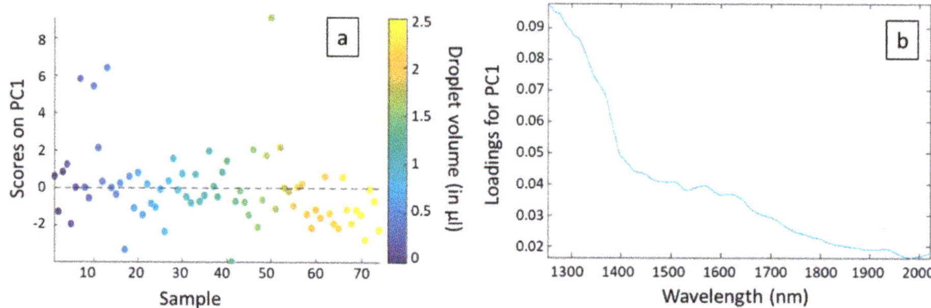

Figure 5. Scores (**a**) and loadings (**b**) related to principal component 1 (PC1). Note that in (**a**) the samples are classified by increasing volume, regardless of the origin of the series (S1, S2, or S3).

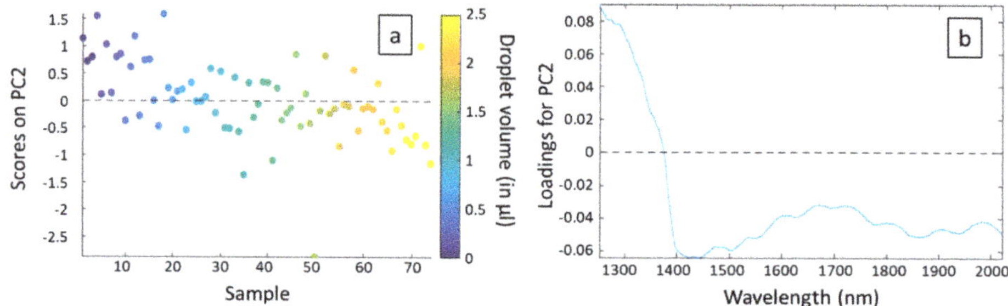

Figure 6. Scores (**a**) and loadings (**b**) related to PC2. Note that in (**a**) the samples are classified by increasing volume, regardless of the origin of the series (S1, S2, or S3).

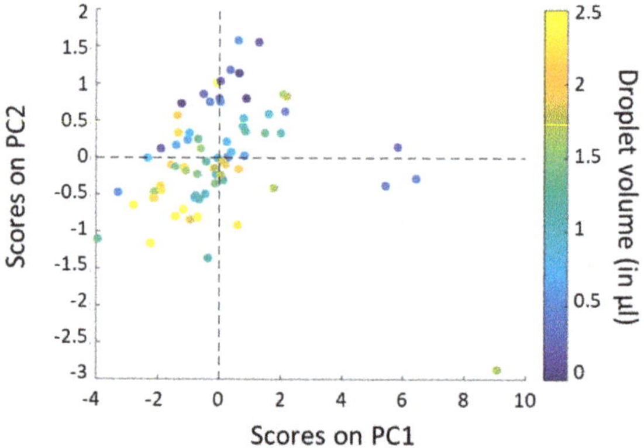

Figure 7. Score plot on PC1 and PC2 generated from the whole dataset.

In Figure 5a we can observe that, except for four points, scores on PC1 are homogeneously distributed around the axis corresponding to a score of 0. It signifies that there is no correlation between PC1 and the droplet volume (color coded). The main variance between the different spectra is thus not due to absorption of water but more to physical phenomena. This is confirmed by Figure 5b. Indeed, we can see that loadings for PC1 are all positive and decrease with the wavelength: PC1 is probably correlated with injection losses (which are supposed to be the lowest at short wavelengths where they were optimized) and propagation losses. Concerning the four points that are in margin of Figure 5a, they correspond to the atypical spectra already observed in Figure 4. They probably correspond to errors in the measurements.

In Figure 6a, we observe a correlation between the scores on PC2 and the droplet volumes. Indeed, even if the effect is not very pronounced, we can notice that scores on PC2 are mainly positive for low volumes, decrease when volumes increase and are mainly negative for high volumes. The principal component PC2 is thus related to the absorption of water. This is confirmed by Figure 6b. Indeed, the loadings for PC2 versus the wavelength form a curve which is totally in agreement with the absorption spectrum of water, with a high absorption peak at around 1450 nm and another one at around 1940 nm. The loadings for PC2 are negative in the spectral range where water absorbs.

Note that the outliers observed in Figure 5a are not observed in Figure 6a. It confirms the fact that these outliers are well correlated to a problem in the measurements and not to the volume of the droplets.

The conclusions that can be drawn from Figures 4 and 5 are therefore that 85.72% of the explained variance between the different spectra is due to physical phenomena, while only 10.09% of the explained variance is due to the presence of water and therefore to the volume of the droplet deposited. Unfortunately, this means that system disturbances have a greater impact on spectra than the measurand. But the positive point is that the effect of volume is measurable.

In Figure 7, corresponding to the score plot on PC1 and PC2, outliers are still present: This is consistent with the previous observations (Figures 4 and 5a). The four spectra corresponding to the four outliers are removed for the prediction model, because a problem of measurement is suspected. Then, we notice that there are no groups formed by the different points (representing the initial spectra): The distribution of the points is indeed quite homogeneous as volume correlated groups overlap. Nevertheless, we find the same trend in PC2 as in Figure 6a: One can see that points corresponding to low volumes are mainly characterized by positive scores whereas points corresponding to high volumes are mainly characterized by negative scores.

3.4. Prediction Model

A PLS model was built with a calibration set containing spectral dataset from S1 + S2, after having removed the outliers. The results of the calibration model are presented in Figure 8a where the predicted volumes obtained are plotted versus the real values. The metrological performance of the system (laser + probe + detector + model) can then be extracted. Repeatability is given by the standard error of cross-validation (SECV) or the variance, accuracy by the bias, and precision by the coefficient of correlation R^2. This latter is equal to 0.803, the bias is 0.00508, and the variance is of 0.299 µL. So, it tends to show that the precision of the model for the calibration set is mainly limited by the variance, since most of it is due to physical perturbation (PC1) and limits the learning process. The low value of the bias may be interpreted as the fact that variance of S1 + S2 is symmetric and has little impact on mean error. The result is quite encouraging.

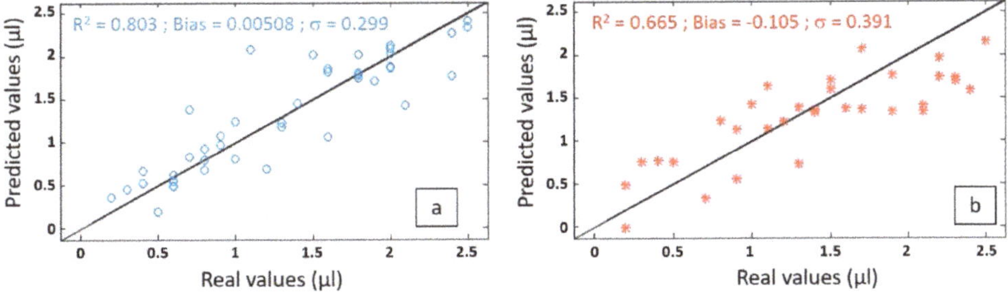

Figure 8. Model performance for the (**a**) cross-validation and (**b**) prediction.

In order to validate the predictive model in real conditions, it was then tested on the independent spectral dataset from the series S3. The predicted volumes obtained are plotted versus the actual volumes in Figure 8b. The coefficient of correlation is now of 0.665 and the standard error of prediction or the variance in the predicted volume increased from 0.299 to 0.391 µL. The bias is now 50 times much stronger (at a value of −0.105), which negatively impacts the R^2.

The failure of the coefficient of determination (66.5%) isn't due to the volume but to the test conditions: Light injection quality, droplet position on the waveguide, spectrometer setting, etc. A first physical improvement is related to light injection, this can be optimized and made robust versus wavelength changes and mechanical vibrations by studying injection alignment impact, pigtailing the fibre from the light source with the waveguide and monitoring directly the output signal of the light source. Also, the temperature may slightly modify the evanescent field extent and thermal monitoring can quantify the temperature explained variance. Finally, droplet lateral position modifies the surface of interaction between water and light and thus varies the output signal; this effect can be studied by imaging droplet position and can be limited by measuring the transmission of a series of close (few microns or tens of microns distance) waveguides on which the droplet is deposited. At the end, by decreasing the physical variance, the percentage of chemically explained variance will increase and improve the learning process to order to obtain a better model. Moreover, the model can be improved from a numerical point of view. Using a bigger calibration test should improve the learning process and, secondly, using a bigger test set should permit better evaluation performance in real conditions. The PLS construction depends on PLS dimension and its error can be minimized by choosing the proper dimensionality.

The 66.5% prediction coefficient of determination can thus be increased, as well as the 0.39 µL precision can be reduced. Nevertheless, it already unveils the possibility of reaching µL or lower resolution, which is more than needed for spray deposit analysis since a less-than-1 µL droplet cannot be deposited as it is suspended in air. The integrated optics approach will permit the addition of functions as power splitting to simplify pigtailing on any number of waveguides in a compact way. Mach–Zehnder or multimode interferometers can be used for on chip temperature monitoring. Additionally, wavelength filters and (de)multiplexers can be used to select wavelengths of interest to reduce the size, complexity, and cost of the system; before this, a study of the reduction of the number of wavelength variables study must be conducted with PCA to feed PLS construction and choose the correct wavelengths depending on prediction performance.

4. Conclusions

The accurate determination of the leaf coverage, a key measurement of the spraying process to maximize spray efficiency, requires specific measuring tools. We have developed the first brick of micro-sensors that can be used directly in the field to quickly and automatically measure leaf coverage surface. Even if this work is still a proof of principle, the first results are encouraging. The probe based on Ge-Se-Te rib optical waveguides is sensitive throughout the whole range of volumes studied, i.e., from 0.1 to 2.5 µL.

A principal component analysis performed thanks to spectral measurements showed that the spectra variance was related at 10.09% to the presence of water onto the waveguide surface and that 85.72% of the variance had to be suppressed or strongly limited. The tests for each volume were repeated three times to apply a prediction model that estimates the droplet volume present on the waveguide. First results show that the prediction, on an independent data set, is 66.5% reliable and the precision on the predicted volume is equal to 0.39 µL.

Two complementary ways can be investigated to increase the percentage of variance explained by the volume of the deposited droplets: (a) To increase the measuring system robustness versus interfering physical quantities, and (b) to deepen data analysis with thanks to the learning process in order to create a better prediction model. The implementation of the technology to develop a sensor for use in real condition environments will then need further studies to determine the right optical waveguide structure to develop, low cost and portable laser sources and spectrometers, as well as the optimized packing and connectors.

Author Contributions: Conceptualization, R.B.; methodology, M.R.; software, R.K.K. and M.R.; validation, A.T.B.; formal analysis, A.T.B. and M.R.; investigation, A.T.B.; resources, R.E. and D.H.; data curation, R.B., R.K.K. and C.V.; writing-original draft preparation, A.T.B. and M.R.; writing-review and editing, R.B., R.K.K. and C.V.; supervision, R.B. and C.V.; project administration, R.B.; funding acquisition, R.B.

Funding: This work was supported by the French National Research Agency under the Investments for the Future Program, referred as ANR-16-CONV-0004.

References

1. Paul, C.; Techen, A.K.; Robinson, J.S.; Helming, K. Rebound effects in agricultural land and soil management: Review and analytical framework. *J. Clean. Prod.* **2019**, *227*, 1054–1067. [CrossRef]

2. Pandey, K.; Dangi, R.; Prajapati, U.; Kumar, S.; Maurya, N.K.; Singh, A.V.; Pandey, A.K.; Singh, J.; Rajan, R. Advance breeding and biotechnological approaches for crop improvement: A review. *Int. J. Chem. Stud.* **2019**, *7*, 837–841.

3. Lee, R.; Uyl, R.; Runhaar, H. Assessment of policy instruments for pesticide use reduction in Europe; Learning from a systematic literature review. *Crop Prot.* **2019**, *126*, 104929. [CrossRef]

4. Lamichhane, J.R.; Aubertot, J.N.; Begg, G.; Birch, A.N.E.; Boonekamp, P.; Dachbrodt-Saaydeh, S.; Hansen, J.G.; Hovmøller, M.S.; Jensen, J.E.; Jørgensen, L.N.; et al. Networking of Integrated Pest Management: A Powerful Approach to Address Common Challenges in Agriculture. *Crop Prot.* **2016**, *89*, 139–151. [CrossRef]

5. Grella, M.; Marucco, P.; Balsari, P. Toward a new method to classify the airblast sprayers according to their potential drift reduction: Comparison of direct and new indirect measurement methods. *Pest Manag. Sci.* **2019**, *75*, 2219–2235. [CrossRef]

6. Hilz, E.; Vermeer, A.W.P. Spray drift review: The extent to which a formulation can contribute to spray drift reduction. *Crop Prot.* **2013**, *44*, 75–83. [CrossRef]

7. Balsari, P.; Grella, M.; Marucco, P.; Matta, F.; Miranda-Fuentes, A. Assessing the influence of air speed and liquid flow rate on the droplet size and homogeneity in pneumatic spraying. *Pest Manag. Sci.* **2019**, *75*, 366–379. [CrossRef]

8. Nuyttens, D.; Baetens, K.; De Schampheleire, M.; Sonck, B. PDPA laser based characterization of agricultural sprays. *Agric. Eng. Int. CIGR J.* **2006**, *8*, 1130–1682.

9. Tröndle, J.; Ernst, A.; Streule, W.; Zengerle, R.; Koltay, P. Non-contact optical sensor to detect free flying droplets in the nanolitre range. *Sens. Actuators A Phys.* **2010**, *158*, 254–262. [CrossRef]

10. Bilsky, A.V.; Lozhkin, Y.A.; Markovich, D.M. Interferometric technique for measurement of droplet diameter. *Thermophys. Aeromech.* **2011**, *18*, 1–12. [CrossRef]

11. Turner, C.R.; Huntington, K.A. The use of a water sensitive dye for the detection and assessment of small spray droplets. *J. Agric. Eng. Res.* **1970**, *15*, 385–387. [CrossRef]

12. Nansen, C.; Ferguson, J.C.; Moore, J.; Groves, L.; Emery, R.; Garel, N.; Hewitt, A. Optimizing pesticide spray coverage using a novel web and smartphone tool, SnapCard. *Agron. Sustain. Dev.* **2015**, *35*, 1075–1085. [CrossRef]

13. Marchal, A.R.S.; Cunha, M. Image processing of artificial targets for automatic evaluation of spray quality, Trans. *Asabe (Am. Soc. Agric. Biol. Eng.)* **2008**, *51*, 811–821. [CrossRef]

14. Fox, R.D.; Derksen, R.C.; Cooper, J.A.; Krause, R.; Ozkan, T.H.E. Visual and image system measurement of spray deposits using water-sensitive paper. *Trans. ASAE* **2003**, *19*, 549–552. [CrossRef]

15. Llop, J.; Gil, E.; Gallart, M.; Contador, F.; Ercilla, M. Spray distribution evaluation of different settings of a hand-held-trolley sprayer used in greenhouse tomato crops. *Pest Manag. Sci.* **2016**, *72*, 505–516. [CrossRef] [PubMed]

16. Cerruto, E.; Manetto, G.; Longo, D.; Failla, S.; Papa, R. A model to estimate the spray deposit by simulated water sensitive papers. *Crop Prot.* **2019**, *124*, 104861. [CrossRef]

17. Wang, L.; Yue, X.; Liu, Y.; Wang, J.; Wang, H. An intelligent vision based sensing approach for spraying droplets deposition detection. *Sensors* **2019**, *19*, 933. [CrossRef] [PubMed]

18. Foque, D.; Dekeyser, D.; Langenakens, J.; Nuyttens, D. Evaluating the usability of a leaf wetness sensor as a spray tech monitoring tool. *Asp. Appl. Biol.* **2018**, *137*, 191–200.

19. Kesterson, M.A.; Luck, J.D.; Sama, M.P. Development and preliminary evaluation of a spray deposition sensing system for improving pesticide application. *Sensors* **2015**, *15*, 31965–31972. [CrossRef]
20. Giles, K.; Crowe, T. Real-Time Electronic Spray Deposition Sensor. U.S. Patent No. US 7,280,047 B2, 9 October 2007.
21. Taleb Bendiab, A.; Bathily, M.; Vigreux, C.; Escalier, R.; Pradel, A.; Kribich, R.K.; Bendoula, R. Chalcogenide rib waveguides for the characterization of spray deposits. *Opt. Mater.* **2018**, *86*, 298–303. [CrossRef]
22. Martens, H.; Naes, T. Multivariate calibration. In *Chemometrics*; Springer: Berlin, Germany, 1984; pp. 147–156.
23. Vigreux, C.; Escalier, R.; Kribich, R.; Pradel, A. Chalcogenide circuits for the realization of CO_2 micro-sensors operating at 4.23 μm. In Proceedings of the IEEE 18th ICTON, Trento, Italy, 10–14 July 2016. [CrossRef]
24. Vigreux, C.; Vu Thi, M.; Escalier, R.; Maulion, G.; Kribich, R.; Pradel, A. Channel waveguides based on thermally co-evaporated Te-Ge-Se films for infrared integrated optics. *J. Non-Cryst. Solids* **2013**, *377*, 205–208. [CrossRef]
25. Vigreux, C.; Vu Thi, M.; Maulion, G.; Kribich, R.; Barillot, M.; Kirschner, V.; Pradel, A. Wide-range transmitting chalcogenide films and development of micro-components for infrared integrated optics applications. *Opt. Mater. Exp.* **2014**, *4*, 1617–1631. [CrossRef]
26. Wold, S.; Esbensen, K.; Geladi, P. Principal Component Analysis. *Chemom. Intell. Lab. Syst.* **1987**, *2*, 37–52. [CrossRef]
27. Wold, S.; Sjöström, M.; Eriksson, L. PLS-regression: A basic tool of chemometrics. *Chemom. Intell. Lab. Syst.* **2001**, *58*, 109–130. [CrossRef]
28. Mark, H.; Workman, J. *Chemometrics in Spectroscopy*, 1st ed.; Elsevier/Academic Press: Cambridge, MA, USA, 2007.
29. Savitzky, A.; Golay, M.J. Smoothing and differentiation of data by simplified least squares procedures. *Anal. Chem.* **1964**, *36*, 1627–1639. [CrossRef]
30. Barnes, R.J.; Dhanoa, M.S.; Lister, S.J. Standard Normal Variate Transformation and De-Trending of Near-Infrared Diffuse Reflectance Spectra. *Appl. Spectrosc.* **1989**, *43*, 772–777. [CrossRef]
31. Wold, S. Cross-Validatory Estimation of the Number of Components in Factor and Principal Components Models. *Technometrics* **1978**, *20*, 397–405. [CrossRef]

Article

Automated Measurement and Control of Germination Paper Water Content

Lina Owino *, Marvin Hilkens, Friederike Kögler and Dirk Söffker

Chair of Dynamics and Control, University of Duisburg-Essen, Lotharstraße 1, 47057 Duisburg, Germany;
marvin.hilkens@de.bertrandt.com (M.H.); friederike.koegler@uni-due.de (F.K.); dirk.soeffker@uni-due.de (D.S.)
* Correspondence: lina.owino@uni-due.de; Tel.: +49-203-37-91866

Received: 21 January 2019; Accepted: 11 May 2019; Published: 14 May 2019

Abstract: Germination paper (GP) is used as a growth substrate in plant development studies. Current studies bear two limitations: (1) The actual GP water content and variations in GP water content are neglected. (2) Existing irrigation methods either maintain the GP water content at fully sufficient or at a constant deficit. Variation of the intensity of water deficit over time for plants grown on GP is not directly achievable using these methods. In this contribution, a new measurement and control approach was presented. As a first step, a more precise measurement of water content was realized by employing the discharging process of capacitors to determine the electrical resistance of GP, which is related to the water content. A Kalman filter using an evapotranspiration model in combination with experimental data was used to refine the measurements, serving as the input for a model predictive controller (MPC). The MPC was used to improve the dynamics of the irrigation amount to more precisely achieve the required water content for regulated water uptake in plant studies. This is important in studies involving deficit irrigation. The novel method described was capable of increasing the accuracy of GP water content control. As a first step, the measurement system achieved an improved accuracy of 0.22 g/g. The application of a MPC for water content control based on the improved measurement results in an overall control accuracy was 0.09 g/g. This method offers a new approach, allowing the use of GP for studies with varying water content. This addressed the limitations of existing plant growth studies and allowed the prospection of dependencies between dynamic water deficit and plant development using GP as a growth substrate for research studies.

Keywords: moisture measurement; Kalman filter; model predictive control; germination paper

1. Introduction

Germination paper (GP) is frequently used as substrate in research concerning plants, mainly in studies involving root phenotyping [1,2], root development in early developmental stages [3], and germination [4,5]. The plants are grown on vertically arranged sheets of GP with the roots growing on the GP surface [1–3]. This allows image acquisition of the root system using flatbed-scanners, and therefore can be implemented in an automated, high-throughput setting as shown in Adu et al. [1]. Irrigation of the GP is applied by a pouch-and-wick system, with the lower edge of the GP immersed in a nutrient solution.

A platform for high-throughput, high resolution root phenotyping is described in Adu et al. [1], as applied to different genotypes of Brassica Rapa. Here the pouch-and-wick system is used with the lower 10 cm section immersed in nutrient solution. It is stated that the plants "showed no symptoms of mineral deficiencies when provided with an appropriate nutrient solution through the wick."

A root phenotyping platform applied in the investigation of root growth dynamics of corn (*Zea mays* L.) is presented in Hund et al. [2]. Here the pouch-and-wick system is used with the lower 2 cm of GP immersed in nutrient solution. The root systems growing on the GP surface were scanned daily. The results showed a linear growth rate of axial roots and an exponential growth rate of lateral roots.

In an examination of the effect of phosphorus availability on root gravitropism for five different genotypes of common bean (*Phaseolus vulgaris* L.), Liao et al. [3] carried out experiments with different substrates (GP, sand, and soil) to "validate existing observations of seedlings in growth pouches with studies of older plants in soil and solid media." The GP was also used in a pouch-and-wick configuration with the lower edge of GP hanging into nutrient solutions with different phosphorous concentrations. The results showed that root gravitropism depends on phosphorous availability and show different characteristics among genotypes. Additionally, the adaption of genotypes to phosphorous deficit is related to their ability to develop roots in the top layers of substrate.

The existing studies demonstrate the applicability of GP in studies related to root growth and development, typically under fully watered conditions. The previous studies state that the available water content is responsible for several effects, and should therefore be controlled if related research questions are to be addressed.

1.1. Induction of Water Deficit in GP

Exploration of water deficit effects on root growth and development with GP as a growth substrate requires the induction of water deficit in GP. Solutions of polyethylene glycol (PEG) as described in [6] have been applied in various studies to induce a water deficit in plants [4,5]. The solutions show different osmotic potentials dependent upon PEG concentration. Plants irrigated with PEG6000-solutions can therefore obtain only a part of the water contained in the solution. The osmotic potential can be measured by thermocouple psychrometry or vapor pressure osmometry [6] and is specified as pressure (measured in bar).

The effect of water deficit on sixteen different wheat genotypes in early developmental stages is described in Rauf et al. [4]. The seeds were placed between two layers of GP moistened with PEG6000-solutions of different concentration to induce water stress of different intensities. It was shown that the tested genotypes differ significantly in water stress tolerance.

In an investigation on the effect of water stress on germination indices, soluble sugar, and proline for five different genotypes of wheat, Qayyum et al. [5] employed petri dishes containing GP moistened with PEG6000-solutions at different concentrations as a growth substrate. The results show that the germination percentage, mean germination time, and coleoptile length decreased with rising water stress while soluble sugar and proline increased.

For studies involving fixed water deficit levels, PEG provides a practical solution. Where dynamic variation of water deficit levels in the course of the growth cycle is necessary, alternative approaches are needed to achieve the required water uptake.

1.2. Current Limitations of GP Use as a Growth Substrate

In some studies using a pouch-and-wick system for irrigation [1–3], a sufficient amount of nutrient solution (not in deficit) is applied to the plants. However, the actual volumes of nutrient solution contained in the GP at different points in time as well as variations of GP water content between different GP sheets are not considered. It was observed in preliminary experiments conducted for this work that GP water content of different GP sheets strongly varied even if irrigated with the same volumes of nutrient solution at the same time, as illustrated in Figure 1. Taking into account the actual water content of each GP used in a study would result in a more accurate evaluation of the effects of plant water uptake in irrigation studies using GP as a growth substrate.

The pouch-and-wick configuration using a constant supply of nutrient solution leads to a maximal distribution of GP water content. This is defined by the equilibrium between absorption of nutrient solution and evapotranspiration. For research tasks concerning the effect of water deficit on root growth and development, improved approaches allowing for supply of water to GP below maximum level would be helpful.

In studies prospecting water stress [4,5], defined water deficit intensities are applied to GP sheets using a fixed concentration of PEG6000. This method maintains a constant water deficit level throughout and is therefore not applicable to studies requiring an application of a sequence of different water deficit intensities to the same GP sheet.

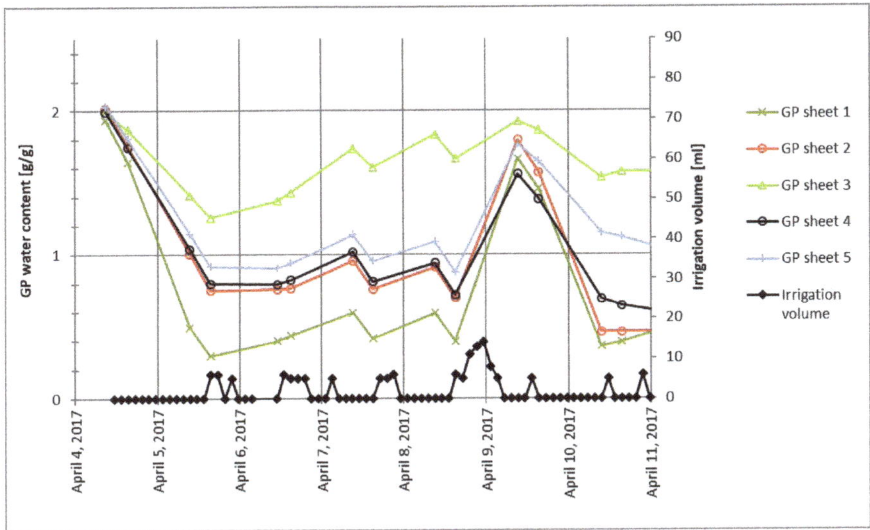

Figure 1. An example of germination paper (GP) water content behavior in sheets with an identical irrigation treatment.

1.3. Resulting Targets for this Research

The contributions of this work are:

(i) The development of a method for an automated measurement of GP water content, which considers the current GP water content and variations of GP water content (between different sheets of GP and over time). Determination of the actual GP water content over time would improve the evaluation of dependencies between water uptake and plant growth and development;

(ii) The development of GP water content control approach using automated measurements and an automated irrigation system. This allows for the application of water deficit with dynamically varying intensities, expanding the scope of applicability of GP, especially for dynamic water deficit studies.

2. Materials and Methods

2.1. Ambient Conditions and Test Rig

A system for measurement and control of GP water content was installed in a test rig for the research of plant dynamics under water stress at the Chair of Dynamics and Control, U Duisburg-Essen, Germany. A partial view of the test rig is shown in Figure 2. The test rig contains up to 20 maize plants (*Zea mays* L., Ronaldinio variety from KWS. divided into four groups of five plants each. Sheets of GP (210 × 297 mm, Hahnemühle, grade 3644, approx. 45 g) were chosen as substrate to allow a scan of the root systems. One plant was placed on each sheet. Each GP sheet was mounted on a sheet of Perspex, which was covered with a second, transparent sheet to allow scanning. Small tanks were put on the bottom right edge of the GP for irrigation, clamped between the two Perspex sheets. Approximately 3 cm of the GP was submerged in the tanks. These tanks could automatically be filled at discrete times with 5–30 mL of nutrient solution via two peristaltic pumps and a distribution system. The system for GP irrigation was described by Sattler [7].

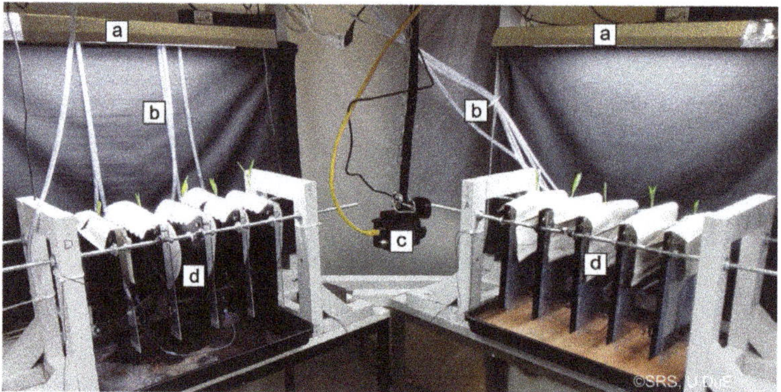

Figure 2. Partial view of the test rig. Two out of four test groups are shown. (a) Artificial illumination; (b) Irrigation tubes; (c) Infrared camera (not relevant in this work); (d) Test group of five specimens each consisting of a Perspex frame, GP, and a corn plant.

Test runs showed that the GP could absorb 30 ml of nutrient solution from the tank in less than 2 h (=sampling time of the measurements), as long as its water content was beneath 0.67 g/g. The maximal water content of GP sheets was between 1.78 and 2.00 g/g.

Preliminary tests to determine the distribution of nutrient solution within the GP were conducted. Four pieces of GP were each marked into eight equal sections measuring 105 mm × 74 mm and submerged in nutrient solution to saturation. In experiments, the GP sheets were dried out until the mass of retained water within the GP was 50, 40, 35, and 20 g respectively, with periodic measurements taken using a precision balance to determine mass of water within the GP. As soon as each GP attained the desired mass, it was divided into the eight pre-marked sections, and water content for each individual section was determined gravimetrically.

Environmental conditions within the test rig were controlled to maintain air temperature of 21–23 °C and artificial illumination for 14 h of simulated daylight. Illumination was provided by four 150 W, 9500 K fluorescent bulbs.

Relative air humidity was not controlled, but was measured twice a day. Relative humidity values ranged between 21–33% during the experiment.

2.2. GP Water Content Measurement

In pre-studies, different methods were evaluated for the measurement of GP water content (Table 1). Methods involving the direct measurement of electrical resistance were eliminated due to inconsistency in values obtained. This was taken to be due to the extremely high resistances exhibited by GP. Indirect measurement of the GP electrical resistance via the discharging of capacitors as described in Hering [8] was applied in the development of the moisture measurement system described in this work.

The relationship between the capacitor charge and electrical resistance can be described as:

$$Q(t) = Q_0^{-\frac{t}{RC}}, \tag{1}$$

with,

$Q(t)$: Current capacitor charge;
Q_0 : Initial capacitor charge;
t : Elapsed time;
R : Electrical resistance, and;
C : Capacitance.

Table 1. Available GP moisture measurement methods.

Measuring Equipment	Measurement Principle	Results
Multimeter	Direct measurement of electrical resistance	Inconsistent values obtained
Air humidity sensor	Measurement of air humidity in a capsule attached to the GP	Inconsistent values obtained
Capacitor with GP as dielectric	Measurement of capacitance	No dependence on GP water content observed
Capacitor circuit with GP as resistance	Measurement of electrical resistance via capacitor discharge	Dependency on GP water content observed

From this relationship, the electrical resistance of the GP is obtained as:

$$R = \frac{t}{C \ln\left(\frac{Q_0}{Q(t)}\right)}. \tag{2}$$

The system for the measurement of GP water content is described in Zimmermann [9]. An electric circuit as described in Hering [8] was implemented. Four capacitors (6300 μF ± 20% each) were connected in a series to facilitate a higher operating voltage as well as to increase the range of measurable resistance. The capacitors were connected to multiple relays, two resistors (100 kΩ ± 5%, 5.3 kΩ ± 10%), a laboratory power supply, and a programmable logic controller (PLC) (ifm CR0403). In Figure 3, the circuit for one sheet of GP is exemplarily shown. For simplification, the relays are drawn as switches connected to the PLC. Connections are denoted by chain-dotted lines. The measurement is realized using the PLC, which is connected to a PC via CAN-BUS. Every measurement is saved on the PC with a time stamp.

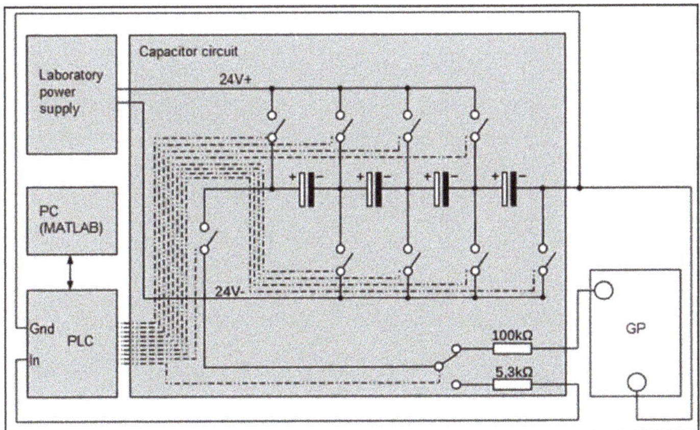

Figure 3. Circuit for the measurement of GP water content.

The measurements were implemented as follows: Each capacitor was consecutively charged to its maximum. The series-wound capacitors were connected to the GP at defined positions, starting the discharging process. The discharging process was terminated after 3 min. The remaining capacitor charge was measured via an analog current measurement via the PLC with an accuracy of ± 0.2 mA. To facilitate the analog-to-digital conversion of the signal by the PLC, a sampling frequency of 10 Hz was selected.

The water content of 8 out of 20 GP sheets (two in each group) was measured successively. A single measurement lasted 15 min and the sampling time was 2 h. Two characteristic curves for the calculation of GP water content from the measured remaining charge were derived for two different positions as shown in Figure 4. The capacitor circuit was discharged through the GP between the indicated measurement positions and the ground. Only one measurement position was connected at a time.

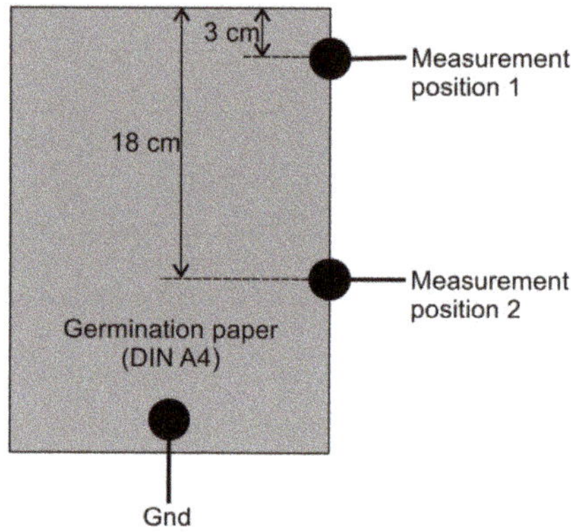

Figure 4. Positions of contacts for measurement of GP water content.

The data were collected by repeated measurements of remaining charge and corresponding gravimetric water contents. The gravimetric water contents were measured using a precision balance. The dedicated measurements and characteristic curves are shown in Figure 5. The assumed characteristic curves were generated by numerical fitting assuming a logarithmic relationship between GP electrical resistance (and, by extension, water content) and remaining charge as suggested by Equation (2) [8].

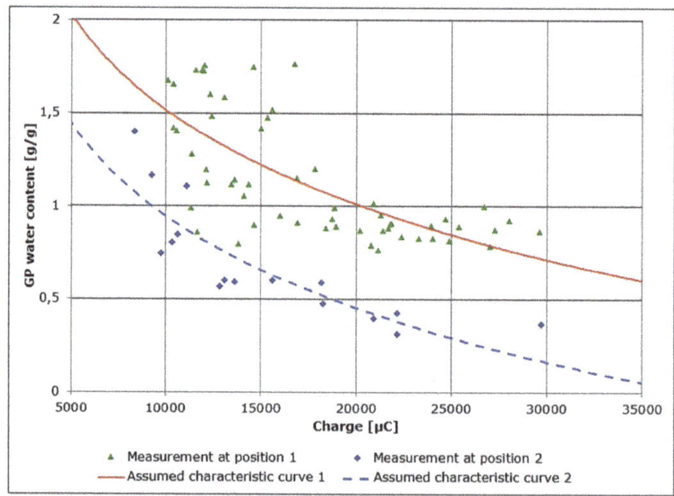

Figure 5. GP water content and remaining charge for the two different measurement positions.

The evaluation of the measurement accuracy shows a mean deviation of automated measurements from gravimetric measurements of more than ± 0.44 g/g for the whole effective range. This initial measurement accuracy, though typical for the implemented circuit, is unfortunately not sufficient enough to be used for the effective control of GP water content. From Figure 5 the practical problem resulting from the use of GP in combination with direct measurements from the automated measurement system can be clearly detected. From a theoretical point of view, the measurements are not accurate or can be assumed as noisy. To overcome this problem using GP, a new (filtered) measurement approach has to be developed. A similar problem is known from studies involving measurement of soil moisture content [10], where filters are also employed as solutions.

2.3. Improved Model-Based Measurement System

Uncertainties in a system measurement often limit the performance of controllers. Observers, filters, and especially Kalman filters were developed in the last few decades to improve estimations of real states resulting from noisy measurements and/or noisy processes/systems (systems with randomly varying system parameters). Kalman filters have been applied in Groenendyk et al. [10] and Rosnay et al. [11], combining model-based predictions of soil moisture content with in-situ measurements to generate more reliable moisture content estimates.

In this work, for measurement improvement, a specific filtering approach was employed. A model for time behavior of GP water content was developed to generate a prediction, which was combined with actual measurements based on correction factors derived from the mean variation of measurements. The model equation:

$$m_{GP}W_k = m_{GP}W_{k-1} + I_{k-1} - E_{GP}T_0, \tag{3}$$

with,

m_{GP} : GP mass;
W_k : Current gravimetric water content;
W_{k-1} : Gravimetric water content at previous time step;
I_{k-1} : Irrigation amount at previous time step;
E_{GP} : GP evaporation rate, and;
T_0 : Sampling time;

is based on the assumption that the evaporation E_{GP} depends on the GP water content. Preliminary tests showed a logarithmic dependency between the water content of GP and the rate of evaporation shown in Figure 6. This is consistent with a previously documented relationship between evaporation and soil water depletion [12]. Measurements were carried out gravimetrically using a precision balance. The two curves show the assumed dependencies between evaporation and GP water content. From these results, a strong (and formalizable) dependency on air humidity can be assumed.

To prospect these dependencies on evaporation the Penman equation as expressed in Ostrowski [13] as:

$$E_{pot} = \frac{s}{s+\gamma}\frac{E_R^{net}}{L} + \frac{\gamma}{s+\gamma}f(v)(e_s - e_a) \tag{4}$$

is employed, where,

$$e_a = \frac{e_s H}{100} \tag{5}$$

is assumed with:

E_{pot} : Potential evaporation;
s : Slope of saturation vapor pressure curve;
γ : Psychrometric constant;
E_R^{net} : Net irradiance;
L : Latent heat of vaporization;

e_s : Saturated vapor pressure;
e_a : Vapor pressure of free flowing air;
H : Relative humidity, and;
$f(v)$: Wind velocity function.

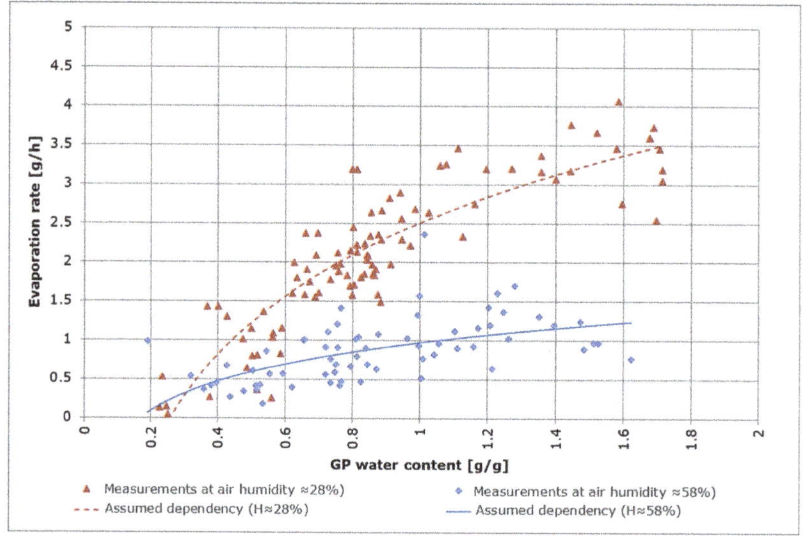

Figure 6. Evaporation and GP water content for two different values of air humidity (28% and 58%).

The parameters saturation vapor pressure gradient s, wind velocity v, net irradiance E_R^{net}, and latent heat equivalent L are assumed to be constant. From this the dependency:

$$E_{pot} = C_1 + C_2(1 - H/100), \tag{6}$$

using constant terms C_1 and C_2 defining the relation between the relative air humidity H and the potential evaporation E_{pot}, the evaporation from the germination paper, E_{GP}, can be derived. The resulting equation for GP evaporation is:

$$E_{GP}(H, W_k) = (c_1 H + c_2) \ln(m_{GP} W_k) - c_3 H + c_4. \tag{7}$$

This model is fitted to the measurements using coefficients c_1 to c_4 (Figure 6) resulting in:

$c_1 = -0.038292758;$
$c_2 = 2.901962527;$
$c_3 = -0.096373808,$ and;
$c_4 = 7.18246835.$

The transpiration influence can be neglected since the surface of plants leaves at the growth stage of interest in the studies related to this work (3 to 5 leaves) was significantly smaller than the surface of GP sheets. Measurements of evapotranspiration from plants at a similar growth stage grown under identical environmental conditions in 200 ml PET pots were found to be lower than 1 g/h, which is significantly lower than the evaporation from the GP sheets.

The automated measurements with model-based correction showed a significantly better correlation to the gravimetrically measured water contents. Implementation of the Kalman filter resulted in the elimination of measurement outliers and consequently a more accurate representation of the system state was obtained as compared to direct measurements taken from the system (Figure 7).

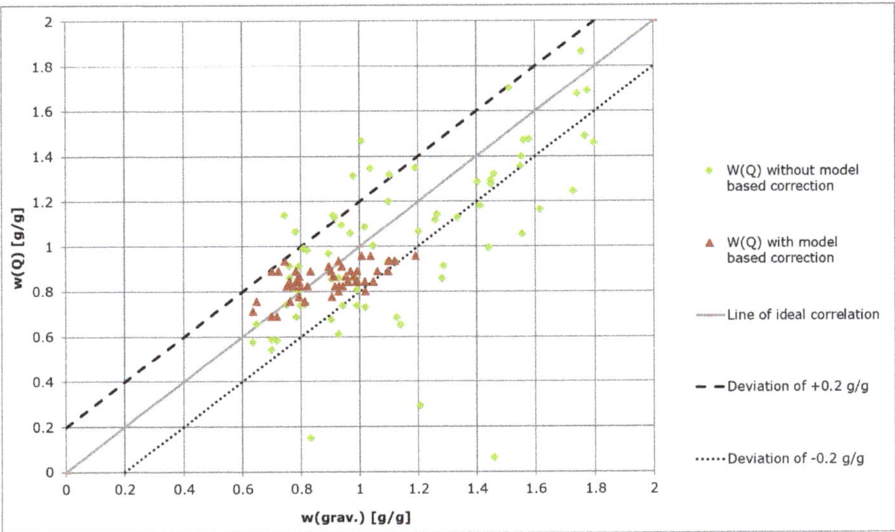

Figure 7. Automated measurements of GP water content W(Q) with and without model-based correction (Kalman filtering).

2.4. GP Water Content Control (MPC and PI)

The control loop for GP water content control is shown in Figure 8. The irrigation element was assumed as a zero order hold element. The measurement circuit was assumed as ideal. The evaporation was considered as a disturbance variable acting directly on the output.

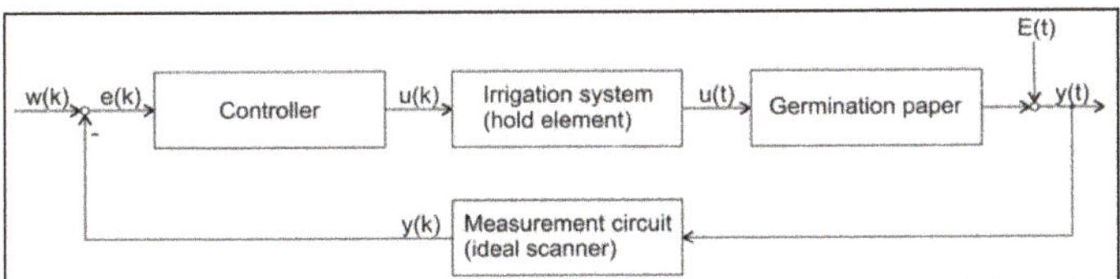

Figure 8. Control loop for the control of GP water content. w(k)—reference variable; e(k)—control deviation; u(k)—calculated irrigation volume; u(t)—irrigation volume; E(t)—evaporation; y(t)—GP water content; y(k)—measured GP water content. (k) denotes a variable in discrete time; (t) denotes a variable in continuous time.

The system to be controlled (GP) is modeled as an integral transfer element. Including the hold element of zero order the z-transfer function:

$$G_1(z) = \frac{T_0}{T_1} \frac{1}{1-z} \tag{8}$$

is deduced. From the transfer function as described by Equation (8), two controllers are derived: A model predictive controller (MPC) based on Equations (3) and (7) [14] and a time discrete PI-Controller [15] described as:

$$G_{R,PI}(z) = \frac{1 - 0.6z^{-1}}{1 - z^{-1}}. \tag{9}$$

The MPC approach was selected because it allowed the quantization of the actuation as well as related limitations. Both controllers were applied to realize two different control tasks, which are:

(i)   An individual control of the water content of a single sheet of GP and;
(ii)   A group control whose implementation is subsequently described:
The mean value of the two automated water content measurements from each group under consideration was calculated. Subsequently a single irrigation volume was calculated by the controller. The irrigation volume was applied to all five GP sheets of the group. Therefore, the GP sheets without water content measurement within the group were controlled by an open-loop control.

2.5. Testing

A test run was carried out:

(i)   To test the accuracy of the water content measurement;
(ii)   To validate the functionality of the described control, and;
(iii) To compare the accuracy of the two controller types defined by the control deviations and the integral of squared error.

Test groups A and C were controlled via individual control mode (two GP sheets each). Groups B and D were controlled via group control mode (five GP sheets each). Groups A and B were controlled using the discrete PI-controller while groups C and D were controlled using MPC. The reference variable was set to be 0.89 g/g for five days and afterwards 0.56 g/g for four days for all groups. The ambient conditions were adjusted as stated above. The GP sheets were weighed twice a day using a precision balance. These measurements were taken as reference for the evaluation of accuracy of the automated measurements.

3. Results

3.1. Preliminary Tests

The investigation of distribution of nutrient solution in GP over time shows a heterogeneous distribution pattern dependent on total GP water content, as qualitatively shown in Figure 9. The distribution gradient increased as the GP water content decreased, indicating an acceleration of evaporation with time.

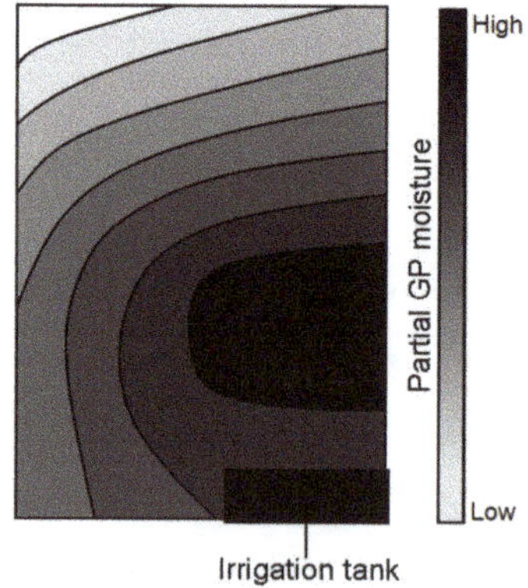

Figure 9. Qualitative distribution of nutrient solution within a GP sheet.

Drying out the GP from saturation and evaluation of the water distribution indicated greater evaporation from the corner samples. This was due to a greater exposure to the environment at the frame boundary. Similarly, the left edge (Figure 9) experienced greater exposure to the atmosphere as a result of facing the interior of the greenhouse. This is assumed to be the reason for the left-right gradient of water distribution obtained. Gravitational effects would also contribute to faster loss of water from the topmost sections.

3.2. Accuracy of Water Content Measurements

The accuracy of the automated water content measurements was defined by the maximal difference between the automated measurements and the gravimetrical measurements as shown in Table 2. An accuracy of ± 0.2 g/g was assumed to be sufficient for most applications within the experimental setup employed. Therefore the effective range was defined to 1.64 g/g $> W > 0.80$ g/g for measurement position 1 and to 0.66 g/g $> W > 0.45$ g/g for measurement position 2. Outside of these ranges, the automatic measurements are not accurate enough to be used for an effective control of GP water content. Variations of 0.2 g/g are large in comparison with a maximum water content of 2.0 g/g. It is assumed that these variations can be minimized by correction using a look-up table and elimination of disturbances within the measurement unit.

Table 2. The results of the measurement accuracy's evaluation. The measurement accuracy is defined as the maximal deviations of automated measurements from the manual measurements . The accuracy is evaluated for different ranges of retained GP moisture.

GP Moisture W	Position 1 Accuracy	Position 2 Accuracy
W > 1.67 g/g	± 0.27 g/g	$>\pm 0.67$ g/g
1.64 g/g > W > 1.02 g/g	± 0.16 g/g	$>\pm 0.67$ g/g
1.01 g/g > W > 0.80 g/g	± 0.16 g/g	± 0.69 g/g
0.79 g/g > W > 0.67 g/g	± 0.31 g/g	± 0.42 g/g
0.66 g/g > W > 0.45 g/g	$>\pm 0.67$ g/g	± 0.27 g/g

3.3. Assessment of Water Content Control

The gravimetric GP water content over time for all groups and for a reference variable of 0.89 g/g is shown in Figure 10.

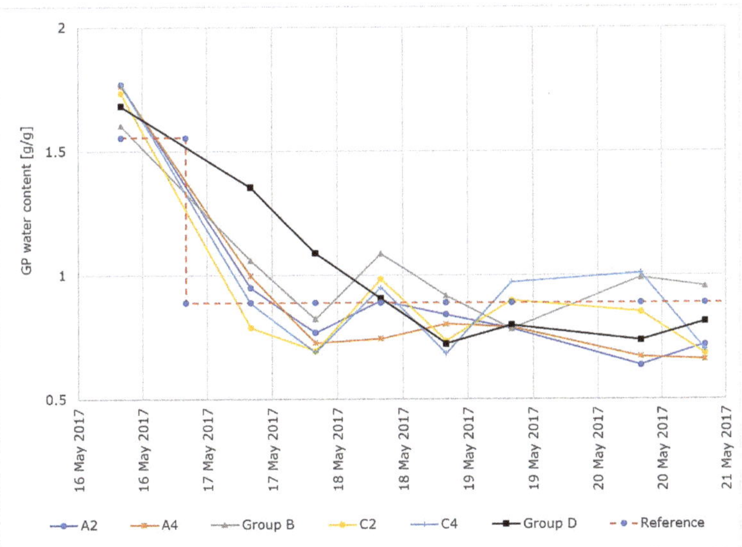

Figure 10. Results of the control test run for a reference variable of 0.89 g/g. GP moisture values over time for all tested GP sheets. Groups A and C were controlled individually and groups B and D were controlled in group control mode.

The control deviation:

$$e = w - W_{grav} \tag{10}$$

with the reference variable w and the gravimetrically measured water content W_{grav} as output variable is evaluated. Additionally, the integral of squared error:

$$A_{quad} = \int_0^{\infty} e^2 dt \tag{11}$$

is evaluated. The maximal control deviations and the integral of squared error of the different groups for $w = 0.89$ g/g are shown in Table 3.

Table 3. Maximal absolute control deviations and integrals of squared error for the different test groups.

Group/Control Applied	Max. Absolute Control Deviation	Integral of Squared Error
A2 (PI-Individual)	0.25 g/g	0.474
A4 (PI-Individual)	0.23 g/g	0.650
B (PI-Group)	0.23 g/g	0.453
C2 (MPC-Individual)	0.20 g/g	0.450
C4 (MPC-Individual)	0.20 g/g	0.565
D (MPC- Group)	0.09 g/g	0.329

For the plants under group control, the means and standard deviations of the control error within the groups are shown in Table 4. Each group consisted of 5 plants, with 7 measurements for each plant taken into consideration.

Table 4. Medians and standard deviations of control error for group-controlled plants.

Group	Median of Control Error	Standard Deviation of Control Error
B	0.08 g/g	0.185
D	0.04 g/g	0.147

4. Discussion

The actual GP water content as well as variations in GP water content between different GP sheets over time have not been considered in past studies [1–3]. One aim of this work was the development of a method for automated measurement of GP water content allowing the consideration of the above mentioned variables. Another aim was the development of GP water content control using automated measurements. This allows the application of defined water deficits with a variable intensity over time to examine related plant growth. This is not possible with common methods in existing studies [1–5].

The main focus of this work was the development of an automatic measurement of GP water content with an adequately effective range and accuracy. The measurements were realized by a capacitor circuit ultimately measuring the electrical resistance of GP. The measured values were additionally processed by a model-based Kalman filter to improve accuracy.

The effective range of measurement was 1.64 g/g > W > 0.80 g/g (for contact position 1) with a maximal GP water content of 2.0 g/g. This effective range could be adjusted by the reduction of distance between the measurement contacts on the GP sheets. The final accuracy of measurements (including model-based correction) within the effective range was smaller than ±0.2 g/g.

The control sytem performance was tested at two different levels of GP water content. For the experiments, a MPC was used to reduce the control deviations of the implemented control loop. The best control result in the test run was realized in a single group of GP sheets with a maximal control deviation of 0.09 g/g.

The application of MPC resulted in smaller maximal control deviations when compared to a time discrete PI-controller. Based on the integral of squared error, a direct advantage of one of the two controllers over the other cannot be ascertained. However, both controllers induced similar dynamical behavior. The best control result was obtained using a MPC applied in group control mode.

The control output in group control was based on the measurements from only 2 plants per group which were used for the control of the water content of all the plants in the group. From the results, it can be concluded that this method could be used to improve the control of GP water content based on measurements taken from a sample of GP sheets rather than from all individuals.

The quality of water content control strongly depends on measurement accuracy. The improvement of measurement accuracy in future could be achieved by improving the measurement circuit and by the reduction of disturbance values acting on the measurement.

The system allows the application of defined time-varying intensities of water deficit on plants grown on GP. Therefore it can be used to prospect the dependencies between dynamically changing water deficit and plant development. The system for measurement of GP water content can also be applied to test rigs using a pouch-and-wick configuration to identify variations in GP water content, which are not considered in many studies [1–3].

Causes of the differences between test groups were not explored for this work. The assumption was that these differences are either caused by differences in GP attributes or by local variations of surrounding conditions like temperature or wind.

5. Summary and Conclusions

Plant experiments involving water deficit require specific knowledge about the moisture available to the plant for control of irrigation. In studies using germination paper as a substrate, moisture measurement and precise watering based on the individual water uptake situation is difficult due to extreme measurement noise.

A novel method for improved measurement and suitable control of GP water content was described in this work. The measurement principle was based on the relationship between electrical resistance of GP, determined via capacitance measurements, and the moisture content of the GP. Kalman filtering was applied for improved accuracy. The control of GP water content was implemented using a model predictive controller.

The measurement system allowed for accurate determination of water content within the 0.8 g/g to 1.64 g/g range for GP with maximal water content of 2.0 g/g. The measurement range could be further extended by adjusting the positioning of the electrical contacts during measurements. The model predictive controller allowed the GP water content to be maintained at predetermined levels, with the ability to vary the desired GP water content over time. System measurement accuracy was ±0.22 g/g and control accuracy was ±0.09 g/g.

The introduced system strongly improved the existing technology and helped to improve sensing and actuation within lab procedures. The application of the system would allow continuous tracking of GP moisture content which is crucial in mapping dependencies between water content and plant behavior under water deficit conditions. The system also allows the maintenance of GP at dynamically varying water content values, which would allow the exploration of controlled variable water deficit sequences in plant growth studies.

Further work on the distribution of water on GP during water uptake would be useful in mapping out the actual water availability at different sections of the GP.

Author Contributions: The contribution of the authors to the work is as follows: Conceptualization, D.S. and F.K.; Methodology, D.S.; Validation, M.H., F.K. and L.O.; Formal Analysis, M.H.; Investigation, M.H.; Resources, D.S.; Data Curation, F.K.; Writing—Original Draft Preparation, M.H. (in 2017) and L.O.; Writing—Review & Editing, L.O. and D.S.; Proofreading—D.S.; Visualization, M.H.; Supervision, D.S.; Project Administration, D.S.; Funding Acquisition, D.S.

Funding: The APC was funded by the Open Access Publication Fund of the University of Duisburg-Essen. The research was partially funded by the German Academic Exchange Service.

Acknowledgments: Support by the Open Access Publication Fund of the University of Duisburg-Essen and German Academic Exchange Service is gratefully acknowledged.

Conflicts of Interest: The authors declare no conflict of interest.

Abbreviations

The following abbreviations are used in this manuscript:

CAN   Controller area network
GP   Germination paper
MPC   Model predictive controller
PEG   Polyethylene glycol
PLC   Programmable logic controller

References

1. Adu, M.O.; Chatot, A.; Wiesel, L.; Bennett, M.J.; Broadley, M.R.; White, P.J.; Dupuy, L.X. A scanner system for high-resolution quantification of variation in root growth dynamics of *Brassica rapa* genotypes. *J. Exp. Bot.* **2014**, *65*, 2039–2048, doi:10.1093/jxb/eru048. [CrossRef] [PubMed]
2. Hund, A.; Trachsel, S.; Stamp, P. Growth of axile and lateral roots of maize: I development of a phenotying platform. *Plant Soil* **2009**, *325*, 335–349, doi:10.1007/s11104-009-9984-2. [CrossRef]
3. Liao, H.; Rubio, G.; Yan, X.; Cao, A.; Brown, K.M.; Lynch, J.P. Effect of phosphorus availability on basal root shallowness in common bean. *Plant Soil* **2001**, *232*, 69–79, doi:10.1023/a:1010381919003. [CrossRef] [PubMed]
4. Rauf, M.; Munir, M.; ul Hassan, M.; Ahmad, M.; Afzal, M. Performance of wheat genotypes under osmotic stress at germination and early seedling growth stage. *Afr. J. Biotechnol.* **2007**, *6*, 971–975.
5. Qayyum, A.; Razzaq, A.; Ahmad, M.; Jenks, M.A. Water stress causes differential effects on germination indices, total soluble sugar and proline content in wheat (*Triticum aestivum* L.) genotypes. *Afr. J. Biotechnol.* **2011**, *10*, 14038–14045, doi:10.5897/ajb11.2220. [CrossRef]
6. Michel, B.E.; Kaufmann, M.R. The osmotic potential of polyethylene glycol 6000. *Plant Physiol.* **1973**, *51*, 914–916. [CrossRef] [PubMed]
7. Sattler, A. SPS-Basierte Regelung und Erweiterung eines Versuchsaufbaus Sowie Experimentelle Modellbildung zur Automatisierten Dosierung von Flüssigkeiten. Master's Thesis, Universität Duisburg-Essen, Duisburg, Germany, 2016.
8. Hering, E.; Martin, R.; Stohrer, M. *Physik für Ingenieure*; Springer: Berlin/Heidelberg, Germany, 2012; doi:10.1007/978-3-642-22569-7.
9. Zimmermann, F. Modellbildung und Validierung eines komplexen Systems: Versuchsstandaufbau für die Erhebung Pflanzenphysiologischer Parameter, Automatisierung der Versuchsdurchführung sowie Experimentelle Bestimmung des Dynamischen Verhaltens. Master's Thesis, Universität Duisburg-Essen, Duisburg, Germany, 2016.
10. Groenendyk, D.G.; Kaleita, A.L.; Thorp, K.R. Assimilating In Situ Soil Moisture Measurements into the DSSAT-CSM Using a Kalman Filter. Master's Thesis, American Society of Agricultural and Biological Engineers, Louisville, KY, USA, 2011; doi:10.13031/2013.37421. [CrossRef]
11. de Rosnay, P.; Drusch, M.; Vasiljevic, D.; Balsamo, G.; Albergel, C.; Isaksen, L. A simplified Extended Kalman Filter for the global operational soil moisture analysis at ECMWF. *Q. J. R. Meteorol. Soc.* **2012**, *139*, 1199–1213, doi:10.1002/qj.2023. [CrossRef]
12. Smith, M.; Kivumbi, D. *Use of the FAO CROPWAT Model in Deficit Irrigation Studies*; Deficit Irrigation Practices (Water Report); Food and Agriculture Organization of the United Nations (FAO): Rome, Italy, 2002.
13. Ostrowski, M.W. *Ingeniuerhydrologie II*. Technische Universität Darmstadt: Darmstadt, Germany. Available online: https://www.ihwb.tu-darmstadt.de/media/fachgebiet_ihwb/lehre/ingenieurhydrologieii/skript-ih-ii.pdf (accessed on 10 December 2018).

14. Dittmar, R.; Pfeiffer, B.M. *Modellbasierte PräDiktive Regelung*; Oldenbourg Wissenschaftsverlag GmbH: München, Germany, 2004; doi:10.1524/9783486594911.
15. Isermann, R. *Digitale Regelsysteme: Band I, Grundlagen Deterministische Regelungen*; Springer: Berlin, Germany, 1987.

Article

Evaluating Soil-Borne Causes of Biomass Variability in Grassland by Remote and Proximal Sensing

Sebastian Vogel [1,*], Robin Gebbers [1], Marcel Oertel [1] and Eckart Kramer [2]

[1]   Leibniz Institute for Agricultural Engineering and Bioeconomy, Max-Eyth-Allee 100, 14469 Potsdam, Germany; rgebbers@atb-potsdam.de (R.G.); moertel@atb-potsdam.de (M.O.)

[2]   Department of Landscape Management and Nature Conservation department, Eberswalde University for Sustainable Development, Schicklerstr. 5, 16225 Eberswalde, Germany; ekramer@hnee.de

*   Correspondence: svogel@atb-potsdam.de; Tel.: +49-331-5699-426

Received: 20 August 2019; Accepted: 18 October 2019; Published: 22 October 2019

Abstract: On a grassland field with sandy soils in Northeast Germany (Brandenburg), vegetation indices from multi-spectral UAV-based remote sensing were used to predict grassland biomass productivity. These data were combined with soil pH value and apparent electrical conductivity (ECa) from on-the-go proximal sensing serving as indicators for soil-borne causes of grassland biomass variation. The field internal magnitude of spatial variability and hidden correlations between the variables of investigation were analyzed by means of geostatistics and boundary-line analysis to elucidate the influence of soil pH and ECa on the spatial distribution of biomass. Biomass and pH showed high spatial variability, which necessitates high resolution data acquisition of soil and plant properties. Moreover, boundary-line analysis showed grassland biomass maxima at pH values between 5.3 and 7.2 and ECa values between 3.5 and 17.5 mS m^{-1}. After calibrating ECa to soil moisture, the ECa optimum was translated to a range of optimum soil moisture from 7% to 13%. This matches well with to the plant-available water content of the predominantly sandy soil as derived from its water retention curve. These results can be used in site-specific management decisions to improve grassland biomass productivity in low-yield regions of the field due to soil acidity or texture-related water scarcity.

Keywords: apparent electrical conductivity (ECa); pH; UAV; boundary-line; quantile regression; law of minimum

1. Introduction

Precision agriculture (PA) technologies are increasingly applied on arable land to improve resource efficiency, reduce environmental impacts, and optimize agricultural productivity (e.g., [1–3]). This can only be achieved by understanding and controlling within-field variability of soil and/or vegetation properties [4–6]. Soil and biomass mapping using on-the-go proximal and remote sensing are a time, cost, and labor effective way to investigate soil characteristics and biomass production, quantify their spatial variability, and delimit homogeneous management zones (HMZ) for variable rate applications [4,7,8].

In grassland research and management, the advent of PA has long lagged behind [9]. One major obstacle was the high variability of soil and crop characteristics, both within-field and between grassland plots, as well as the lack of knowledge in explaining the causes of such heterogeneity [9,10]. However, over the past decade numerous studies focused on that matter to close that knowledge gap (e.g., [11–15]). They monitored grassland biophysical parameters by means of optical remote sensing technologies [16–20], or mapped soil properties using proximal soil sensors, most of all applying geophysical methods to measure the apparent electrical conductivity (ECa) [11,13,21]. Especially soil ECa is recently being used for precision grassland management to characterize within-field variability of soil properties and identify HMZ. This is because ECa correlates to a series of yield effecting soil properties such as soil texture or moisture availability [22,23], while, on the other hand, ECa sensors are relatively easy to use and cause only minor damage to the grass sward [11,24].

However, only a few studies exist that combine soil data from proximal sensing with crop data from remote sensing (e.g., [14]). This is of interest because vegetation mapping by remote sensing may indicate spatial variation in a very efficient way but often does not tell the reasons for the observed variability. However, for an appropriate site-specific grassland management it is indispensable to understand the cause-effect relationship between soil properties and grassland productivity. Several soil properties such as pH, nutrient content, bulk density, water content, and soil texture can affect plant growth considerably. A couple of them are manageable (e.g., by fertilization) while others cannot be changed (e.g., soil texture) but have to be regarded in the decision making.

The overarching objective of this study was to investigate the relationship between soil characteristics and biomass production to identify high and low yielding regions within the field and their possible soil-borne causes. Specific aims were to: (i) Map a grassland field on-the-go for the pH value and the apparent electrical conductivity (ECa) using proximal soil sensors, (ii) quantify grassland biomass production from unmanned aerial vehicles (UAV)-based remote multi-spectral imagery, (iii) evaluate the spatial variability of pH, ECa, and grassland biomass using semivariance modelling, and (iv) investigate the correlation among pH, ECa, and grassland biomass by means of boundary-line analysis to evaluate causes of grassland biomass variability with special emphasis on the yield-limiting effect of pH and ECa.

2. Materials and Methods

2.1. Site Description

The grassland field Königsfeld is situated northwest of the city of Potsdam, in the federal state of Brandenburg (Germany; 5811800N, 365600E; ETRS89). According to the Koeppen-Geiger climate classification system, Potsdam belongs to the humid continental climate (Dfb). It has an average temperature of 9.2 °C with January and July being the coldest (−0.6 °C) and hottest (18.6 °C) month, respectively. The mean annual total precipitation is 566 mm with February being the driest (35 mm) and June the wettest (67 mm) month (German Meteorological Service [DWD], 1981–2010).

The soil cover consists mainly of Gleyic Cambisols of sandy soil textures [25]. Whereas Cambisols, as pedogenetically young soils, are associated with the glacial and periglacial origin of the soil's parent material. The principal qualifier Gleyic derives from the vicinity to the Sacrow-Paretz Canal in the northeast and its related groundwater influence. However, during the field campaigns in 2016 and 2017, no water logging was observed at the soil surface.

The Königsfeld has a total area of 15 ha and was covered by a mixture of grass and alfalfa. The field was used as grassland since 2011, predominantly as grazing land for cattle. In the past, a fence parted the field into two halfs. Despite removal, the growing patterns of the grassland are still affected at its former location. The same applies to a former watering trough for the cattle that was situated at the western corner of the field. For that reason, both areas were disregarded in the following analyses (Figure 1A).

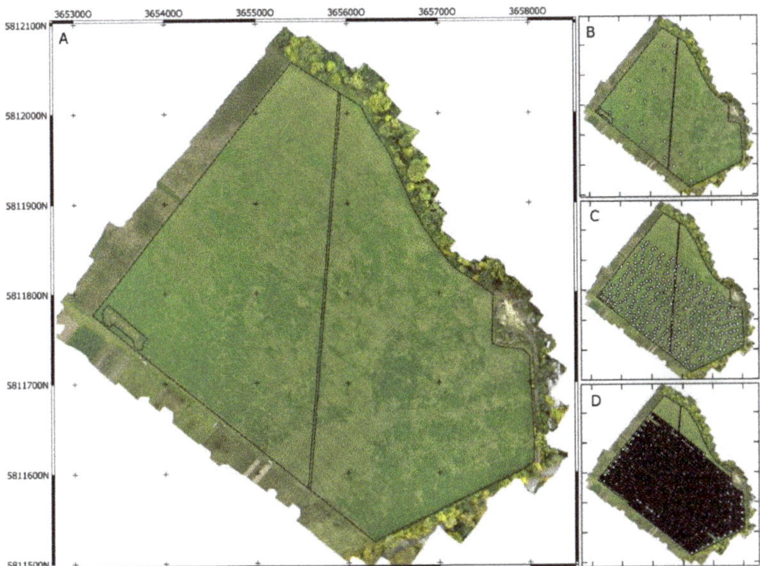

Figure 1. Unmanned aerial vehicles (UAV)-based orthophoto of (**A**) the Königsfeld with the locations of (**B**) biomass sampling sites, (**C**) sensor-based pH, and (**D**) apparent electrical conductivity (ECa) measurements.

2.2. Methods

2.2.1. Proximal Soil Sensing of Grassland Soil

The Königsfeld was mapped in March 2017 for pH and apparent electrical conductivity (ECa) using the Mobile Sensor Platform (MSP) by Veris technologies (Salinas, KS, USA) as described by Lund et al. [26] and Schirrmann et al. [27] (Figure 2A).

The pH value was measured on-the-go by two ion selective antimony electrodes on naturally moist soil material (Figure 2B). While driving across the field, a sampler is lowered into the soil to about 12 cm depth and a soil core flowed through the sampler's orifice. When the soil sampler is raised out of the soil it pressed the sample against two pH electrodes for measurement. Measurements were stopped if sufficiently stable or if a maximum time of 20 s was reached. A logger recorded the raw potential data along with Global Navigation Satellite System (GNSS) coordinates. Additionally, an on-line conversion of the voltage data into pH values was conducted based on a preceding calibration with pH 4 and 7 standard solutions. After each measurement, the sampler was pushed into the soil again and the old soil core was replaced by new material that entered the sampler trough. In the meantime, the pH electrodes were cleaned with tab water from two spray nozzles to prepare them for the next measurement cycle. Typically, pH values were recorded every 10 to 12 s. GNSS coordinates were recorded when the sampler shank was raised out of the soil.

ECa was measured every second with a galvanic coupled resistivity instrument using six parallel rolling coulter electrodes (Figure 2A). This electrode configuration provided readings over two depths with a median depth of exploration of 0.12 and 0.37 m [28]. This enables the identification of significant soil textural and/or soil moisture changes between soil horizons.

Figure 2. (**A**) Veris mobile sensor platform, (**B**) Veris pH-Manager, and (**C**) furrow cut by the pH sampler trough. Numbers indicate (1) rolling ECa coulter electrodes, (2) pH sampler trough, (3) pH electrodes, and (4) cleaning nozzles.

2.2.2. UAV-Based Remote Sensing of Grassland Biomass

The UAV flight was performed at the end of October 2016. The UAV platform was a hexa-copter system HP-Y6 (HEXAPILOTS AHLtec, Dresden, Germany) and with a flight control software based on DJI Wookong M (DJI Innovations, Shenzhen, China). The camera was mounted on a two-axis gimbal underneath the copter. A Sony α6000 (ILCE-6000) (Sony corporation, Tokyo, Japan) RGB camera was used for image acquisition. It had a 23.5 × 15.6 mm sensor (APS-C format) with 4000 × 6000 pixels. The Sony SEL-16F28 lens had a fixed focal length of 16 mm and a maximum aperture of F2.8. The copter was navigated along a predefined route at an altitude of about 100 m. The camera produced a sequence of nadir images at fixed points along the route with an overlap of 70%. From these images an orthophoto mosaic with 10 cm resolution was generated with Agisoft PhotoScan photogrammetry software (Agisoft LLC, St. Petersburg, Russia). Before the flight, white 30 ∗ 30 cm landmarks were placed as ground control points in the field and were georeferenced with an RTK GNSS.

To estimate grassland biomass from the RGB ortho image, the following primary data and derivatives were used:

- Reflectance (R) of red, green, and blue (R_{Red}, R_{Green}, R_{Blue}),
- Hue,
- Saturation,
- Value,
- Normalized difference vegetation index,

$$NDVI = \frac{R_{NIR} - R_{Red}}{R_{NIR} + R_{Red}} \tag{1}$$

- Visible atmospheric resistant index,

$$VARI = \frac{R_{Green} - R_{Red}}{R_{Green} + R_{Red} - R_{Blue}} \qquad (2)$$

2.2.3. Reference Sampling and Laboratory Analysis

To relate the ECa data to soil texture and/or soil moisture, 25 reference soil samples were taken at locations that meet the following three requirements [29]:

i.   To represent extreme values of the target parameter,

ii.   to be spatially homogeneous,

iii.   to be well distributed throughout the area of investigation.

At each reference sampling point, five subsamples were taken with an auger from 0 to 30 cm depth within a radius of 0.5 m. The subsamples were mixed and filled in a plastic bag.

In the laboratory, the soil samples were oven-dried at 105 °C. Afterwards, the particle distribution of the fraction <2 mm was determined according to the German standard in soil science (DIN ISO 11277) by wet sieving and sedimentation after removal of organic matter with H_2O_2 and dispersal with 0.2 N $Na_4P_2O_7$.

To calibrate the pH sensor data 13 reference samples were analyzed in the laboratory by the German standard method (DIN ISO 10390). The soil was dried and sieved (see above) and 10 g of it were mixed with 25 ml of a 0.01 M $CaCl_2$ solution. The pH was measured with a glass electrode after 30 min.

For relating the multispectral aerial photograph on the grassland biomass, at the beginning of November 2016, aboveground biomass from 1 m^2 was cut at 20 points along four transects (Figure 1B). The biomass samples were weighted before and after oven-drying at 75 °C to obtain fresh and dry matter weight.

2.2.4. Data Analysis

Spatial Data Alignment and Visualization

Spatial data alignment and visualization was accomplished with QGIS (QGIS Geographic Information System, QGIS Development Team, Open Source Geospatial Foundation). Coordinates were transformed to a common Cartesian ETRS89 system. Image data at sampling locations and sensor measurement points were extracted by spatial queries.

Analysis of Spatial Variability and Spatial Interpolation by Geostatistics

Spatial variability of the sensor data was quantified by variogram modeling [30–32]. The variogram can provide information about the range of spatial autocorrelation (range parameter) and the disparity between observations beyond the range of autocorrelation, which is called the sill parameter. Additionally, the nugget parameter summarizes the measurement error and sample micro-variability. For variogram modeling, firstly, the method of moments [32] was used to obtain the empirical semivariogram, which relates the average squared differences between observed values to their respective distance class (lag interval). Secondly, a theoretical semivariogram model was fitted to the empirical semivariogram. Final variogram models were established after outlier removal (see below). Calculation and visualisation was carried out with own MATLAB (The MathWorks, Inc., Natick, MA, USA) codes based on [30–33].

Outlier Removal by Spatial Cross-Validation

Since ECa and pH sensor measurements can be prone to error, data were checked for spatial outliers, which strongly deviate from the surrounding observations. After an initial fit of a variogram model (see above) a leave-one-out cross validation procedure was applied [32], which used the variogram model for estimating the values by kriging at each measurement location after excluding the sample value there. A linear regression was calculated between the measured and the estimated values and data points having a Cooks distance larger than 0.033 were excluded from the data set.

Calibration of Sensor Data

Veris pH data: Even though the electrode readings in mV were referenced by pH 4 and 7 standard solutions, the measurements of actual soil samples need further calibration to match them with the standard laboratory method. This was necessary because the standard laboratory method extracts more protons from the soil due to the longer extraction time and because glass electrodes show higher sensitivity/responsiveness compared to antimony electrodes, which results in a trade-off between the laboratory method and the mobile sensor measurements, in particular at lower pH values. Since outlier removal (see above) reduced the number of samples that were co-located with sensor measurements, also samples which fall in a four-meter distance to sensor measurements were included in the calibration. A spatially weighted linear regression was applied which yielded a slope of 0.35, an intercept of 4, a RMSE of 0.2, and a Pearson's r of 0.66.

RGB image: Data from the primary RGB image and derived indices were picked with the zonal statistics procedure of QGIS within a four-meter radius around the biomass sampling locations. Biomass (fresh and dry) was related to image data by linear regression.

Modeling of Growth Response by Boundary-Line Analysis

Under outdoor conditions, biomass development can be restricted by various environmental factors in space and time. Consequently, the bivariate correlation between environmental factors and biomass of a whole field very often show only weak relationships and statistical distributions of non-constant variances. This is because a particular factor controls biomass development only in a subset of field observations, namely when all other growth factors are in their optimum range. This fact was first noted by Carl Sprengel in 1828 and later popularized by Justus von Liebig as the "Law of the Minimum" or "Liebig's Law". Consequently, for quantifying the effect of a single factor on biomass development in observational studies (uncontrolled experiments), where several factors can be limiting in different situations, the assumption of a symmetric error distribution in classical regression analyses is not valid. Instead, Webb [34] suggested to use the observable upper boundary of the point cloud in a bivariate scatter plot to describe the cause-and-effect relationship (with the limiting factor on the abscissa and the biomass or yield on the ordinate). While this early approach was mainly biologically motivated, mathematical solutions were presented later by Koenker and Bassett [35], Kaiser and Speckman [36], and Lark and Milne [37]. In particular, quantile regression is a powerful tool to detect relationships between variables when an asymmetric distribution of errors (residuals) is assumed [35]. When estimating an upper quantile (e.g., 90th, 95th, or 99th percentile) of the conditional distribution of a dependent variable, quantile regression can model the effect of limiting factors by masking the effect of other unknown or unmeasured limiting factors [38–40]. From controlled experiments (with all factors but one being optimum) it was learned that growth-controlling factors often vary between too low, optimum, and too high [40]. This is particularly true for pH: Many crops show increasing growth depressions if pH values sink below 5.5 and rise above 7.5 while pH values around 6.5 are optimum. This effect can be modeled by a piecewise linear function with a trapezoidal shape. Fitting of trapezoidal models of growth response by quantile regression was implemented in MATLAB based on the loss function as described in Koenker and Bassett [35].

3. Results and Discussion

3.1. Descriptive Statistics and Correlation Between Grassland Biomass and Multi-Spectral Indices

Descriptive statistics of the sensor and laboratory data are shown in Table 1. During the field survey at 7 November 2016 as well as by examination of the UAV-based aerial photograph from 24 October 2016, variation in stand height and species composition was clearly visible. In areas of taller and denser vegetation, alfalfa was dominant, whereas zones of smaller and rather sparse vegetation were characterized by grasses (Figures 1A and 3).

Table 1. Descriptive statistics of the sensor and laboratory data.

	Minimum	Mean	Median	Maximum	Standard Deviation	Coefficient of Variation
Veris ECa$_{\text{deep}}$ [mS m^{-1}]	0.5	4.0	3.3	23.8	2.4	0.6
Veris pH	4.9	6.0	6.0	6.9	0.5	0.1
NDVI	0.08	0.23	0.23	0.29	0.02	0.09
Fresh biomass [g m^{-2}]	219.1	595.7	678.1	1114.9	308.4	0.5
Dry biomass [g m^{-2}]	71.2	146.2	148.3	251.2	51.0	0.3
Clay [%]	0.0	3.3	3.0	7.0	1.6	0.5
Silt [%]	8.0	11.3	11.0	16.0	2.1	0.2
Sand [%]	77.0	85.3	86.0	89.0	2.6	0.03
Soil moisture [%]	4.8	8.1	7.9	11.7	1.3	0.2

Figure 3. Visible small-scale variation of the grassland vegetation at Königsfeld. Alfalfa is dominant in the fore- and background while mainly graminaceous plants can be seen in the midst of the image.

To estimate area-wide grassland biomass of Königsfeld, multi-spectral indices derived from the UAV-based orthophoto mosaic were checked for correlation with fresh and dry matter (Figure 4). Hue, saturation, normalized difference vegetation index (NDVI) and visible atmospheric resistant index (VARI) correlated well with fresh and dry matter weight of the biomass (Figure 4). Despite the risk of insufficient estimation of high biomass due to saturation of vegetation indices at high densities [41,42] the correlation between NDVI and dry matter weight was strongest having a Pearson correlation coefficient (Pearson's r) of 0.812. The NDVI was used to generate a regression model for the quantification of dry matter (DM) grassland biomass for the entire field (Figure 5). To derive DM, the regression model was rearranged (Equation (3)). The RMSE was 36.1 g.

$$DM = \frac{(NDVI - 0.028)}{0.001} \tag{3}$$

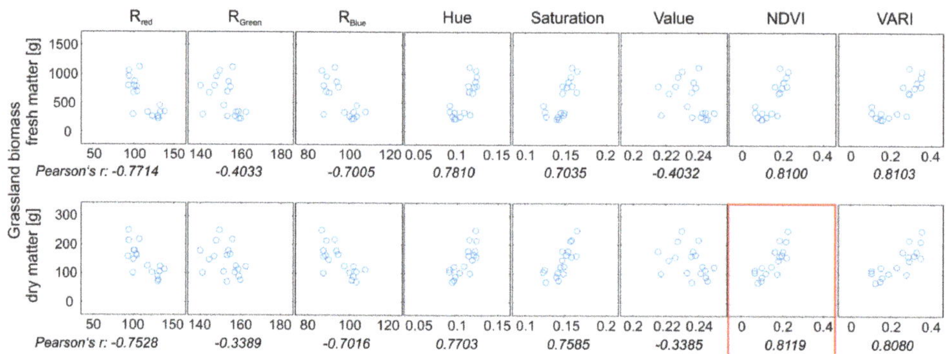

Figure 4. Scatterplots and correlations between fresh and dry matter weight of the grassland biomass and multi-spectral indices deduced from the UAV-based aerial photographs.

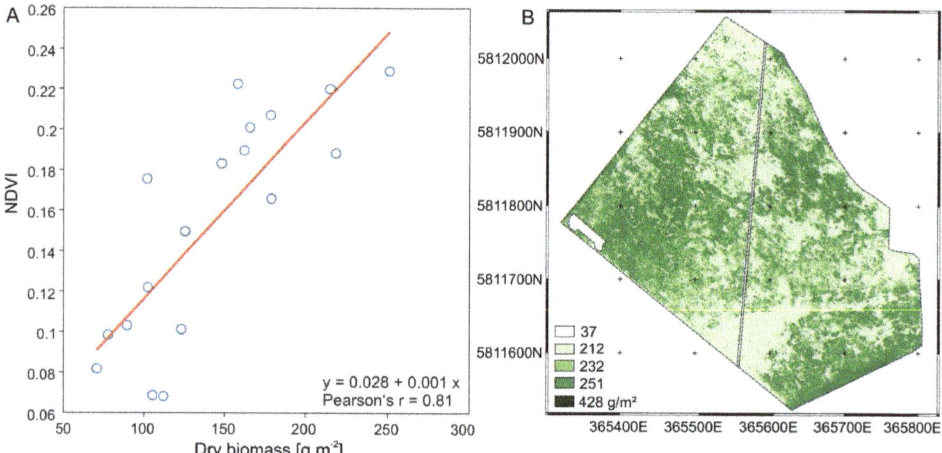

Figure 5. (**A**) Linear regression model of grassland biomass (dry) and normalized difference vegetation index (NDVI) and (**B**) grassland biomass estimation for the entire field.

3.2. Spatial Variability of Grassland Biomass, pH, and ECa

The results of spatial variability assessment of the sensor data by means of semivariance modeling are shown in Figure 6. As for the grassland biomass, pH and ECa, two nested models plus the nugget effect were fitted. That means that there were at least two spatial processes with different ranges. Variography of grassland biomass, derived from NDVI, result in a very low nugget effect as the first component of spatial correlation. This indicates a very low spatial micro-variance and a very low random error from the measurement. Two spherical models were fitted components of spatial (macro) correlation. The range of the first spherical model is about 18 m and that of the second is 245 m (Figure 6A). From the first range, it can be deduced that about half of the spatial variance of grassland biomass already occurs at distances smaller than 18 m. This corresponds to the visual impression of the UAV-based grassland biomass map (Figures 1A and 3) showing pronounced small-scale disparities.

The empirical semivariogram of the pH value also exhibits a nested structure (Figure 6B). The nugget effect is relatively high which can be attributed (a) to microvariance, which is partly to the small sensor footprint and (b) to measurement errors. The first spatial macro process was described by a spherical model with a range of 41 m indicating a high small-scale pH variance of the soil. This can be generally attributed to the high degree of soil variability in Brandenburg due to the Pleistocene and Holocene origin of its parent material and is especially amplified by decades of agricultural use [43–45]. Finally, the third component of spatial correlation within the pH semivariogram was described a Gaussian model showing a range of 295 m.

As with the biomass, the semivariogram model of the apparent electrical conductivity (ECa) was also characterized by a very low nugget effect (Figure 6C). Two nested exponential model components are showing ranges of 89 and 900 m.

The semivariance models have shown that most of the spatial variability of grassland biomass, pH, and ECa occurs at relatively low distances of 18, 41, and 89 m. This illustrates the necessity of small-scale data acquisition of soil and plant properties for precision agriculture applications.

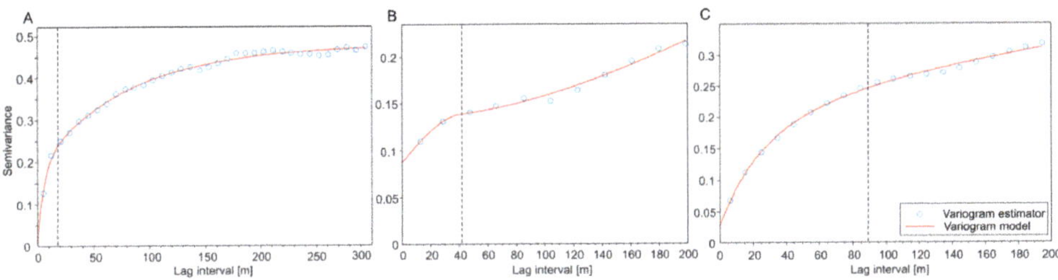

Figure 6. Empirical semivariograms and fitted models of (**A**) NDVI derived grassland biomass [dry matter weight, g m^{-2}], (**B**) pH value, and (**C**) ECa[mS m^{-1}]. Dashed vertical lines indicate the range of the first component of the nested theoretical models.

3.3. Relationship Between ECa, pH, and Grassland Biomass

To investigate the relationship between sensor-based pH value, ECa data, and grassland biomass, in a first step scatterplots were created (Figure 7). At first sight, the scatterplots show no correlation between the dependent and the independent variable and, especially visible for the ECa data, a non-constant variance (Figure 7B). In contrast, the variance of ECa clearly increases with increasing grassland biomass which violates one of the key assumptions of linear regression. This is generally due to the complex nature of factors effecting biomass production. In the context of Liebig's law of the minimum, the variability in the biomass data cannot be entirely explained by ECa and the pH value alone since many other unmeasured factors are a potentially limiting resource for plant growth [46]. Thus, instead of linear regression analyzing the correlation of the means, quantile regression was applied estimating the 95th percentile of the conditional distribution of grassland biomass. This masks the effect of other unknown or unmeasured yield-limiting factors (e.g., [36,38,39]) and the limiting effect of ECa and the pH value can be visualized by means of the boundary line.

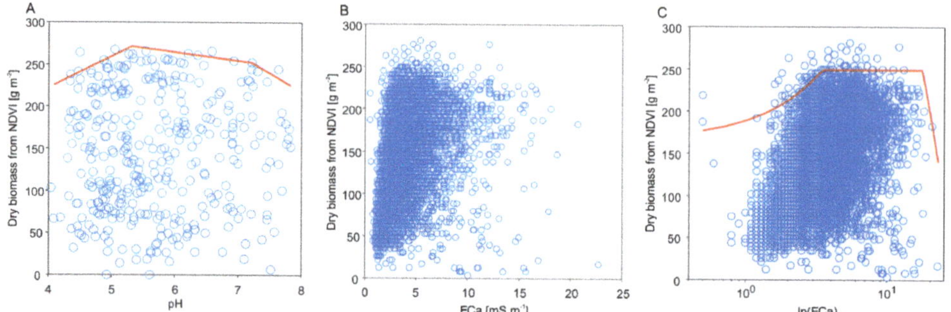

Figure 7. Association between pH, (**A**) ECa, (**B**) log-transformed ECa, (**C**) and grassland biomass of a alfalfa-grass mixture and modelling of their yield-limiting effect by means of quantile regression.

A trapezoidal boundary-line model fitted best to the 95th percentile of the pH value (Figure 7A). It characterizes three stages of response of alfalfa-grass mixture productivity to soil pH [47]: A linear yield increase until a pH value of 5.3, a middle stage of yield stability consisting of a light negative slope until pH 7.2, and a linear yield decline with increasing pH values. This corresponds with several publications stating a pH optimum for alfalfa between 5.8 and 7.2. However, some of these references [48–50] indicate that there might be an interaction between pH and soil texture: The pH optimum shifts to a higher range on fine-textured soils and to a lower range on sandy soils. As soil texture at Königsfeld is strongly dominated by sand (86% sand, 11% silt, and 3% clay) this could be a reason for observing a lower threshold of optimum pH at 5.3. For management of the field it can be considered to raise the pH value by liming at sites with pH lower than 5.3. In contrast, lowering too high pH values (>7.2) is more difficult, in particular if the high pH is due to carbonaceous parent material. The farmer might, at least, avoid any liming in these areas and can reduce other inputs as well.

Before applying quantile regression to ECa, the data were log-transformed ($\log_{10}$) to improve visibility of patterns in the data (Figure 8C). Additionally, for the ECa data a trapezoidal boundary-line model matched best with the 95th percentile. The model can be interpreted as follows: Biomass increases until 3.5 mS m^{-1}, no limits related to ECa are imposed until 17.5 mS m^{-1}, and biomass decreases with increasing ECa above 17.5 mS m^{-1}. In fact, by visually comparing the grassland biomass with the ECa$_{deep}$ map, it is striking that especially low (light green) and high (dark green) values of grassland biomass coincide well with low and high ECa$_{deep}$ values, respectively (Figure 8A,B).

As statistical analyses often show correlations between ECa and clay or sand content, geoelectrical methods are commonly accepted for the estimation of the soil texture [28,51,52]. However, at Königsfeld, the analysis of correlation between ECa$_{deep}$ and soil texture of 25 lab-analyzed reference soil samples showed only poor results. Pearson coefficients of −0.45, 0.43, and 0.25 were obtained for the sand, silt, and clay fraction, respectively. The observation of sometimes weak and inconsistent relationships between ECa and soil properties were also reported by Corwin and Lesch [53] and Sudduth et al. [54]. One reason is that ECa is a cumulative parameter which is affected by a series of soil characteristics such as clay content, soil moisture, salinity, temperature, bulk density, or organic matter content [53,55]. If these parameters are not correlated and if a single parameter is not dominating the influence on ECa statistical analysis becomes difficult. A low range of variation in the control variable can be another reason for poor correlations. At Königsfeld, soil texture varied only within three soil texture classes, namely slightly silty sand (Su2), slightly loamy sand (Sl2), and pure sand (Ss) (Figure 9A). However, the correlation between ECa$_{deep}$ and soil moisture (θ) showed much better results, receiving a Pearson's r of 0.84 (Figure 9B). Thus, in the next step, the ECa$_{deep}$ data were calibrated on the soil moisture following Equation (4):

$$\theta \; [\%] = 6.0 + 0.4 \; ECa_{deep} \tag{4}$$

where θ is the soil moisture in % and ECa_{deep} the sensor-based apparent electrical conductivity in mS m^{-1} at a depth of 0.37 m. Similarly, Cousin et al. [56] and Besson et al. [57] used the relationship between geoelectrical methods and water content to map the spatial distribution of soil water content at the field scale.

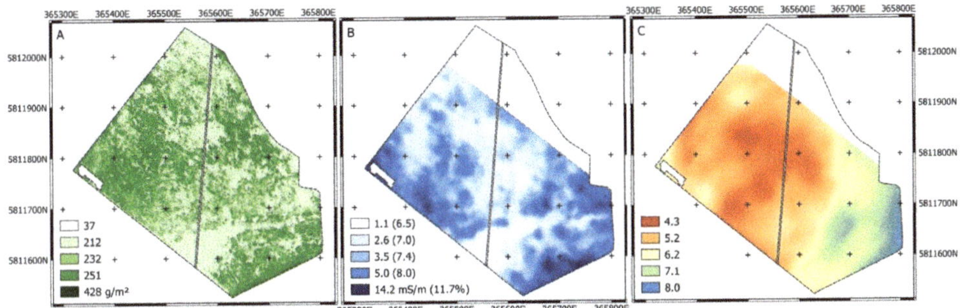

Figure 8. (**A**) Sensor-based estimate of dry grassland biomass, (**B**) soil ECa_{deep} with soil moisture in parentheses, and (**C**) pH value at Königsfeld.

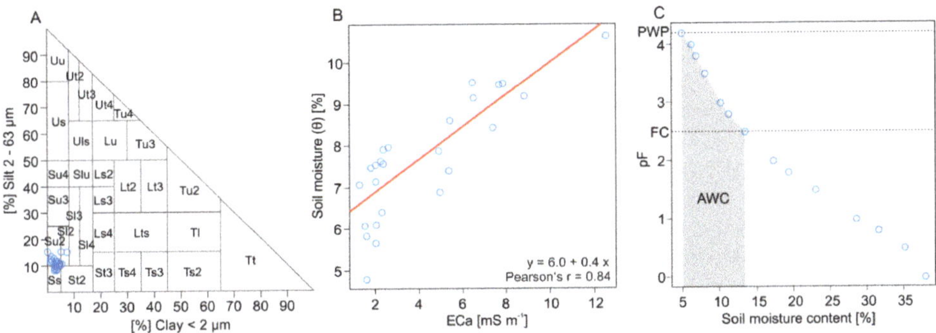

Figure 9. (**A**) Soil texture classes of the 25 lab-analysed reference soil samples. (**B**) Linear regression analysis of apparent electrical conductivity (deep) and soil moisture. (**C**) Mean water retention curve of Königsfeld based on soil texture of the samples and the soil physical data base of [37]. (PWP: Permanent wilting point, FC: Field capacity, AWC: Available water content).

Using the calibration model in Equation (4), soil moisture optimum of the alfalfa-grass mixture was derived from the ECa_{deep} and biomass data. It ranges between 7% and 13%. To better understand the ecological significance and plant availability of the two soil moisture values, we calculated a mean water retention curve on the basis of the mean soil texture data of Königsfeld and the soil physical data base of Wessolek et al. [58] (Figure 9C). Special emphasis was put on the two ecologically important soil moisture states field capacity (FC, pF 2.5) and permanent wilting point (PWP, pF 4.2). FC is the maximum amount of soil water that a soil is able to hold in the root zone against gravity at a matrix potential of −0.33 kPa (pF 2.5). In contrast, the PWP is the amount of soil water that is so strongly retained by the smallest soil pores at a matrix potential of −15 kPa (pF 4.2) [59]. As a consequence, plants are not able to absorb that water and start to wilt. The difference between FC and PWP describes the plants available water content of a soil (AWC) [60]. According to Figure 9C the PWP and FC at Königsfeld is at a soil moisture content of 5% and 13%, respectively. At low soil moistures, biomass production of the alfalfa-grass mixture is already decreasing at soil moistures below 7% even though it is still within the range of the usable field capacity of the soil. This is probably caused by an uneven soil water distribution within the range of AWC especially at low moisture contents [61]. Moreover, plant roots are not uniformly distributed in the soil and the soil water potential that can be overcome by roots differ with different plant species [62]. To increase the water holding capacity (WHC) of these sandy and dry sites, showing ECa below 3.5 mS m^{-1}, the farmer can try to ameliorate by spreading organic fertilizers. However, a significant improvement of WHC will require huge amounts of organic matter. Alternatively, an adaptive strategy could include the seeding of grassland species that are more drought-tolerant.

At high soil moistures, on the other hand, grassland yield begins to decline when soil moisture exceeds field capacity (>13%). Below a pF value of 2.5, water is not held by the soil matrix potential anymore. Instead, it starts to slowly percolate through the soil following the gravitational potential and may contribute to the groundwater below the root zone or is retained by soil layers with poor drainage in the root zone. The latter situation creates water logging with adverse effects on crop growth due to the lack of oxygen. High ECa values were observed at elevated areas, in particular top-slopes, of the field. We assume that at these terrain positions soils were eroded and glacial until it can be found in shallow depths. This is in line with information from the national soil survey [63].

4. Conclusions

A grassland field on sandy soils in Northeast Germany was mapped by proximal soil sensors and remote sensing. The UAV-based orthophoto was used to estimate spatial variation in grassland biomass. Soil pH value and apparent electrical conductivity (ECa) from proximal sensing were investigated as indicators for soil-borne causes of biomass variation. The spatial variability and correlation between pH, ECa, and grassland biomass was analyzed by means of semivariance modelling and boundary-line analysis focusing on the yield-limiting behavior of pH and ECa.

Area-wide mapping of biomass provided means to obtain a sound description of spatial variability by variography based on a large data set. However, the RGB imagery alone does not explain the reasons of spatial variation in biomass. Knowledge about cause-effect relationships is fundamental for taking the appropriate measures in grassland management. Proximal soil sensing in conjunction with boundary line analysis helped to understand some of the drivers of biomass variability.

Comparing the two proximal soil sensors, the pH meter offers the benefit of direct assessment of a key soil fertility factor. The agronomic interpretation of pH data is straight forward and many recommendation tables exist that relate pH values to decisions on liming or selection of crops or varieties. In contrast, the ECa values are affected by several soil properties, including water, salinity, texture, temperature, and compaction. It requires additional efforts to interpret the ECa data and to transfer them into information that is meaningful in agricultural decision making. Technically, however, the EC sensor showed a better performance than the pH system. The ECa data were less noisy (as derived from the nugget effect parameter in the variogram model), they provided a high spatial resolution due to the higher measurement frequency (1 Hz), and the coulter electrodes did not seriously harm the turf. The marks of the pH sampler, however, were visible even after one year. This can promote the expansion of unwanted plant species in the grassland or initiate soil erosion. Compared to the ECa sensor, the pH data were relatively noisy and much sparser. To improve the pH measurements, better solutions for sampling and/or sample preparation have to be found and measurement time has to be reduced.

As one of our reviewers pointed out, this study has neglected the temporal aspect of vegetation dynamics. In particular, seasonal variation of plant growth, due to changes in temperature and photoperiod, could modify the spatial distribution of biomass and probably even vegetation composition. For example, above ground plant parts can die off due to frost over winter. When the plants regrow in the following spring, spatial variation of biomass might not be very strong or may even show different spatial patterns over time. Thus, the relationship between biomass and soil properties (ECa and pH) could change within the season. On the other hand, spatial patterns of vegetation composition and biomass might evolve over several vegetation periods. The vegetation might be spatially more uniform after seeding while more pronounced patterns will appear in subsequent years. This evolution of stable and pronounced spatial patterns can not only be attributed to abiotic environmental factors but also to interspecies competition and to species persistence due to accumulation of root systems and/or seeds of certain plant species at certain spots. Unfortunately, we had not enough resources to thoroughly monitor temporal dynamics of spatial vegetation patterns over time in this study. From qualitative observation (field scouting) we had the impression that the patterns of Königsfeld were relatively stable five years after sowing. With regard to the interaction of spatial and temporal dynamics of alfalfa-grass mixtures we suggest that biomass should be assessed a few years after sowing and preferentially later in the vegetation period. However, further studies are required to substantiate this assumption.

Author Contributions: Conceptualization: R.G. and S.V.; Methodology: R.G.; Formal Analysis: R.G.; Investigation: R.G., S.V., M.O., and E.K.; Data Curation: R.G., S.V., and M.O.; Writing—Original Draft Preparation: S.V.; Writing—Review and Editing: R.G., M.O., and E.K.; Visualization: S.V.; Supervision: R.G.; Project Administration: R.G. and E.K.; Funding Acquisition: R.G. and E.K.

Funding: The study was mainly funded by the Federal Ministry of Food and Agriculture of Germany within the research project pHGreen (Grants 2812NA116 and 2812NA031).

Acknowledgments: The authors thank Jörn Selbeck for UAV flights, Antje Giebel for photogrammetry, Christiane Oertel, Harald Kohn, and Michael Facklam for laboratory analysis as well as two anonymous referees for their constructive comments that considerably improved the manuscript.

Conflicts of Interest: The authors declare no conflict of interest. The sponsors had no role in the design, execution, interpretation, or writing of the study.

References

1. Zhang, N.; Wang, M.; Wang, N. Precision agriculture—A worldwide overview. *Comput. Electron. Agric.* **2002**, *36*, 113–132. [CrossRef]
2. Bongiovanni, R.; Lowenberg-Deboer, J. Precision Agriculture and Sustainability. *Precis. Agric.* **2004**, *5*, 359–387. [CrossRef]
3. Lindblom, J.; Lundström, C.; Ljung, M.; Jonson, A. Promoting sustainable intensification in precision agriculture: Review of decision support systems development and strategies. *Precis. Agric.* **2017**, *18*, 309–331. [CrossRef]
4. Castrignanò, A.; Buttafuoco, G.; Quarto, R.; Vitti, C.; Langella, G.; Terribile, F.; Venezia, A. A Combined Approach of Sensor Data Fusion and Multivariate Geostatistics for Delineation of Homogeneous Zones in an Agricultural Field. *Sensors* **2017**, *17*, 2794. [CrossRef]
5. Peralta, N.R.; Costa, J.L.; Balzarini, M.; Angelini, H. Delineation of management zones with measurements of soil apparent electrical conductivity in the southeastern pampas. *Can. J. Soil Sci.* **2013**, *93*, 205–218. [CrossRef]
6. Xu, H.-W.; Wang, K.; Bailey, J.; Jordan, C.; Withers, A. Temporal stability of sward dry matter and nitrogen yield patterns in a temperate grassland. *Pedosphere* **2016**, *16*, 735–744. [CrossRef]
7. Nawar, S.; Corstanje, R.; Halcro, G.; Mulla, D.; Mouazen, A.M. Delineation of soil management zones for variable-rate fertilization: A review. *Adv. Agron.* **2017**, *143*, 175–245.
8. Moral, F.J.; Terrón, J.M.; Marques da Silva, J.R. Delineation of management zones using mobile measurements of soil apparent electrical conductivity and multivariate geostatistical techniques. *Soil Tillage Res.* **2010**, *106*, 335–343. [CrossRef]
9. Schellberg, J.; Hill, M.J.; Gerhards, R.; Rothmund, M.; Braun, M. Precision agriculture on grassland: Applications, perspectives and constraints. *Eur. J. Agron.* **2008**, *29*, 59–71. [CrossRef]
10. Geypens, M.; Vanongeval, L.; Vogels, N.; Meykens, J. Spatial Variability of Agricultural Soil Fertility Parameters in a Gleyic Podzol of Belgium. *Precis. Agric.* **1999**, *1*, 319–326. [CrossRef]
11. Bernardi, A.C.C.; Bettiol, G.M.; Ferreira, R.P.; Santos, K.E.L.; Rabello, L.M.; Inamasu, R.Y. Spatial variability of soil properties and yield of a grazed alfalfa pasture in Brazil. *Precis. Agric.* **2016**, *17*, 737–752. [CrossRef]
12. Wang, D.; Xin, X.; Shao, Q.; Brolly, M.; Zhu, Z.; Chen, J. Modeling Aboveground Biomass in Hulunber Grassland Ecosystem by Using Unmanned Aerial Vehicle Discrete Lidar. *Sensors* **2017**, *17*, 180. [CrossRef] [PubMed]
13. Serrano, J.M.; Shahidian, S.; Marques da Silva, J. Spatial variability and temporal stability of apparent soil electrical conductivity in a Mediterranean pasture. *Precis. Agric.* **2017**, *18*, 245–263. [CrossRef]

14. Serrano, J.; Shahidian, S.; Marques da Silva, J.; de Carvalho, M. A Holistic Approach to the Evaluation of the Montado Ecosystem Using Proximal Sensors. *Sensors* **2018**, *18*, 570. [CrossRef] [PubMed]

15. Knoblauch, C.; Watson, C.; Berendonk, C.; Becker, R.; Wrage-Mönnig, N.; Wichern, F. Relationship between Remote Sensing Data, Plant Biomass and Soil Nitrogen Dynamics in Intensively Managed Grasslands under Controlled Conditions. *Sensors* **2017**, *17*, 1483. [CrossRef]

16. Rueda-Ayala, V.P.; Peña, J.M.; Höglind, M.; Bengochea-Guevara, J.M.; Andújar, D. Comparing UAV-Based Technologies and RGB-D Reconstruction Methods for Plant Height and Biomass Monitoring on Grass Ley. *Sensors* **2019**, *19*, 535. [CrossRef]

17. Jin, Y.; Yang, X.; Qui, J.; Li, J.; Gao, T.; Wu, Q.; Zhao, F.; Ma, H.; Yu, H.; Xu, B. Remote Sensing-Based Biomass Estimation and its Spatio-Temporal Variations in Temperate Grassland, Northern China. *Remote Sens.* **2014**, *6*, 1496–1513. [CrossRef]

18. Pullanagari, R.R.; Yule, I.J.; Tuohy, M.P.; Hedley, M.J.; Dynes, R.A.; King, W.M. Proximal sensing of the seasonal variability of pasture nutritive value using multispectral radiometry. *Grass Forage Sci.* **2013**, *68*, 110–119. [CrossRef]

19. Jia, W.; Liu, M.; Yang, Y.; He, H.; Zhu, X.; Yang, F.; Yin, C.; Xiang, W. Estimation and uncertainty analyses of grassland biomass in Northern China: Comparison of multiple remote sensing data sources and modeling approaches. *Ecol. Indic.* **2016**, *60*, 1031–1040. [CrossRef]

20. Xu, D.; Guo, X. Some insights on grassland health assessment based on remote sensing. *Sensors* **2015**, *15*, 3070–3089. [CrossRef]

21. Moral, F.J.; Serrano, J.M. Using low cost geophysical survey to map soil properties and delineate management zones on grazed permanent pastures. *Precis. Agric.* **2019**, *20*, 1000–1014. [CrossRef]

22. Kitchen, N.R.; Sudduth, K.A.; Drummond, S.T. Soil electrical conductivity as a crop productivitymeasure for claypan soils. *J. Prod. Agric.* **1999**, *12*, 607–617. [CrossRef]

23. Luchiari, A.; Shanahan, J.; Francis, D.; Schlemmer, M.; Schepers, J.; Liebig, M. Strategies for establishing management zones for site specific nutrient management. In Proceedings of the 5th International Conference on Precision Agriculture, Madison, WI, USA, 16–19 July 2001.

24. Kitchen, N.R.; Drummond, S.T.; Lund, E.D.; Sudduth, K.A.; Buchleiter, G.W. Soil electrical conductivity and topography related to yield for three contrasting soil–crop systems. *Agron. J.* **2003**, *95*, 483–495. [CrossRef]

25. Soil Map of Germany. Bodenübersichtskarte der Bundesrepublik Deutschland 1:200.000 (BÜK200). Available online: https://www.bgr.bund.de/DE/Themen/Boden/Projekte/Informationsgrundlagen-laufend/BUEK200/BUEK200.html (accessed on 11 June 2019).

26. Lund, E.D.; Adamchuk, V.I.; Collings, K.L.; Drummond, P.E.; Christy, C.D. Development of soil pH and lime requirementmaps using on-the-go soil sensors. In Proceedings of the Precision Agriculture: Papers from the Fifth European Conference on Precision Agriculture, Uppsala, Sweden, 9–12 June 2005.

27. Schirrmann, M.; Gebbers, R.; Kramer, E.; Seidel, J. Soil pH Mapping with an On-The-Go Sensor. *Sensors* **2011**, *11*, 573–598. [CrossRef] [PubMed]

28. Gebbers, R.; Lück, E.; Dabas, M.; Domsch, H. Comparison of instruments for geoelectrical soil mapping at the field scale. *Near Surf. Geophys.* **2009**, *7*, 179–190. [CrossRef]

29. Adamchuk, V.I.; Viscarra Rossel, R.A.; Marx, D.B.; Samal, A.K. Using targeted sampling to process multivariate soil sensing data. *Geoderma* **2011**, *163*, 63–73. [CrossRef]

30. Deutsch, C.V.; Journel, A.G. *GSLIB Geostatistical Software Library and User's Guide*, 2nd ed.; Oxford University Press: New York, NY, USA, 1998.

31. Goovaerts, P. *Geostatistics for Natural Resources Evaluation*; Oxford University Press: New York, NY, USA, 1997.

32. Webster, R.; Oliver, M.A. *Geostatistics for Environmental Scientists*, 2nd ed.; Wiley: Chichester, UK, 2007; p. 315.

33. Christakos, G.; Bogaert, P.; Serre, M. *Temporal GIS. Advanced Functions for Field-Based Applications*; Springer: Berlin, Germany, 2002.

34. Webb, R.A. Use of the boundary line in analysis of biological data. *J. Hortic. Sci. Biotechnol.* **1972**, *47*, 309–319. [CrossRef]

35. Koenker, R.; Bassett, G.J. Regression quantiles. *Econometrica* **1978**, *46*, 33–50. [CrossRef]

36. Kaiser, M.S.; Speckman, P.L.; Jones, J.R. Statistical models for limiting nutrient relations in inland water. *J. Am. Stat. Assoc.* **1994**, *89*, 410–423. [CrossRef]

Sensors 2019, 19, 4593

37. Lark, R.M.; Milne, A.E. Boundary line analysis of the effect of water-filled pore space on nitrous oxide emission from cores of arable soil. *Eur. J. Soil. Sci.* **2016**, *67*, 148–159. [CrossRef]
38. Thomson, J.D.; Weiblen, G.; Thomson, B.A. Untangling multiple factors in spatial distributions: Lilies, gophers and rocks. *Ecol.* **1996**, *77*, 1698–1715. [CrossRef]
39. Cade, B.S.; Terrell, J.W.; Schroeder, R.L. Estimating effects of limiting factors with regression quantiles. *Ecology* **1999**, *80*, 311–323. [CrossRef]
40. Cade, B.S.; Noon, B.R. A gentle introduction to quantile regression for ecologists. *Front. Ecol. Environ.* **2003**, *1*, 412–420. [CrossRef]
41. Mutanga, O.; Skidmore, A.K. Hyperspectral band depth analysis for a better estimation of grass biomass (Cenchrus ciliaris) measured under controlled laboratory conditions. *Int. J. Appl. Earth Obs. Geoinf.* **2004**, *5*, 87–96. [CrossRef]
42. Chen, J.; Gu, S.; Shen, M.; Tang, Y.; Matsushita, B. Estimating aboveground biomass of grassland having a high canopy cover: An exploratory analysis of in situ hyperspectral data. *Int. J. Remote Sens.* **2009**, *30*, 6497–6517. [CrossRef]
43. Gebbers, R. *Fehleranalyse im System der Ortsspezifischen Grunddüngung. Forschungsbericht Agrartechnik des Arbeitskreises Forschung und Lehre der Max-Eyth-Gesellschaft Agrartechnik im VDI, 474*; Leibniz-Inst. für Agrartechnik Potsdam-Bornim eV: Potsdam, Germany, 2008.
44. MLUV. *Landnutzung ändert Böden—Steckbriefe Brandenburger Böden. Ministerium für Ländliche Entwicklung, Umwelt und Verbraucherschutz des Landes Brandenburg (MLUV)*; Landesvermessung und Geobasisinformation Brandenburg: Frankfurt, Germany, 2011.
45. Walker, D.B.; Haugen-Kozyra, K.; Wang, C. Effects of Long-term Cultivation on a Morainal Landscape in Alberta, Canada. In Proceedings of the Third International Conference on Precision Agriculture, Madison, WI, USA, 23–26 June 1996.
46. Cade, B.S.; Guo, Q. Estimating effects of constraints on plant performance with regression quantiles. *Oikos* **2000**, *91*, 245–254. [CrossRef]
47. Van Dorp, J.R.; Kotz, S. Generalized trapezoidal distributions. *Metrika* **2003**, *58*, 85–97. [CrossRef]
48. Schmidt, L.; Märtin, B. *Produktionsanleitung und Richtwerte für den Anbau von Luzerne und Luzernegras*; Landwirtschaftsausstellung der DDR: Markkleeberg, Germany, 1978.
49. Kreil, W.; Simon, W.; Wojahn, E. *Futterpflanzenbau, Empfehlungen, Richtwerte, Normative. Band 2—Ackerfutter*; VEB Deutscher Landwirtschaftsverlag: Berlin, Germany, 1983.
50. Bundessortenamt. *Beschreibende Sortenliste, Futtergräser, Esparsette, Klee, Luzerne*; Bundessortenamt: Hannover, Germany, 2009.
51. McBratney, A.B.; Minasny, B.; Whelan, B.M. Obtaining 'useful' high-resolution soil data from proximally-sensed electrical conductivity/resistivity (PSEC/R) surveys. *Precis. Agric.* **2005**, *5*, 503–510.
52. Viscarra Rossel, R.A.; Adamchuk, V.I.; Sudduth, K.A.; McKenzie, N.J.; Lobsey, C. Proximal Soil Sensing: An Effective Approach for Soil Measurements in Space and Time. *Adv. Agron.* **2011**, *113*, 237–282.
53. Corwin, D.L.; Lesch, S. Application of Soil Electrical Conductivity to Precision Agriculture. *Agron. J.* **2003**, *95*, 455–471. [CrossRef]
54. Sudduth, K.A.; Kitchen, N.R.; Wiebold, W.J.; Batchelor, W.D.; Bollero, G.A.; Bullock, D.G. Relating apparent electrical conductivity to soil properties across the north-central USA. *Comput. Electron. Agric.* **2005**, *46*, 263–283. [CrossRef]
55. Lück, E.; Spangenberg, U.; Rühlmann, J. Comparison of different EC-mapping sensors. In Proceedings of the Precision agriculture '09, 7th European Conference on Precision Agriculture, Wageningen, The Netherlands, 6–8 July 2009.
56. Cousin, I.; Besson, A.; Bourennane, H.; Pasquier, C.; Nicoullaud, B.; King, D.; Richard, G. From spatial-continuous electrical resistivity measurements to the soil hydraulic functioning at the field scale. *C. R. Geosci.* **2009**, *341*, 859–867. [CrossRef]
57. Besson, A.; Cousin, I.; Bourennane, H.; Nicoullaud, B.; Pasquier, C.; Richard, G.; Dorigny, A.; King, D. The spatial and temporal organization of soil water at the field scale as described by electrical resistivity measurements. *Eur. J. Soil Sci.* **2010**, *61*, 120–132. [CrossRef]
58. Wessolek, G.; Kaupenjohann, M.; Renger, M. *Bodenphysikalische Kennwerte und Berechnungsverfahren für die Praxis*; Technische Universität Berlin: Berlin, Germany.

59. Richards, L.A.; Weaver, L.R. Moisture retention by some irrigated soils as related to soil-moisture tension. *J. Agric. Res.* **1944**, *69*, 215–235.
60. Veihmeyer, F.J.; Hendrickson, A.H. Soil moisture in relation to plant growth. *Annu. Rev. Plant Physiol.* **1950**, *1*, 285–305. [CrossRef]
61. Richards, L.A.; Wadleigh, C.H. Soil water and plant growth. In *Soil Physical Conditions and Plant Growth*; Byron, T.S., Ed.; Academic Press: London, UK, 1952; Volume 2, p. 491.
62. Brillante, L.; Mathieu, O.; Bois, B.; van Leeuwen, C.; Lévêque, J. The use of soil electrical resistivity to monitor plant and soil water relationships in vineyards. *SOIL* **2015**, *1*, 273–286. [CrossRef]
63. BGR (Federal Institute for Geosciences and Natural Resources). *Borehole Profile Backend Service*; Version 1.0.1; Section Geodata Management: Hannover, Germany, 2016.

Article

High Speed Crop and Weed Identification in Lettuce Fields for Precision Weeding

Lydia Elstone *, Kin Yau How, Samuel Brodie, Muhammad Zulfahmi Ghazali, William P. Heath and Bruce Grieve

School of Electrical and Electronic Engineering, The University of Manchester, Oxford Rd, Manchester M13 9PL, UK; hkyhow12@gmail.com (K.Y.H.); brodie667@hotmail.com (S.B.); ghazalizulfahmi@gmail.com (M.Z.G.); william.heath@manchester.ac.uk (W.P.H.); bruce.grieve@manchester.ac.uk (B.G.)
* Correspondence: lydia.elstone@postgrad.manchester.ac.uk

Received: 7 November 2019; Accepted: 25 December 2019; Published: 14 January 2020

Abstract: Precision weeding can significantly reduce or even eliminate the use of herbicides in farming. To achieve high-precision, individual targeting of weeds, high-speed, low-cost plant identification is essential. Our system using the red, green, and near-infrared reflectance, combined with a size differentiation method, is used to identify crops and weeds in lettuce fields. Illumination is provided by LED arrays at 525, 650, and 850 nm, and images are captured in a single-shot using a modified RGB camera. A kinematic stereo method is utilised to compensate for parallax error in images and provide accurate location data of plants. The system was verified in field trials across three lettuce fields at varying growth stages from 0.5 to 10 km/h. In-field results showed weed and crop identification rates of 56% and 69%, respectively. Post-trial processing resulted in average weed and crop identifications of 81% and 88%, respectively.

Keywords: precision weeding; multispectral imaging; kinetic stereo imaging; plant detection

---

1. Introduction

Traditionally, weed management has been achieved through broadcast application of selective herbicide. A large proportion of herbicide used in broadcast spraying is released into the environment through run-off and drift [1]. Increasing environmental and public health concerns have resulted in stricter regulation of herbicide use [2,3]. Traditional selective herbicides also have to contend with a growing number of resistant weed species [4]. Weeds represent a loss potential of 32%, with an effect comparable to that of pathogens and parasites combined [5]. Therefore, effective weed control is essential to maintaining and increasing worldwide food productivity to provide for a growing global population [6]. Speciality crops, such as vegetables, are disproportionally affected by herbicide resistance, as very few herbicides are registered for use in the sector [7]. Such farmers have been forced to use hand-weeding methods, which are expensive, inefficient, and made more difficult by an industry-wide labour shortage [8,9].

Several approaches have attempted to tackle weed management issues through the introduction of new technology. Herbicide waste can be moderately reduced through the use of variable-rate spraying and modifying existing spray booms [10]. Herbicide inputs may be completely eliminated through robotic weed removal [11,12]. These approaches can be implemented using relatively simple detection techniques, estimating overall plant coverage [10,13] or crop position [12,14]. Individual weed targeting for mechanical removal [15] or herbicidal micro-dosing [16,17] has also been investigated in recent years. Non-selective herbicides less effected by herbicide resistance can be utilised for micro-dosing systems, which can reduce herbicide requirements by up to 99.3% according to [18]. In order to achieve individual weed targeting, plant identification and accurate target position information are essential.

Weed detection methods most often use a combination of a colour index followed by image segmentation and feature extraction, to locate weeds in the crop bed [19]. Other methods have been considered, but not so extensively, such as the use of LIDAR (light detection and ranging) to discriminate between crops, plants, and soil using height information [20]. Colour indices use some combination of visible and near infra-red (NIR) reflectance to create a single greyscale image [19,21]. Some methods require the transformation into different colour spaces [22], such as normalised RGB [23] or HSV (hue, saturation, and value) [24]. Common indices include ExG (excess green), ExR (excess red), NDI (normalised difference index), and NDVI (normalised difference vegetative index), which discriminate between plants and soil with varying effectiveness [19]. Images are segmented using a variation of thresholding techniques, but most commonly Otsu's thresholding technique [21]. To discriminate crops from weeds, other information is needed; shape [25] or textural features [26] are most commonly used, which can be computationally expensive [3].

Machine learning has been used extensively in the discrimination of vegetation and soil and between vegetation types. Support vector machines proposed in [27–29] have been utilised, in combination with a selection of features, such as colour or shape information, for plant identification. Fuzzy decision making methods were considered in [25] for the identification of monocot and dicot weeds, using a set of shape features achieving up to a 92.9% classification accuracy. Each of these approaches for decision making is applied following the segmentation of the image into crop and weed areas and the calculation of a selection of features for plant areas. Convolutional neural networks (CNN) have also been utilised successfully in [30,31] to locate weed patches, without the need to provide feature sets prior to processing. While the method increases the training time for the algorithm compared to, for example, support vector machines, it may provide a more universally applicable system [31]. These methods show an important next step in the identification process, but are complex and relatively computationally expensive. They are most notably being used to identify patches of weeds within a field, not individual weeds in real-time.

This study focuses on the development of a low cost, high-speed detection of individual plants for use with a herbicide micro-dosing system, assuming a targeting area 20 mm in diameter. As such, our detection system was designed with the target of operating at 5 km/h with the capacity for cost-effective retrofitting to existing tractor tool-bar mounts, and minimal equipment cost. The approach should be robust to variations in ambient lighting and to inadvertent wind dispersion on the injected herbicidal products. The assembly, therefore, is enclosed under a tractor-mounted hood (Figure 1) and controlled illumination is provided. The research aimed to determine the effectiveness of using spectral reflectance in the red, green, and NIR wavebands using controlled illumination, combined with size information in the discrimination between crops, weeds, and soil in horticultural crops. Accurate position information for individual weed targets is required whilst maintaining a large field-of-view from the camera to ensure high-speed operation. A system to mitigate the effect of parallax error on location information of plants has been proposed through a kinematic stereo method. Where possible, the detection method should provide flexibility for use across a range of scenarios. As such, a modular design was utilised to allow for customisable width up to a maximum of seven modules (3.78 m wide), which may be implemented concurrently for each PC.

2. Materials and Methods

2.1. Hardware Design

Each camera-lighting module (Figure 1b) covers a width of 0.54 m, allowing for customisation according to application. Weather, time, and location may cause significant variations in the imaging environment, and the detection system must be robust to these changes. Therefore, a closed canopy houses the imaging system to provide illumination control. For easy integration with current farming equipment, the canopy is mounted to the rear of a tractor (Figure 1a) and covers one crop bed (1.6 m, 3 camera-lighting modules).

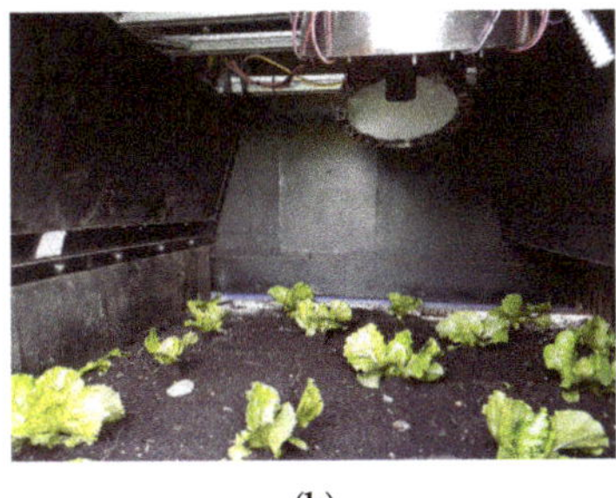

(a)   (b)

Figure 1. The system in field trials. (**a**) The system mounted to the rear of the tractor. (**b**) The sensor system in-field.

Three main components are used for crop and weed identification: a PC, RGB camera, and lighting rig; a 12 V lead-acid battery powers the system in conjunction with an inverter to supply the PC; LED arrays, 650 (red), 850 (NIR), and 525 nm (green) (Justar Electronic Technology Co. LTD). 5×5 arrays at 525–530 nm, 650 nm, 850 nm, 25 W, 1.75 A are used for illumination.

An RGB machine vision camera (PointGrey, Grasshopper 3, GS3-U3-41C6C-C), with a wide angle lens (Fujinon CF12.5HA-1), is used for image capturing. The camera is mounted between two lamps to provide shadowless illumination. Each lamp consists of three LED arrays of each wavelength placed at 120° intervals within a metal hemisphere (Figure 2). The camera is positioned 0.6 m above the ground resulting in a field of view of 0.54×0.54 m and a resolution of 0.9481 px/mm. This height was chosen to balance coverage and precision requirements, whilst minimising image distortion.

Figure 2. Illumination system containing nine LED arrays per lamp, positioned at regular intervals.

Illumination control is facilitated through two circuit boards: the LED driver and microcontroller boards. The camera receives signals from the PC to capture images, which are relayed to the microcontroller board. The microcontroller generates PWM signals to control the intensity of the LED arrays of different wavelengths, and multiplexes these with the ON/OFF signals received from the camera. The LED drivers take this control signal and a 12 V supply to illuminate the area when required by the camera. The duty cycles for each of the LED arrays are 50%, 50%, and 20% for the NIR, green, and red, respectively. Comparisons were made between a selection of images taken at varying duty cycles (between approximately 10% and 70% for each wavelength). Where the duty cycle for each wavelength was equal, the red response appeared to dominate the image. The combination of duty cycles chosen allowed for a balanced response in each channel and resulted in an image with clearly differing responses between plants and soil, and to a lesser extent between plant types.

A custom built PC provides the required processing power for high speed identification, including a dedicated GPU. The full PC component list can be seen in Table 1.

Table 1. PC components.

Component	Name
Motherboard	MSI Z370-A PRO ATX LGA1151
CPU	Intel Core i7-8700
GPU	Gigabyte GeForce® GTX 1080
RAM	G.Skill Ripjaws 4 Series 16 GB
Storage	Samsung PM961
Power Supply	EVGA 600B

2.2. Plant Identification

Our approach uses red, green, and NIR reflectivity, combined with an assumption that crops are generally larger than weeds, to identify crops and weeds in-field. In [32] a modified RGB camera is used to implement NDVI using multiple filters to allow the green and blue channels to detect NIR wavelengths. Our approach similarly modifies an RGB camera through the removal of the NIR filter and introduction of a 515 nm long-pass filter to remove the unwanted response to blue light in the blue channel. However, in addition to utilising red and NIR wavelengths, the green response is maintained from the camera. The green response is intended to obtain minor tonal differences between plant types to be included in the index calculation (RGNIR—Equation (1)). This may be useful where leaves of crops and weeds overlap or large bodies of multiple weeds appear to be a single larger object.

A single shot method is used, flashing all LEDs simultaneously and capturing the red, green and NIR reflectance in one frame. The RGB channels of the camera overlap, and as such, all respond to the wavelengths used in the system to a varying extent. From the datasheet for the CMOSIS CMV400 chip, an estimation of the quantum efficiency of the RGB channels at the system wavelengths is shown in Table 2.

Table 2. Estimated quantum efficiency of camera channels.

RGB Channel	Wavelength		
	Red (650 nm)	NIR (850 nm)	Green (525 nm)
Red Channel	45%	15%	5%
Green Channel	3%	15%	45%
Blue Channel	5%	15%	15%

The removal of the NIR filter led to colour distortion of the captured image; to compensate, a background image is taken in a black box condition and then removed from the image. An example of each of the captured RGB frames is shown in Figure 3. The channel responses are combined in Equation (1) to produce a single greyscale image for segmentation. This equation is similar to the normalised green index proposed by [33], although in our case the blue channel is a combined response to NIR and green wavelengths. The RGNIR intensity is calculated as

$$RGNIR = \frac{\beta(G_c - g_c)}{\alpha(R_c + r_c) + \beta(G_c - g_c) + \gamma(B_c - b_c) + L'} \tag{1}$$

where G_c, R_c, and B_c are the green, red, and blue channels, respectively; g_c, r_c, and b_c are the background corrections for each channels (0–255); α, β, and γ are the channel weights (0–1); L is the soil adjusted constant.

The histogram of the RGNIR image (Figure 4a) is taken and segmentation is performed using Otsu's multi-level (in this case three) thresholding method [34,35]. The features from the segmented image are identified as either crop or weeds using a size exclusion method (Figure 4b).

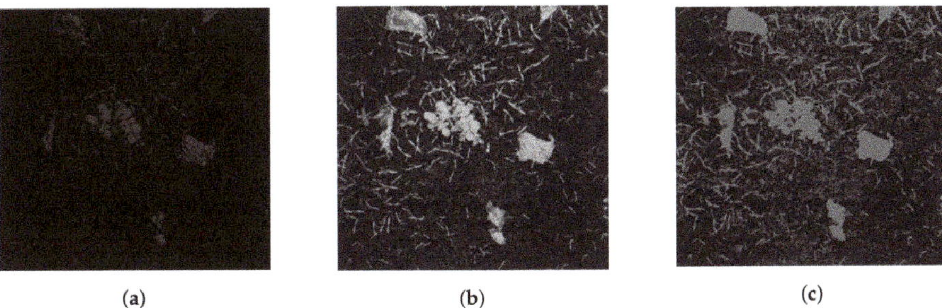

(a)   (b)   (c)

Figure 3. Responses of the red (**a**), green (**b**), and blue (**c**) camera channels, in controlled illumination.

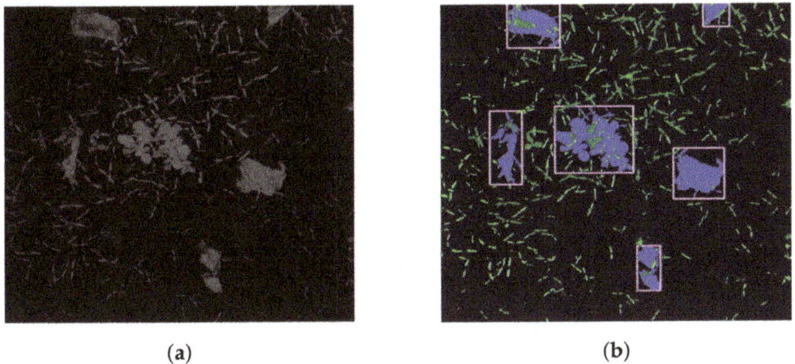

(a)   (b)

Figure 4. Image processing steps. (**a**) Greyscale output of the RGNIR function. (**b**) Image following thresholding.

2.3. Height Estimation

In order to image a field efficiently at high speed, it is essential that each frame captures the largest feasible area while maintaining sufficient detail to accurately identify and locate small plants. However, where plants are not located directly below the camera, variations in their height can result in an inaccurate determination of their position due to parallax error. In our system, the usable frame area (assuming a parallax error of less than 5% is acceptable) is approximately a 19% slice across the centre (Figure 5).

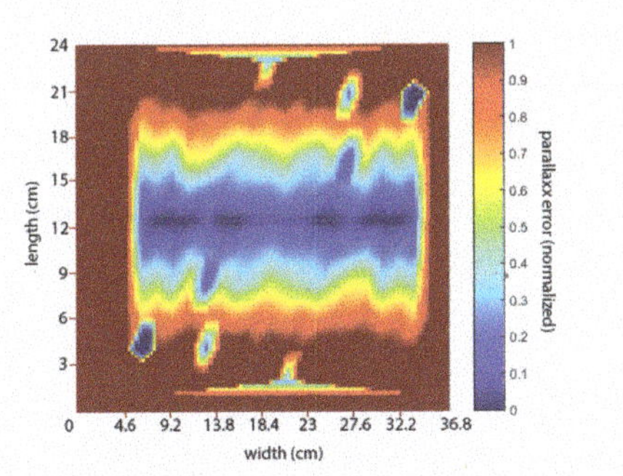

Figure 5. Heat map of the normalised position error across a frame.

While it is possible to disregard all but this area within the image, the camera would need to capture five times more images in order to cover the same area. This effectively reduces the camera frame rate from 90 fps to ≈17 fps, limiting the maximum speed of the system and the ability to incorporate additional wavelengths in the future. A stereo imaging technique using a single camera in motion [36,37] was chosen to compensate for the error, eliminating the need to purchase additional equipment.

Given that the imaging system is in constant motion, and the camera has a high frame rate, it is possible to use a single camera to achieve kinetic depth estimation. As the camera moves between two consecutive images, taller objects will appear to move faster than their shorter counterparts. Therefore the height disparity between the objects can be calculated using a dense optical flow algorithm. The system utilises the Farneback optical flow algorithm [38], chosen because of the low processing time and availability of the GPU kernel in the OpenCV library.

Tests were carried out to verify the reliability of the algorithm using objects of known height imaged twice with 2.1 cm camera displacement between images (chosen from the 90 fps camera frame rate and target 5 km/h system speed). The item on the left in Figure 6a is 2.5 cm tall; the object on the right is 3.5 cm tall at its lowest point and 11 cm at its highest. A height disparity map was produced from these images (Figure 6b), which shows significant error in the height estimation algorithm. The error was attributed to the small number of features, and thus, low spatial frequency of the image. Given that neighbouring pixels had very similar values, the algorithm was not able to accurately calculate the pixel shift.

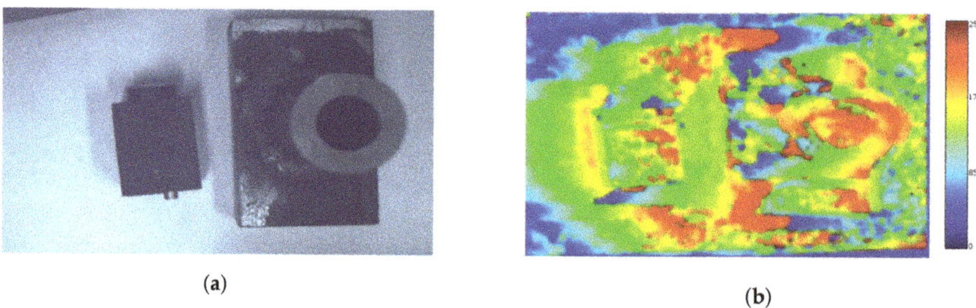

(a)   (b)

Figure 6. Height estimation using kinematic stereo method—preliminary test. (**a**) The image prior to processing containing two objects of different heights. (**b**) The height disparity map produced using the optical flow algorithm on the first attempt.

The error in the disparity values can be reduced by applying a Laplacian filter to the image and summing the filtered and non-filtered images (Figure 7). This can improve the texture of the image, preserving the high spacial frequency components. As can be seen in Figure 7, the error in disparity values across the image is reduced. Approximately 1 cm in height corresponds to 3 units in the height disparity map (Figure 7), showing a good estimation of the object height by the algorithm. However, some inconsistencies still exist at object edges, which can be mitigated by taking the disparity value in the middle of any given object.

Using Equation (2) the real distance can be calculated, allowing the system to provide accurate position information of individual plants across an entire frame. The distance between two frames is calculated using a speed estimation algorithm and the time between images, as:

$$x - x' = \frac{bf}{d} \tag{2}$$

where x and x' are the point of interest in frame one and the one in frame two respectively; b is the distance between frames; f is the focal length of the camera; and d is the disparity map (object height).

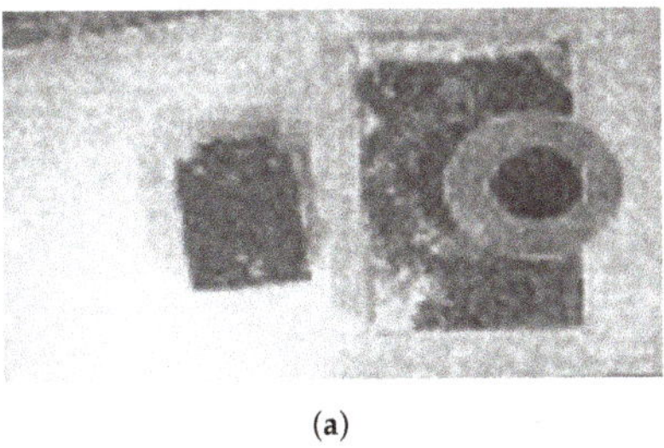
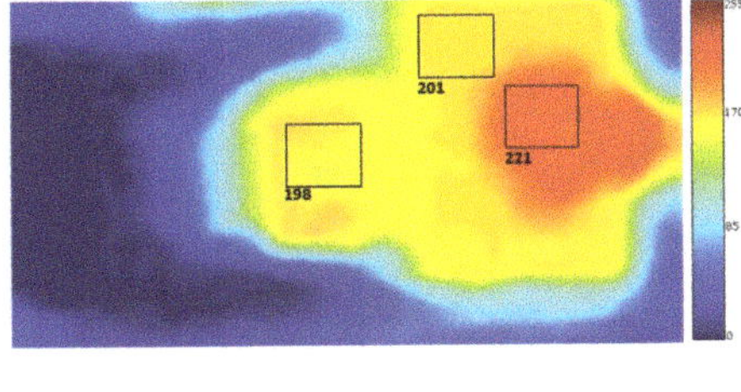

(**a**)   (**b**)

Figure 7. Height estimation using the kinematic stereo method with the addition of a Laplacian filtered image. (**a**) Sum of the original and Laplacian filtered images. (**b**) The height disparity map using the optical flow algorithm with addition of laplacian filtered image to reduce error.

2.4. System Optimisation

The system uses a NVIDIA GTX1080 GPU with a parallel implementation of the software developed in NVIDIA Compute Unified Device Architecture (CUDA)—Toolkit 9.0. A parallel prefix sum algorithm is used to reduce the run time of the multi-thresholding algorithm by several orders of magnitude. Three images were processed 1000 times, and the average speed was taken for the implementations on each the CPU and GPU to establish the time advantage of introducing multithreading. The results show the algorithm is up to 10,000 times faster using the GPU method, with the same threshold values found. A trial run of 50 images gave a total average loop time of 100 ms and maximum of 175 ms per frame with one detection module in use.

An in-field calibration method allows the user to optimise the system for different conditions. A background image is taken with no illumination, as are two further images in different positions to allow the user to tune parameters in Equation (1) and the height estimation algorithm. Figure 8 shows the calibration process with the "ideal" processed images output from the tuned parameters.

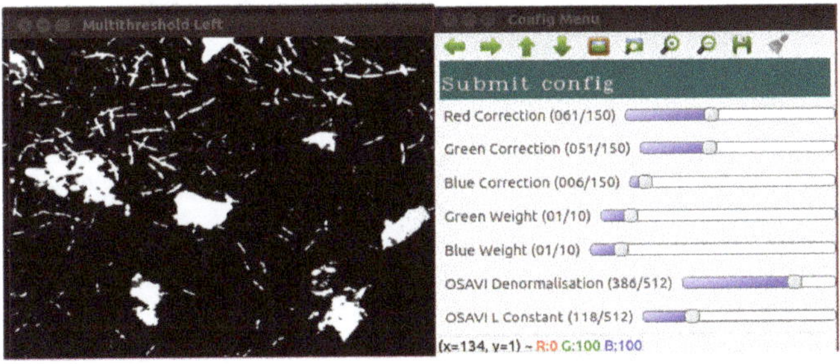

Figure 8. Screenshot of the system during calibration process for plant identification, showing a desirable output following thresholding.

2.5. Field Trial Procedure

Trials were performed in three separate iceberg lettuce fields at three growth stages in Ely, Cambridgeshire, United Kingdom on the 17 August 2018.

The tractor speed was varied from 0.5 to 10 km/h. The average weed density varied across each of the fields (150, 69, 31 weeds/m^2 for fields 1, 2 and 3 respectively). A maximum density of 364 weeds/m^2 was found in those frames analysed. Weeds found in the fields were between the one and three leaf stage and were predominantly broadleaf type.

Over 20,000 unprocessed images were captured from the trials for further testing. Each image file name contains the timestamp of capture, such that any lag in the system may be detected. The time between frame capture is fixed regardless of system speed, resulting in increased overlap between frames at a lower speed. This variable could be optimised to reduce overlap at lower speeds for future trials. Each trial run produced a database entry for each crop detected, including its size and location. These databases were produced to log information which may be of use to the farmer in other activities.

The in-field calibration files were also saved to reproduce the processed images after the trial; there were two files, one used for field 1 and part of field 2, the other for the remainder of field 2 and field 3. A random sample of 50 images was considered for further analysis. Each of the raw images was evaluated, and weeds and crops were identified by eye. This was then compared to the processed images to identify which plants had been correctly identified and where false negatives and positives had occurred according to the colour code described in Table 3. The process was completed first with the trial calibration and then with the post-trial calibration to establish the limitations of the calibration process and the capabilities of the system with optimal calibration (Figure 9).

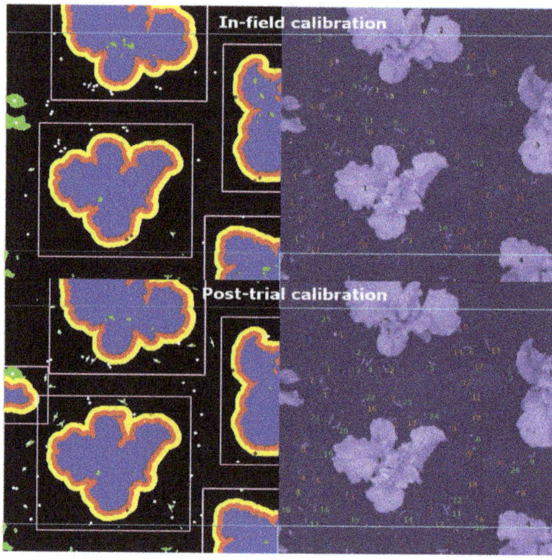

Figure 9. Images following processing (**left**) using in-field (**top**) and post-trial (**bottom**) calibration. Blue areas were designated as crops, and weeds are shown in green, with the centre of each target identified by a star. Unprocessed images are labelled by hand (**right**) according to the accuracy of the system identification using the key as defined in Table 3.

Crops are encircled with yellow and red regions to indicate the area within 2 and 4 cm of a crop respectively. Targets within the red-zone are ignored, as they are considered likely to result in crop damage, those within the yellow-zone are flagged to be handled with care, as are those within the pink box. Blue boxes at the top and bottom of each image identify the edge of the area for targets to be generated within each frame. The post-trial set used also used two calibrations, one for field 1 and another for field 2 and 3; this is mainly due to the significant size difference between the crops in these fields.

Table 3. Key for labelling of system targets.

Actual Object	Identified By System As:				
	Weed	**Crop**	**Debris**	**Missed**	**Multiple Targets**
Weed	Green	Blue	Orange	Red	Yellow
Crop	Pink	Black	N/A	N/A	N/A

3. Results and Discussion

Figure 10 shows the proportion of weeds per frame correctly identified taken from a sample of 50 of the processed images. Post-trial calibration (C2) results are consistently better than their in-field (C1) counterparts. With increasing field number, positive weed identification decreased for both in-field and post-trial calibration, with a particularly poor result (18%) for field 3 in-field calibration.

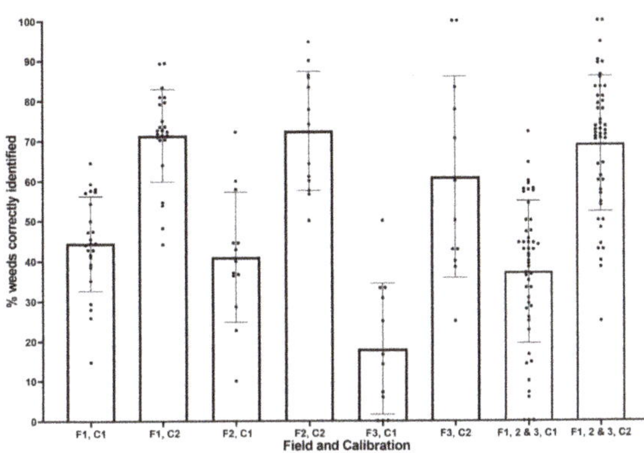

Figure 10. Proportion of weeds correctly identified for each field (F1–3) and calibration (C1, C2).

Field 3 had the smallest crops, and therefore, the fewest weeds/frame on average (nine) which may have resulted in the smaller peak in the histogram, a change in the threshold values, and a worse detection rate. Field 3 and part of field 2 also used a second, different calibration in-field due to the substantial change in crop size from those in field 1. This in-field calibration may have been less effective than that used for field 1. Post-trial calibration results are an improvement due to a process of trial and error not possible in-field and due to a lack of screen visibility in-field making it difficult to finely tune as required.

As can be seen from Figure 11, there was a wide range of values for the percentage of weeds identified as multiple targets, particularly in field 3. For increasing field number, there was a decreasing trend of multiple target weeds. Larger weeds with several leaves are most often misidentified as multiple targets, where each leaf is mistakenly identified as an individual plant. Those fields with fewer and smaller weeds are less likely to have weeds identified as multiple targets. Overall, the post-trial calibration reduces this error in identification (from an average 19% to 12%). Post-trial results in field 3 do not appear to reduce the error as expected; however, the wide range in results makes it difficult to draw a definitive conclusion from these results. Although the addition of extra targets to the system may decrease efficiency, these weeds can still be effectively managed, and the overall effect is determined by the resolution of any actuator system. As such, it is clear the system performs well in identifying weeds, with total average identification rates of 56% and 81% for the in-field and post-trial calibrations respectively (Figure 12).

The system has a significant problem with false positives, where debris in the field (e.g., stones and twigs) are incorrectly identified as a weeds. Whereas weed identification improves in the post-trial calibrations, false positives worsen, and there is a clear trade-off when calibrating the system between improved weed identification and increasing false positives. Figure 13 shows the proportion of generated targets which are in fact debris.

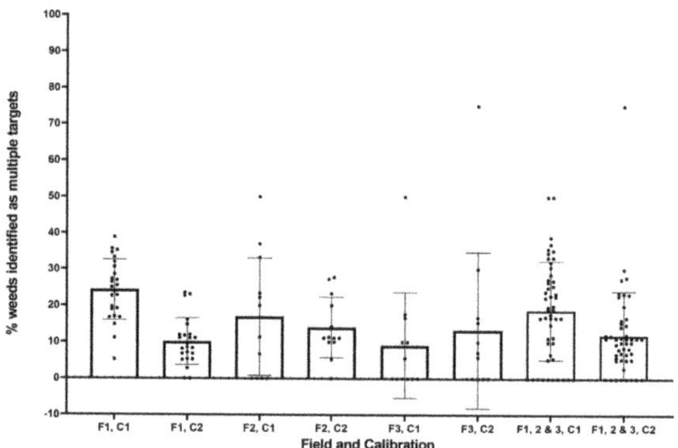

Figure 11. Proportion of weeds identified as multiple targets for each field (F1–3) and calibration (C1, C2).

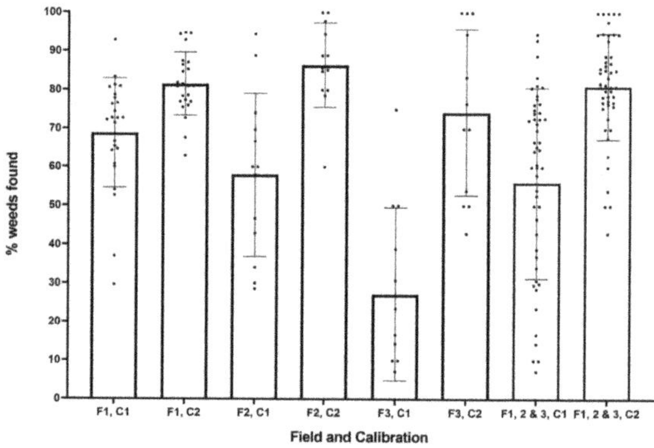

Figure 12. Proportion of weeds identified for each field (F1–3) and calibration (C1, C2).

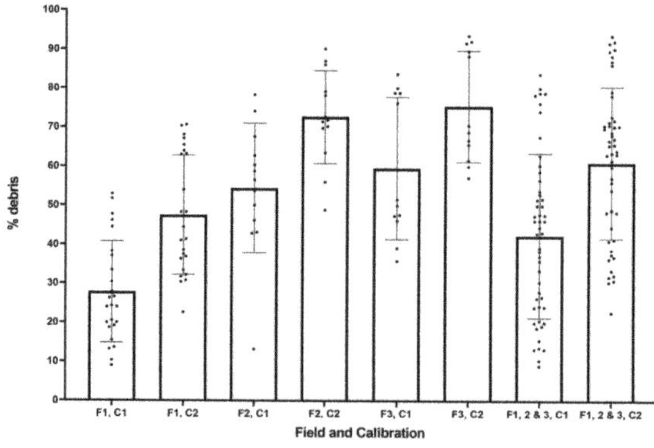

Figure 13. Proportion of objects identified which are debris for each field (F1–3) and calibration (C1, C2).

Field 1 had the lowest proportion of misidentified targets, due to its higher number of weeds/frame (average of 43 weeds/frame). The larger number of weeds, if average debris per image is approximately constant, results in an apparent reduction of debris identified. Where there is higher leaf coverage in an image, more of the debris will also be obscured. Finally, where there is a very high debris-to-weed ratio, it may result in a false peak in the histogram and error in identification.

Approximately 31% and 12% of crops were misidentified as weeds in the in-field and post-trial calibrations respectively, as shown in Figure 14. The vast majority of these false positives were at the edge of images where the crop is not fully visible, and as such the size exclusion method identified the object as a weed (see Figure 9). The effect was most pronounced in field 3, where crops were smallest. This issue could be mitigated in future work using image stitching when more than one module is in use so the full crop is visible to the system. Alternatively, as in [39], plants not fully contained in the image may be excluded from the classification algorithm to avoid crop damage. The increase in crop misidentification in in-field calibration is due to the threshold for size exclusion being set too high. The size exclusion threshold is currently taken from approximate crop size provided during calibration. It may be more effective to derive the value instead from weed size, as there is considerable variation in crop size within each field.

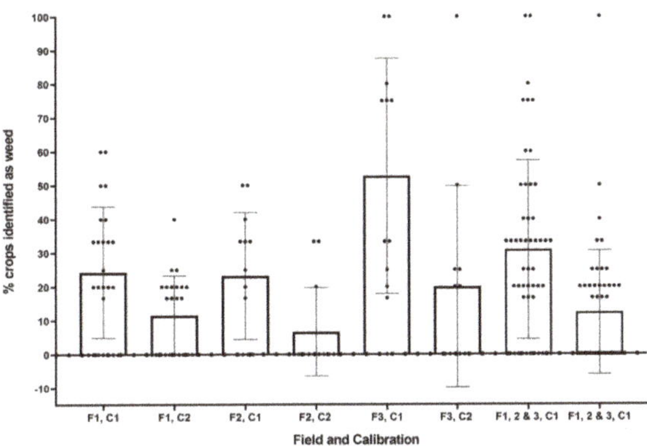

Figure 14. Percentage of crops misidentified for each field (F1-3) and calibration (C1, C2).

Figures 15 and 16a show the debris misidentification and percentage of weeds correctly identified respectively, with respect to weed count. No clear correlation is shown between weed count and correct weed identification or speed and weed identification (Figure 16b). It should be noted that at the highest speeds results may deteriorate as images become more blurry, and identification by eye of objects in raw images becomes more difficult. However percentage of debris and weed count are clearly correlated for the reasons discussed previously.

It should be noted that some lag in the capture of a very small minority of images occurred during the field trial; the reason for this is unclear and should be considered in future studies. Due to the significant overlap of each frame in most of the trial speeds, this did not result in the failure to identify objects in the areas were this error occurred. Individual weed and debris identification may have resulted in the introduction of some human error to the determination of the effectiveness of the system.

These results are promising but further development should be considered to increase classification rates and reduce false positives. The system provides a relatively simple approach for crop and weed identification, but more complex approaches have shown better classification performance. In [28], plants were divided first into three groups of monocot, dicot, and barley with 97.7% accuracy, and weeds further classified by species with varying success using a support vector machine and shape features. Hyperspectral imaging is suggested for species identification in [40], providing 100% crop recognition, and weed species identification (31%–98% correctly identified depending on species and classifier method). This method is implemented at low resolution and operating speeds (average ground speed = 0.09 m/s). Plants were identified in maize fields using a selection of nine colour indices combined with support vector data description in [41] achieving up to 90.79% classification, but with significant variation in results due to weather and time of day. By comparison, our system has a lower classification rate, but provides real-time identification at high resolution for individual plant treatment, whilst correcting target position data through the height estimation algorithm. The system currently has sufficient computational redundancy such that future work may introduce further discriminating features and processes to improve plant recognition and reduce false positives. The need for this increased complexity to improve recognition rates may be required if it offers an economic advantage to total weed management beyond re-running the system periodically and capturing those undetected weeds from the earlier passes.

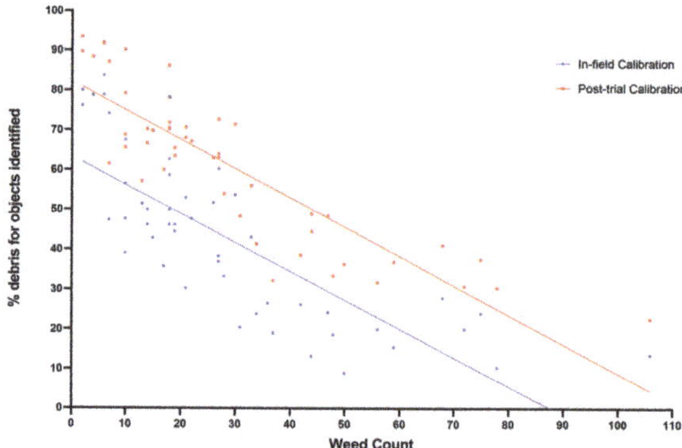

Figure 15. Proportion of objects identified which are debris versus weed count.

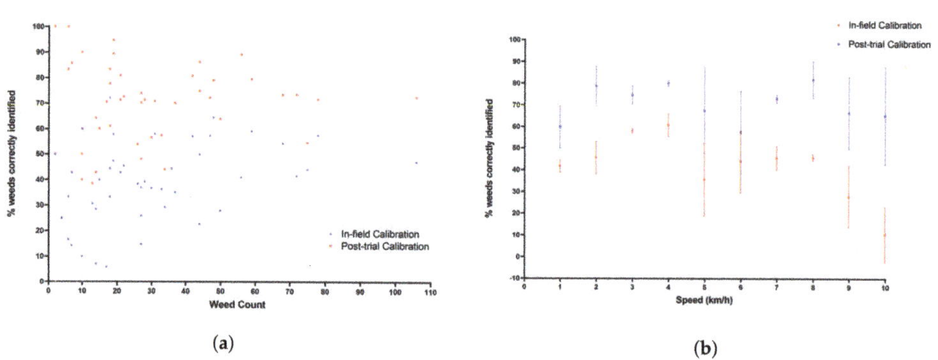

(a)   (b)

Figure 16. System weed identification capability with varying speed and weed count. (**a**) Percentage of correctly identified weeds versus total weed count. (**b**) Percentage of correctly identified weeds versus system speed in-field

4. Conclusions

This paper reports a full-scale, on-tractor, field demonstration of a single-shot, multispectral imaging system for autonomous identification of emergent weeds at forward velocities of up to 10 km/h (2.8 m/s). The method successfully identified an average of 81% of weeds and 88% of crops in field trials, showing promise for the future development of this approach. The approach combines controlled spectrum artificial illumination alongside a short-wavelength (nominal cut-off wavelength, 515 nm), optically filtered, and modified (removal of NIR filter) Bayer-array (colour) digital imaging sensor. Single shot imaging of the red, green, and NIR reflectivity is delivered to achieve plant-soil discrimination.

Weeds are assumed to have a leaf area significantly less than the lettuce crop, and so a size-exclusion approach has been used to separate the target weeds from the neighbouring crop plants. This method of processing relies on the nature of lettuce production and similar horticultural crops where soil pre-treatment is used prior to the growing season to remove emergent weeds, before the transplantation of young plants. The green wavelengths (centred around 525 nm), are intended to handle more difficult weed detection cases associated with overlapping small weeds, which are then interpreted as a larger single plant. Further system testing to establish the performance at high weed density is required to verify the effectiveness of this approach.

A kinematic stereo method has been used to estimate the height of plants in the image and correct the parallax error to allow for accurate targeting of plants by a herbicide micro-dosing system. The method allows the full frame to be processed without increasing the system cost through the addition of expensive equipment, such as stereo vision cameras or LIDAR sensors. As a follow-on from this research programme, tests to ensure the proper functionality of the height-estimation system, and thereby the accuracy of the plant location data in-field are required.

Future developments of the design may include more sophisticated image processing, and taking advantage of the tonal differences between the crop and weed types or the morphologies of the various plant shapes. The significant improvement in plant identification using the post-trial calibration indicates the calibration system should be improved to ensure the system can be operated more effectively in-field.

Author Contributions: Conceptualization, investigation, and methodology K.Y.H., L.E., S.B. and M.Z.G.; software, K.Y.H. and S.B.; validation, L.E., S.B. and K.Y.H.; resources, B.G.; writing—original draft preparation, L.E. and K.Y.H.; writing—review and editing, L.E., B.G. and W.P.H.; visualisation, L.E., K.Y.H. and S.B.; supervision, B.G. and W.P.H.; project administration, K.Y.H. All authors have read and agreed to the published version of the manuscript.

Funding: This research received no external funding.

Acknowledgments: Technical support and materials for experiments were provided by Syngenta and G's fresh. The work was funded by the University of Manchester.

Conflicts of Interest: No conflict of interest to declare.

References

1. Pimentel, D. Amounts of pesticides reaching target pests: Environmental impacts and ethics. *J. Agric. Environ. Ethics* **1995**, *8*, 17–29. [CrossRef]
2. Rüegg, W.T.; Quadranti, M.; Zoschke, A. Herbicide research and development: Challenges and opportunities. *Weed Res.* **2007**, *47*, 271–275. [CrossRef]
3. Slaughter, D.; Giles, D.; Downey, D. Autonomous robotic weed control systems: A review. *Comput. Electron. Agric.* **2008**, *61*, 63–78. [CrossRef]
4. Heap, I. Herbicide Resistant Weeds. In *Integrated Pest Management: Pesticide Problems*; Pimentel, D., Peshin, R., Eds.; Springer: Dordrecht, The Netherlands, 2014; Volume 3, pp. 281–301. [CrossRef]
5. Oerke, E.C.; Dehne, H.W. Safeguarding production—Losses in major crops and the role of crop protection. *Crop Prot.* **2004**, *23*, 275–285. [CrossRef]
6. Godfray, H.C.J.; Beddington, J.R.; Crute, I.R.; Haddad, L.; Lawrence, D.; Muir, J.F.; Pretty, J.; Robinson, S.; Thomas, S.M.; Toulmin, C. Food security: The challenge of feeding 9 billion people. *Science* **2010**, *327*, 812–818. [CrossRef] [PubMed]
7. Fennimore, S.A.; Doohan, D.J. The Challenges of Specialty Crop Weed Control, Future Directions. *Weed Technol.* **2008**, *22*, 364–372. [CrossRef]
8. Fennimore, S.A.; Cutulle, M. Robotic weeders can improve weed control options for specialty crops. *Pest Manag. Sci.* **2019**, *75*, 1767–1774. [CrossRef] [PubMed]
9. Lati, R.N.; Siemens, M.C.; Rachuy, J.S.; Fennimore, S.A. Intrarow Weed Removal in Broccoli and Transplanted Lettuce with an Intelligent Cultivator. *Weed Technol.* **2016**, *30*, 655–663. [CrossRef]

10. Dammer, K.H. Real-time variable-rate herbicide application for weed control in carrots. *Weed Res.* **2016**, *56*, 237–246. [CrossRef]

11. Pérez-Ruiz, M.; Slaughter, D.; Gliever, C.; Upadhyaya, S. Automatic GPS-based intra-row weed knife control system for transplanted row crops. *Comput. Electron. Agric.* **2012**, *80*, 41–49. [CrossRef]

12. Tillett, N.; Hague, T.; Grundy, A.; Dedousis, A. Mechanical within-row weed control for transplanted crops using computer vision. *Biosyst. Eng.* **2008**, *99*, 171–178. [CrossRef]

13. Gonzalez-de Soto, M.; Emmi, L.; Perez-Ruiz, M.; Aguera, J.; Gonzalez-de Santos, P. Autonomous systems for precise spraying—Evaluation of a robotised patch sprayer. *Biosyst. Eng.* **2016**, *146*, 165–182. [CrossRef]

14. Nørremark, M.; Griepentrog, H.; Nielsen, J.; Søgaard, H. The development and assessment of the accuracy of an autonomous GPS-based system for intra-row mechanical weed control in row crops. *Biosyst. Eng.* **2008**, *101*, 396–410. [CrossRef]

15. Michaels, A.; Haug, S.; Albert, A. Vision-based high-speed manipulation for robotic ultra-precise weed control. In Proceedings of the 2015 IEEE/RSJ International Conference on Intelligent Robots and Systems (IROS), Hamburg, Germany, 28 September–2 October 2015; pp. 5498–5505. [CrossRef]

16. Midtiby, H.S.; Mathiassen, S.K.; Andersson, K.J.; Jørgensen, R.N. Performance evaluation of a crop/weed discriminating microsprayer. *Comput. Electron. Agric.* **2011**, *77*, 35–40. [CrossRef]

17. Utstumo, T.; Urdal, F.; Brevik, A.; Dørum, J.; Netland, J.; Overskeid, Ø.; Berge, T.W.; Gravdahl, J.T. Robotic in-row weed control in vegetables. *Comput. Electron. Agric.* **2018**, *154*, 36–45. [CrossRef]

18. Søgaard, H.; Lund, I. Application Accuracy of a Machine Vision-controlled Robotic Micro-dosing System. *Biosyst. Eng.* **2007**, *96*, 315–322. [CrossRef]

19. Wang, A.; Zhang, W.; Wei, X. A review on weed detection using ground-based machine vision and image processing techniques. *Comput. Electron. Agric.* **2019**, *158*, 226–240. [CrossRef]

20. Andújar, D.; Rueda-Ayala, V.; Moreno, H.; Rosell-Polo, J.R.; Escolá, A.; Valero, C.; Gerhards, R.; Fernández-Quintanilla, C.; Dorado, J.; Griepentrog, H.W. Discriminating Crop, Weeds and Soil Surface with a Terrestrial LIDAR Sensor. *Sensors* **2013**, *13*, 14662–14675. [CrossRef]

21. Hamuda, E.; Glavin, M.; Jones, E. A survey of image processing techniques for plant extraction and segmentation in the field. *Comput. Electron. Agric.* **2016**. doi:10.1016/j.compag.2016.04.024. [CrossRef]

22. Bawden, O.; Kulk, J.; Russell, R.; McCool, C.; English, A.; Dayoub, F.; Lehnert, C.; Perez, T. Robot for weed species plant-specific management. *J. Field Robot.* **2017**, *34*, 1179–1199. [CrossRef]

23. Romeo, J.; Pajares, G.; Montalvo, M.; Guerrero, J.; Guijarro, M.; de la Cruz, J. A new Expert System for greenness identification in agricultural images. *Expert Syst. Appl.* **2013**, *40*, 2275–2286. [CrossRef]

24. Hamuda, E.; Mc Ginley, B.; Glavin, M.; Jones, E. Automatic crop detection under field conditions using the HSV colour space and morphological operations. *Comput. Electron. Agric.* **2017**, *133*, 97–107. [CrossRef]

25. Herrera, P.J.; Dorado, J.; Ribeiro, Á. A Novel Approach for Weed Type Classification Based on Shape Descriptors and a Fuzzy Decision-Making Method. *Sensors* **2014**, *14*, 15304–15324. [CrossRef] [PubMed]

26. McCool, C.; Sa, I.; Dayoub, F.; Lehnert, C.; Perez, T.; Upcroft, B. Visual detection of occluded crop: For automated harvesting. In Proceedings of the 2016 IEEE International Conference on Robotics and Automation (ICRA), Stockholm, Sweden, 16–21 May 2016; pp. 2506–2512. [CrossRef]

27. Ahmed, F.; Al-Mamun, H.A.; Bari, A.S.; Hossain, E.; Kwan, P. Classification of crops and weeds from digital images: A support vector machine approach. *Crop Prot.* **2012**, *40*, 98–104. [CrossRef]

28. Rumpf, T.; Römer, C.; Weis, M.; Sökefeld, M.; Gerhards, R.; Plümer, L. Sequential support vector machine classification for small-grain weed species discrimination with special regard to Cirsium arvense and Galium aparine. *Comput. Electron. Agric.* **2012**, *80*, 89–96. [CrossRef]

29. Tellaeche, A.; Pajares, G.; Burgos-Artizzu, X.P.; Ribeiro, A. A computer vision approach for weeds identification through Support Vector Machines. *Appl. Soft Comput. J.* **2011**, *11*, 908–915. [CrossRef]

30. Sa, I.; Chen, Z.; Popovic, M.; Khanna, R.; Liebisch, F.; Nieto, J.; Siegwart, R. WeedNet: Dense Semantic Weed Classification Using Multispectral Images and MAV for Smart Farming. *IEEE Robot. Autom. Lett.* **2018**, *3*, 588–595, [CrossRef]

31. dos Santos Ferreira, A.; Matte Freitas, D.; Gonçalves da Silva, G.; Pistori, H.; Theophilo Folhes, M. Weed detection in soybean crops using ConvNets. *Comput. Electron. Agric.* **2017**, *143*, 314–324. [CrossRef]

32. Dworak, V.; Selbeck, J.; Dammer, K.H.; Hoffmann, M.; Zarezadeh, A.; Bobda, C. Strategy for the Development of a Smart NDVI Camera System for Outdoor Plant Detection and Agricultural Embedded Systems. *Sensors* **2013**, *13*, 1523–1538. [CrossRef]

33. Åstrand, B.; Baerveldt, A. An Agricultural Mobile Robot with Vision-Based Perception for Mechanical Weed Control. *Auton. Robots* **2002**, *13*, 21–35. [CrossRef]
34. Otsu, N. A Threshold Selection Method from Gray-Level Histograms. *IEEE Trans. Syst. Man Cybern.* **1979**, *9*, 62–66. [CrossRef]
35. Huang, D.Y.; Lin, T.W.; Hu, W.C. Automatic Multilevel Threshold Based on Two-stage Otsu's Method with Cluster Determination By Valley Estimation. *Int. J. Innov. Comput.* **2011**, *7*, 5631–5644.
36. Dalmia, A.K.; Trivedi, M. Depth extraction using a single moving camera: An integration of depth from motion and depth from stereo. *Mach. Vis. Appl.* **1996**, *9*, 43–55. [CrossRef]
37. Seal, J.; Bailey, D.G.; Gupta, G.S. Depth Percpetion with a Single Camera. In Proceedings of the International Conference on Sensing Technology, Palmerston North, New Zealand, 21–23 November 2005; pp. 96–101.
38. Farnebäck, G. Two-frame motion estimation based on polynomial expansion. In *Scandanavian Conference on Image Analysis*; Bigun, J., Gustavsson, T., Eds.; Springer: Berlin/Heidelberg, Germany, 2003; Volume 2749, pp. 363–370. [CrossRef]
39. Jeon, H.Y.; Tian, L.F.; Zhu, H. Robust Crop and Weed Segmentation under Uncontrolled Outdoor Illumination. *Sensors* **2011**, *11*, 6270–6283. [CrossRef]
40. Pantazi, X.E.; Moshou, D.; Bravo, C. Active learning system for weed species recognition based on hyperspectral sensing. *Biosyst. Eng.* **2016**, *146*, 193–202. [CrossRef]
41. Zheng, Y.; Zhu, Q.; Huang, M.; Guo, Y.; Qin, J. Maize and weed classification using color indices with support vector data description in outdoor fields. *Comput. Electron. Agric.* **2017**, *141*, 215–222. [CrossRef]

Article

An Autonomous Fruit and Vegetable Harvester with a Low-Cost Gripper Using a 3D Sensor

Tan Zhang, Zhenhai Huang, Weijie You, Jiatao Lin, Xiaolong Tang and Hui Huang *

Guangdong Laboratory of Artificial Intelligence and Digital Economy (SZ), Shenzhen University, Shenzhen 510000, China; tan.zhang@utoronto.ca (T.Z.); zhenhaihuang0@gmail.com (Z.H.); youweijiee@gmail.com (W.Y.); zeup300@gmail.com (J.L.); tangtxl80@gmail.com (X.T.)
* Correspondence: huihuang@szu.edu.cn

Received: 13 October 2019; Accepted: 19 December 2019; Published: 22 December 2019

Abstract: Reliable and robust systems to detect and harvest fruits and vegetables in unstructured environments are crucial for harvesting robots. In this paper, we propose an autonomous system that harvests most types of crops with peduncles. A geometric approach is first applied to obtain the cutting points of the peduncle based on the fruit bounding box, for which we have adapted the model of the state-of-the-art object detector named Mask Region-based Convolutional Neural Network (Mask R-CNN). We designed a novel gripper that simultaneously clamps and cuts the peduncles of crops without contacting the flesh. We have conducted experiments with a robotic manipulator to evaluate the effectiveness of the proposed harvesting system in being able to efficiently harvest most crops in real laboratory environments.

Keywords: harvesting robot; gripper; segmentation; cutting point detection

---

1. Introduction

Manual harvesting of fruits and vegetables is a laborious, slow, and time-consuming task in food production [1]. Automatic harvesting has many benefits over manual harvesting, e.g., managing the crops in a short period of time, reduced labor involvement, higher quality, and better control over environmental effects. These potential benefits have inspired the wide use of agricultural robots to harvest horticultural crops (the term crops is used indistinctly for fruits and vegetables throughout the paper, unless otherwise indicated) over the past two decades [2]. An autonomous harvesting robot usually has three subsystems: a vision system for detecting crops, an arm for motion delivery, and an end-effector for detaching the crop from its plant without damaging the crop.

However, recent surveys of horticultural crop robots [3,4] have shown that the performance of automated harvesting has not improved significantly despite the tremendous advances in sensor and machine intelligence. First, cutting at the peduncle leads to higher success rates for crop detachment, and detachment at the peduncle reduces the risk of damaging the flesh or other stems of the plant and maximizes the storage life. However, peduncle detection is still a challenging step due to varying lighting conditions and occlusion of leaves or other crops [5], as well as similar colors of peduncle and leaves [6,7]. Second, existing manipulation tools that realize both grasping and cutting functions usually require a method of additional detachment [8,9], which increases the cost of the entire system.

In this paper, we present a well-generalized harvesting system that is invariant and robust to different crops in an unstructured environment aimed at addressing these challenges. The major contributions of this paper are the following: (i) development of a low-cost gripper (Supplementary Materials) that simultaneously clamps and cuts peduncles in which the clamper and cutter share the same power-source drive, thus maximizing the storage life of crops and reducing the complexity and cost; (ii) development of a geometric method to detect the cutting point of the crop peduncle based on the Mask Region-based Convolutional Neural Network (Mask R-CNN) algorithm [10].

The rest of this paper is organized as follows: In Section 2, we introduce related work. In Section 3, we describe the design of the gripper and provide a mechanical analysis. In Section 4, we present the approach of cutting-point detection. In Section 5, we describe an experimental robotic system to demonstrate the harvesting process for artificial and real fruits and vegetables. Conclusions and future work are given in Section 6.

2. Related Work

During the last few decades, many systems have been developed for the autonomous harvesting of soft crops, ranging from cucumber [11] and tomato-harvesting robots [12] to sweet-pepper [2,13] and strawberry-picking apparatus [14,15]. The work in [16] demonstrated a sweet-pepper-harvesting robot that achieved a success rate of 6% in unstructured environments. This work highlighted the difficulty and complexity of the harvesting problem. The key research challenges include manipulation tools along with the harvesting process, and perception of target fruits and vegetables with the cutting points.

2.1. Harvesting Tools

To date, researchers have developed several types of end-effectors for autonomous harvesting, such as suction devices [17], swallow devices [18], and scissor-like cutters [15]. The key component of an autonomous harvester is the end-effector that grasps and cuts the crop. A common approach for harvesting robots is to use a suction gripping mechanism [16,19,20] that picks up fruit by pushing the gripper onto the crop to generate air to draw a piece of crop into the harvesting end and through the tubular body to the discharge end. These types of grippers do not touch the crops during the cutting process, but physical contact occurs when they are inhaled and slipped into the container, potentially causing damage to the crops, especially to fragile crops. Additionally, the use of additional pump equipment increases the complexity and weight of the robotic system, resulting in difficulties for compact and high-power density required in autonomous mobile robots [21,22]. Some soft plants, such as cucumbers and sweet peppers, must be cut from the plant and thus require an additional detachment mechanism and corresponding actuation system. The scissor-like grippers [16,23] that grasp the peduncles or crops and cut peduncles can be used for such soft plants. However, these grippers are usually designed at the expense of size. A custom harvesting tool in [16] can simultaneously envelop a sweet pepper and cuts the peduncle with a hinged jaw mechanism; however, size constraints restrict its use for crops with irregular shapes [24].

To overcome these disadvantages, we have designed a novel gripper that simultaneously clamps and cuts the peduncles of crops. In addition, the clamping and detachment mechanism share the same power-source drive rather than using separate drives. The benefits of sharing the same power source for several independent functions are detailed in our previous work [25,26]. Since the gripper clamps the peduncle, it can be used for most fruits that are harvested by cutting the peduncle. Since the gripper does not touch the flesh of the crop, it can maximize the storage life and market value of the crops. As the clamping and cutting occurs simultaneously at the peduncle, the design of the gripper is relatively simple and therefore cost-effective.

2.2. Perception

Detection in automated harvesting determines the location of each crop and selects appropriate cutting points [1,27,28] using RGB cameras [29,30] or 3D sensors [13,15]. The 3D localization allows a cutting point and a grasp to be determined. Currently, some work has been done to detect the peduncle based on the color information using RGB cameras [6,7,29,30]; however, such cameras are not capable of discriminating between peduncles, leaves, and crops if they are in the same color [7]. The work in [31] proposed a dynamic thresholding to detect apples in viable lighting conditions, and they further used a dynamic adaptive thresholding algorithm [32] for fruit detection using a small set of training images. The work in [27] facilitates detection of peduncles of sweet peppers using multi-spectral imagery; however, the accuracy is too low to be of practical use. Recent work in deep neural networks has led to the development of a state-of-the-art object detector, such as Faster Region-based CNN (Faster R-CNN) [33] and Mask Region-based CNN (Mask R-CNN) [10]. The work in [34] uses a Faster R-CNN model to detect different fruits, and it achieves better performance with many fruit images; however, the detection of peduncles and cutting points has not been addressed in this work. To address these shortcomings, we use a Mask R-CNN model and a geometric feature to detect the crops and cutting points of the peduncle.

3. System Overview and the Gripper Mechanism

3.1. The Harvesting Robot

The proposed harvesting robot has two main modules: a detection module for detecting an appropriate cutting point on the peduncle, and a picking module for manipulating crops (see Figure 1). Specifically, the detection module determines the appropriate cutting points on the peduncle of each crop. The picking module controls the robot to reach the cutting point, clamps the peduncle, and cuts the peduncle. The clamping and cutting actions are realized by the newly designed gripper.

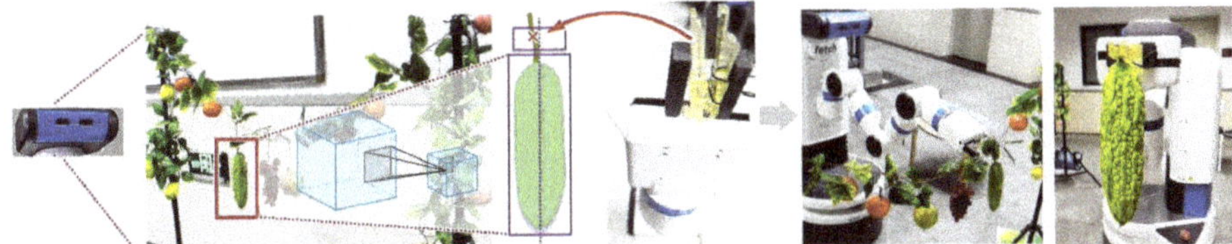

Figure 1. Architecture of the proposed harvesting robot.

3.2. The Gripper Mechanism

The proposed gripper is illustrated in Figure 2 from a 3D point of view. Inspired by skilled workers who use fingers to gently hold the peduncles and use their fingernails to cut the peduncles, this gripper is designed with both clamping and cutting functions with three key parts only: a top plate with a cutting blade, a middle plate to press the peduncle, and a bottom plate to hold the peduncle. Two torsion springs and an axle are included.

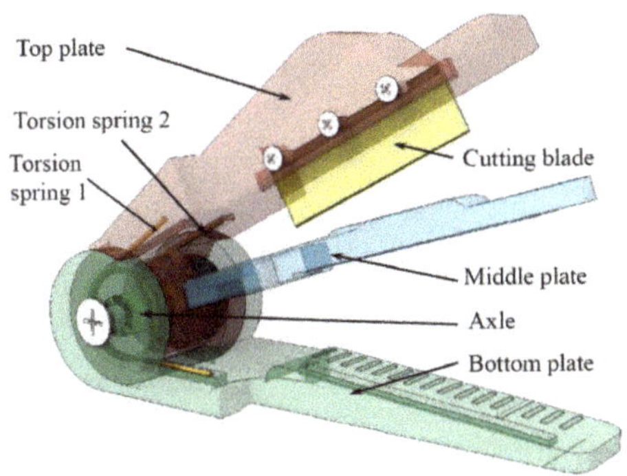

Figure 2. The 3D view of the proposed gripper [35,36].

The clamping and cutting processes work as follows:

1. Clamping the peduncle: pressing the top plate, the middle plate contacts the bottom plate based on *"torsion spring 2"*, and thus the peduncle is clamped, as shown in Figure 3b.

2. Cutting the peduncle: continuing to press the top plate, the cutting blade contacts the peduncle on the bottom plate, thereby cutting the peduncle, as shown in Figure 3c.

After releasing the top plate, the top and middle plates are automatically restored to the rest position (Figure 3a) by *"torsion springs 1 and 2"*, respectively.

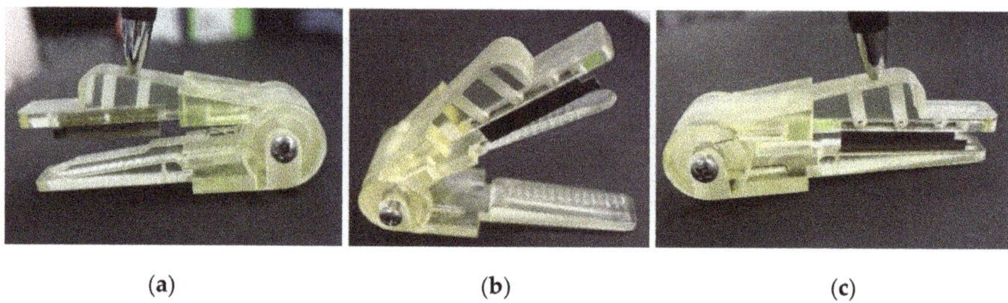

(a)   (b)   (c)

Figure 3. Cutting process: (**a**) rest position; (**b**) clamping state; (**c**) cutting state.

3.3. Force Analysis of the Gripper

As shown in Figure 4, when the cutting blade touches the bottom plate, the cutting force on the peduncle is calculated using the following equations:

$$M_{cut} = M_{push} - M_{spring\_1} - M_{spring\_2} \tag{1}$$

$$M_{spring\_1} = k_1 \cdot \theta_1 \tag{2}$$

$$M_{spring\_2} = k_2 \cdot \theta_2 \tag{3}$$

where M_{cut}, M_{push}, M_{spring_1}, M_{spring_2} are the moments of cutting force, pushing force, spring 1 and spring 2. k_1 and k_2 are the torsion coefficients of "torsion springs 1 and 2", respectively. θ_1 and θ_2 are the angles of deflection from the equilibrium position of "torsion springs 1 and 2", respectively.

"*Torsion spring 1*" is the spring between the top plate and bottom plate, while "*torsion spring 2*" is the spring between the top plate and middle plate; see also Figures 2 and 4d. With the moment M_{cut}, the cutting force F_{cut} can be described as:

$$M_{cut} = F_{cut} \cdot L_{cut} \tag{4}$$

where F_{cut} is the force applied on the peduncle from the cutting blade, L_{cut} is the length between the axis of the coil and the line of the cutting force, see Figure 4b. P is the contacting point between the cutting blade and the peduncle.

Thus, F_{cut} is derived from Equations (1)–(4) as follows:

$$F_{cut} = \frac{M_{push} - k_1 \cdot \theta_1 - k_2 \cdot \theta_2}{L_{cut}} \tag{5}$$

For further investigating the cutting force on peduncle, we analyzed the stress:

$$F_{cut} = \tau_{cut} \cdot A \tag{6}$$

$$\tau_{cut} = \frac{M_{push} - k_1 \cdot \theta_1 - k_2 \cdot \theta_2}{A \cdot L_{cut}} \tag{7}$$

where τ_{cut} is the stress, and A is the surface area of the peduncle. The two variables determine if the peduncle can be cut off. Equation (7) is derived from Equations (5) and (6).

As shown in Figure 5, when pressing the top plate and the middle plate touches the bottom plate, "*torsion spring 2*" creates the pressing force on the middle plate, and thus generates force on the bottom plate. This force ensures that the gripper clamps the peduncle before the cutting blade touches the peduncle:

$$M_{clamp} = M_{spring\_2} \tag{8}$$

$$M_{clamp} = F_{clamp} \cdot L_2 \tag{9}$$

$$M_{spring\_2} = k_2 \cdot \theta_2 \tag{10}$$

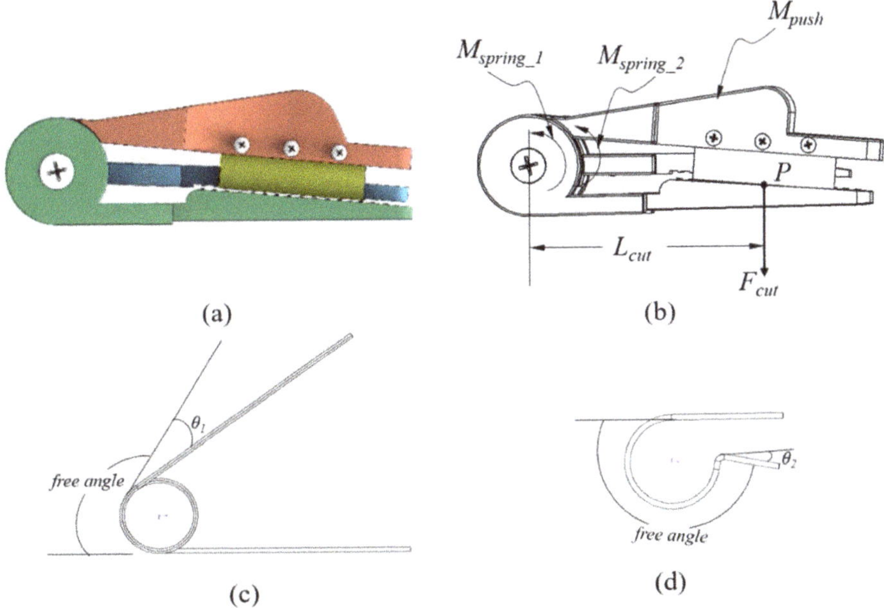

Figure 4. Force diagram of the cutting blade touching the bottom plate. (**a**) The diagram of the gripper. The force analysis of the gripper (**b**), *Torsion spring 1* (**c**), and *Torsion spring 2* (**d**).

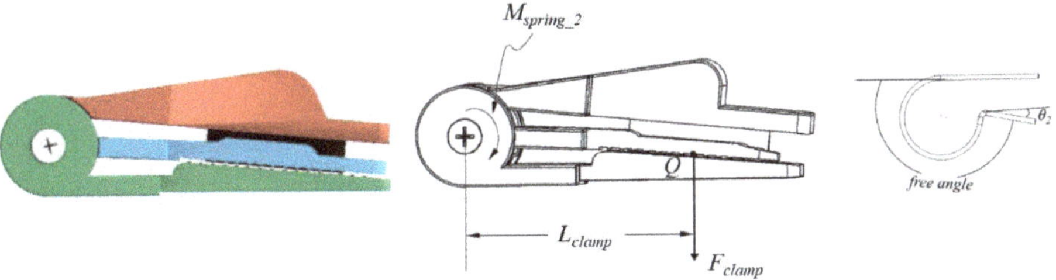

Figure 5. Force diagram of the middle plate touching the bottom plate.

After the peduncle is cut, the crop is detached from the peduncle if the clamping force is smaller than the gravity of the crop. To ensure that the crop is clamped, the minimum clamping force overcoming the gravity is expressed as:

$$F_{clamp} \geq \frac{m \cdot g}{\mu} \tag{11}$$

where m is the weight of the crop, g the gravity, and μ the coefficient of friction for a crop on the gripper plate surface. Based on the Equations (8)–(11), we obtain the minimum coefficient of "*torsion spring 2*":

$$k_2 \geq \frac{m \cdot g \cdot L_{clamp}}{\mu \cdot \theta_2} \tag{12}$$

4. Perception and Control

4.1. Cutting-Point Detection

The approach to detecting the cutting points of the peduncles consists of two steps: (i) the pixel area of crops is obtained using deep convolutional neural networks; (ii) the edge images of fruits are extracted, and then a geometric model is established to obtain the cutting points. Here, we obtain the external rectangle size of each pixel area to determine the region containing the peduncle for each crop. The architecture of the cutting-point-detection approach is shown in Figure 6.

Accurate object detection requires detecting not only which objects are in a scene, but also where they are located. Accurate region proposal algorithms thus play significant roles in the object-detection task. Mask R-CNN [10] makes use of edge information to generate region proposals by using the Region Proposal Network (RPN). Mask R-CNN uses color images to perform general object detection in two stages: region proposal, which proposes candidate object bounding boxes, and a region classifier that predicts the class and box offset with a binary mask for each region of interest (ROI). To train Mask R-CNN for our task, we perform fine-tuning [37], which requires labeled bounding-box information for each of the classes to be trained. After extracting the bounding box for each fruit, the geometric information of each region (e.g., external rectangle) is extracted by geometric morphological calculation. The cutting region of the interest of the targeted crop can thus be determined. A RGB-D camera like the Kinect provides RGB images along with per-pixel depth information in real time. With the depth sensor, the 2D position of the cutting point is mapped to the depth map, and thus the 3D information of the cutting point (x,y,z) is obtained.

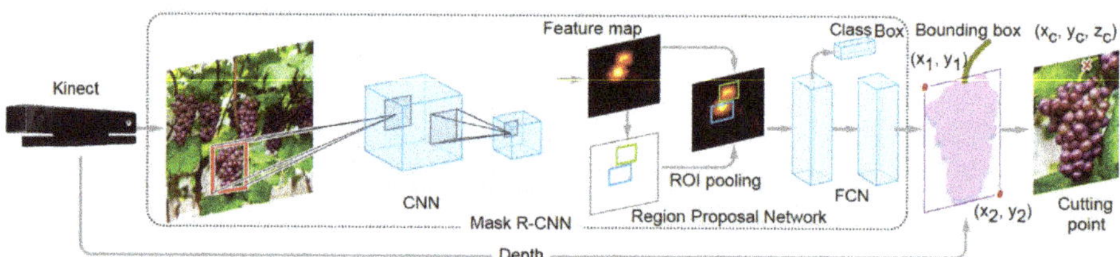

Figure 6. Illustration of the vision system that detects and localizes the crop and its cutting point.

The peduncle is usually located above the fruit. Thus, we set the ROI of the peduncle above the fruit, as shown in Figure 7.

The directions of the crops can be downward or tilted. We also adapted the minimum bounding rectangle to Mask R-CNN, thus to obtain the bounding boxes for crops in tilted direction. Other cases such as the direction of the crop parallel to the camera or the peduncle being "occluded" by its flesh will be discussed in the future work.

We then obtain the bounding box of the fruit, i.e., a top point coordinate (x_t, y_t), and a bottom point coordinate (x_b, y_b). The coordinates of the cutting point (x_c, y_c) can be determined as follows:

$$Roi_L = 0.5 \cdot L_{max} \tag{13}$$

$$Roi_H = 0.5 \cdot H_{max} \tag{14}$$

$$L_{max} = |x_b - x_t| \tag{15}$$

$$H_{max} = |y_b - y_t| \tag{16}$$

$$x_c = 0.5(x_b - x_t) \pm 0.5Roi_L \tag{17}$$

$$y_c = y_t - 0.5Roi_H \tag{18}$$

Figure 7. A schematic diagram of the cutting-point-detection approach.

4.2. System Control

A common method of motion planning for autonomous crop harvesting is open-loop planning [16,38]. Existing work on manipulation planning usually requires some combination of precise models, including manipulator kinematics, dynamics, interaction forces, and joint positions [39]. The reliance on manipulator models makes transferring solutions to a new robot configuration challenging and time-consuming. This is particularly true when a model is difficult to define, as in custom harvesting robots. In automated harvesting, a manipulator is usually customized to a special crop; thus, they are not as accurate as modern manipulators and it is relatively difficult to obtain accurate models of the robots. In addition, requiring high position sensing becomes challenging in manipulators given the space constraints.

In this paper, we demonstrate a learning-based kinematic control policy that maps target positions to joint actions directly, using convolutionally neural network (CNN) [40]. This approach makes no assumptions regarding the robot's configuration, kinematic, and dynamic model, and has no dependency on absolute position sensing available on-board. The inverse-kinematics function is learned using CNNs, which also considers self-collision avoidance, and errors caused by manufacturing and/or assembly tolerances.

As shown in Figure 8, the network comprises three convolutional layers and three fully connected layers. The input is the spatial position of the end-effector, and the output is the displacement of all of the joints. The fully connected layers will serve as a classifier on top of the features extracted from the convolutional layers and they will assign a probability for the joint motion being what the algorithm predicts it is.

We collected data by sending a random configuration request $q = (q_1, q_2, \ldots, q_n)$ to a robot with n joints and observing the resulting end-effector position $x = (x_1, x_2, \ldots, x_n)$ in 3D space. n is the number of joints. This results in a kinematic database with a certain number of trails:

$$X = (<q_1, x_1>, <q_2, x_2>, \ldots, <q_n, x_n>) \tag{19}$$

We use the neural network shown in Figure 8 to estimate the inverse kinematics from the collected data. Then we generate a valid configuration q_n for a given target position x_n. q_n and x_n are the n_{th} set of training data. We use this approach for planning to reach a cutting point of the peduncle.

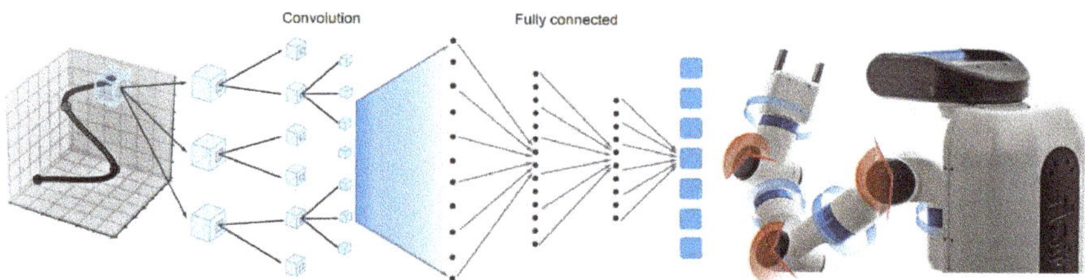

Figure 8. Network architecture for the kinematic control of a robotic manipulator.

5. Hardware Implementation

The following experiments aim at evaluating the general performance of the harvesting system. As shown in Figure 9, the proposed gripper was mounted on the end-effector of the Fetch robot (see Figure 9c). The Fetch robot consists of seven rotational joints, a gripper, a mobile base, and a depth camera. The depth camera was used to detect and localize the crops. The robot is based on the open source robot operating system, ROS [41]. The transformation between the image coordinate frame and the robot coordinate frame are obtained with ROS packages. The three plates and the shaft of the proposed gripper are created using Polylactic Acid (PLA) with a MakerBot Replicator 2 printing machine. The length, width, and height of modules are taken as 11, 5, and 9 cm, respectively.

Since the Fetch robot is not suitable for working in the field, the experiments were conducted in a laboratory environment. In order to verify the effectiveness of the proposed system in identifying and harvesting a variety of crops, we use both real and plastic ones, as shown in Figure 9a,b. Two experiments are presented in the following section aimed at validating the perception system and the harvesting system. First, we present the accuracy of the detection of crops and the cutting points. Second, we present the experiment of the full harvesting platform, demonstrating the harvesting performance of the final integrated system.

(a)   (b)   (c)

Figure 9. Setup for the experiments. (**a**) The experiment setup; (**b**) The plastic crops with plastic peduncles hung on a rope; (**c**) The proposed gripper mounted on the end-effector of the Fetch robot.

5.1. Detection Results

Seven types of crops were tested, including lemons, grapes, common figs, bitter melon, eggplants, apples, and oranges. For each type of crop, we used 60 images for training and noises were introduced in the images. The number of images was inspired by Sa et al. [34]. We collected images from Google and set labelled bounding box information for each crop. The images taken with the Kinect sensor were used for testing only. In this paper, we used 15 testing images for each type of crop. These images are taken from different viewpoints. We have one plant for each type of the real fruits.

An example is shown in Figure 10 where the real crops were on the plants (see Figure 9a). For a convenient harvesting, all the plastic crops were hung on a rope, as shown in Figure 9b. We had three attempts and the crops were repositioned the crops every time. Therefore, we used 27 plastic crops (six apples, three lemons, six oranges, three bitter melons, three clusters of grapes, and six eggplants) and 31 real crops (seven grape bunches, 14 lemons, and 10 common figs).

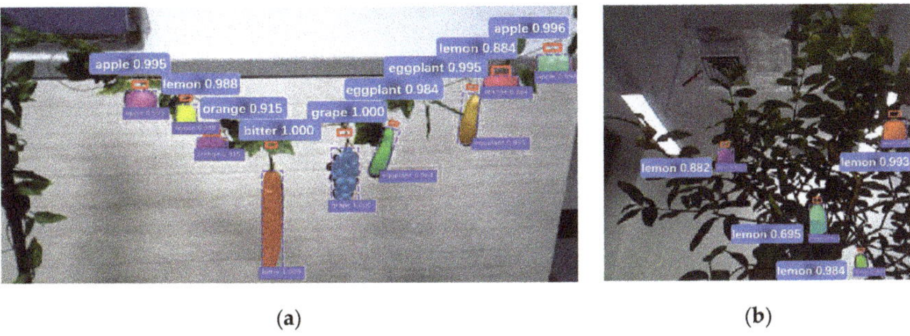

(a)   (b)

Figure 10. Results of the cutting-point-detection for some plastic (**a**) and real crops (**b**).

We utilize the precision-curve [42] as the evaluation metric for the detection of cutting points of lemons. The performance of the proposed detection method and the baseline method Conditional Random Field (CRF) [43] was compared. Figure 11 shows that the two methods have similar performance. It can be seen that there is a gradual linear decrease in precision with a large increase in recall for the proposed approach.

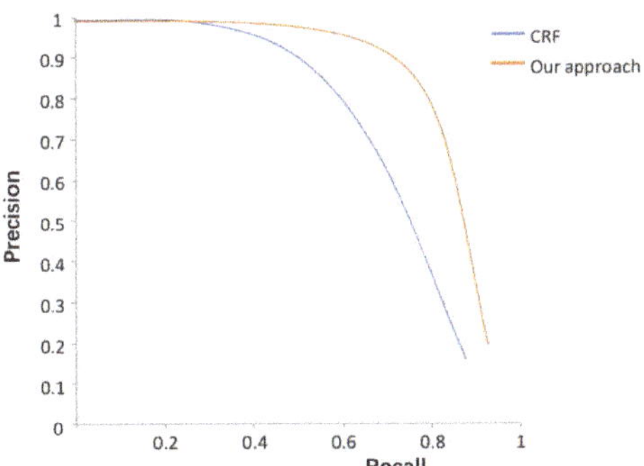

Figure 11. Precision-recall curve for detecting lemons using the CRF-based approach and the proposed approach.

Some of the detection results for real and plastic crops were shown in Figure 10. The results show that the detection success rates for plastic and real crops are 93% (25/27) and 87% (27/31), respectively. The detection success rates for the cutting points for plastic and real crops are 89% (24/27) and 71% (22/31), respectively. The result can be seen in Figure 12.

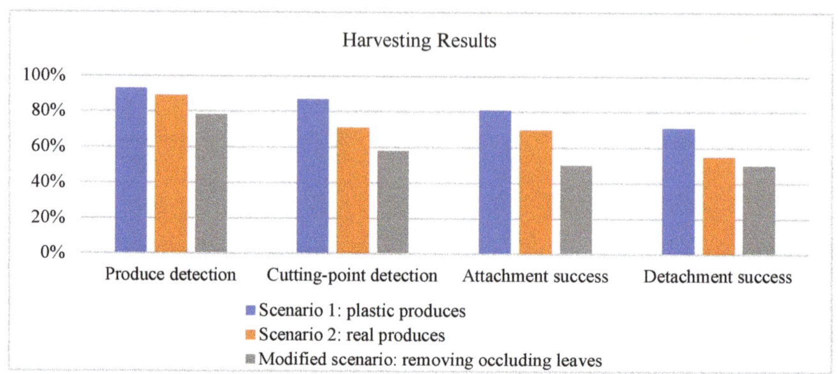

Figure 12. Harvesting rates for autonomous harvesting experiment.

5.2. Autonomous Harvesting

Motion planning is performed for autonomous attachment and detachment. The attachment trajectory starts at a fixed offset from the target fruit or vegetable determined by the bounding box. Then, the trajectory makes a linear movement towards the cutting point. The whole motion was computed using the control policy that maps target positions to joint actions introduced in Section 4.2. Self-collisions of the robot were included within the motion planning framework to ensure the robot safely planned. Once the attachment trajectory has been executed attaching the peduncle, the end-effector is moved horizontally to the robot from the cutting point.

A successful harvest consists of a successful detection of the cutting point and a successful clamping and cutting of the peduncle. The crop that was successfully picked was replaced with a new one hung on the rope. We repositioned the crop for repicking if it was not harvested successfully. For real crops, the robot harvests each crop once, including seven grape bunches, 14 lemons, and 10 common figs. Similar to plastic fruits, reaching the crop's cutting point is counted as a successful attachment, and clamping and cutting of the peduncle is counted as a successful detachment. Results show that most plastic fruits and vegetables were successfully harvested. The result is shown in Figure 12. The detailed success and failure cases can be seen in Table 1. The attachment success, i.e., the success rate of the robot's end-effector reaching the cutting point is 81% and 61% for plastic and real crops, respectively. The detachment success rates are 67% and 52%, respectively.

Table 1. Harvesting results for plastic and real crops.

Success Cases	Plastic Crops							Real Crops			
	Apple (6)	Lemon (3)	Orange (6)	Bitter melon (3)	Grapes (3)	Eggplants (6)	Total	Grapes (7)	Melons (14)	Common figs (10)	Total
Attachment	5	2	5	3	2	5	81%	5	11	6	71%
Detachment	5	2	4	1	2	5	70%	4	8	5	55%

Figure 13 is the video frames showing the robot reaching a cluster of plastic, cutting the peduncle, and placing them onto the desired place. Figure 14 is the video frames showing the robot harvesting real lemons, grapes and common figs in the laboratory environment. A video of the robotic harvester demonstrating the final experiment by performing autonomous harvesting of all crops is available at http://youtu.be/TLmMTR3Nbsw.

For cases where the crop's cutting point is occluded by leaves, the robot can remove the leaves to make the crop's cutting point visible. First, the overlap between crop's cutting area and the leaf is calculated to see if it exceeds a given threshold. Then, the robot detects the leaf's cutting point and cuts it using the same method as that for cutting crops. Figure 15 shows a modified scenario in which the robot removed leaves that occluded the fruits. We made four attempts and failed twice.

The figure shows that the leaf stuck onto the cutting blade and it did not detach even if the gripper is open.

Figure 13. Video frames showing the robot reaching a cluster of plastic grapes, cutting the peduncle, and placing them onto a table.

Figure 14. Video frames showing robot harvests real lemons (top), grapes (middle), and common figs (bottom).

Figure 15. Video frames for the modified scenario in which the robot picked an occluded lemon (bottom) by removing an occluding leaf (top).

5.3. Discussion

A failure analysis was conducted to understand what were the major causes of harvesting failures within the system proposed, and thus to improve the system in the future work. The mode of failures includes: (1) fruits or vegetables not detected or detected inaccurately due to occlusions; (2) cutting point not detected due to the irregular shapes of the crops; (3) peduncle not attached due to motion planning failure; (4) peduncle partially cut since the cutting tool stuck into the peduncle; (5) crop dropped since the gripper does not clamp the crop tightly; (6) detachment fails since the crops or leaves stays on the blade when the gripper is open.

Figure 11 shows the success rates for three scenarios: Scenario 1 with plastic crops, Scenario 2 with real crops, and a modified scenario in which the robot picked crops by removing occlusions. The detection of the fruits and the cutting-points, attachment, and detachment harvest success rate were included. The detachment success rate reflects the overall harvesting performance.

The results showed that Scenario 1 with plastic crops has a higher harvest success. The most frequent failure mode was (2) cutting point not detected and (5), occurring 30% harvest failure. For instance, the two attachment failures (orange 1 and eggplant 1) occurred in the first nine crops. For orange 1, the cutting blade missed the peduncle as the peduncle is short and the cutting blade missed it. Eggplant 1 was irregularly shaped, which resulted in a poor cutting-pose estimation. We then adjusted the directions of the orange and eggplant for the following two rounds, and there was no detection error. The lower detachment success rate is most likely attributed to the grapes and bitter melon since they have a larger weight and fall.

The harvest success rate (55%) was lower for Scenario 2 with real crops. The most frequent failure modes were (4), and (5) which represented a significant portion of the failure cases and showed that attachment and detachment process is challenging due to the lower stiffness of the cutting mechanism, as well as the sophisticated environment. For instance, the common figs have shorter peduncles and the gripper is thick and wider, resulting in a detachment failure. The grape peduncle is thicker and difficult to be cut off.

The results also showed a lower harvest success rate (50%) for the modified scenario. The most failure modes were (2) and (6). For instance, the leaves of the lemon tree did not fall when the gripper is open since the leaves were thin and light. The cutting points of the leaves were difficult to be detected due to the random poses. Thus, the leaves may not be removed completely once. In the future work, we will improve the detection of the occluded crops and the leaves, as well as the robot grasping pose.

6. Conclusions and Future Work

In this paper, we developed a harvesting robot with a low-cost gripper and proposed a cutting-point detection method for autonomous harvesting. The performance of the proposed harvesting robot was evaluated using a physical robot in different scenarios. We demonstrated the effectiveness of the harvesting robot on different types of crops. System performance results showed that our system can detect the cutting point, cutting the fruit at the appropriate position of the peduncle. Most failures occurred in the attachment stage due to irregular shapes. This problem was avoided by adjusting the directions of crops. In the future, we plan to devise a new approach to detect the cutting pose for irregularly shaped crops.

There are some detachment issues related to the fabrication of the gripper. For instance, the crop did not detach from the peduncle when the gripper is open, while the clamped crop falls due to gravity. Thus, we are working to improve the gripper using metal materials with a better cutting blade, so the gripper would be sufficiently stiff to successfully clamp different types of crops, including one with a thicker peduncle.

In this paper, we mainly focused on the cases where the directions of the crops are downward or tilted. The hypothesis posed in this paper is that a robot capable of reaching targets or leaves without considering obstacles. Human approaches a fruit by pushing a leaf aside or going through the stems without causing any damage to the plant since a damage can seriously hamper plant growth and affect future production of fruit. Inspired by this, we demonstrated how a robot removes leaves to reach the target crop. However, the proposed crop picking process by removing leaves is still in an ideal situation. In the future, we will improve the detection of obstacles, such as unripen crops, leaves and stems. For cases where the direction of a crop parallel to the camera or the peduncle being "occluded" by its flesh, the robot needs to rotate the camera to the next best view [44] to make the cut point of the crop visible. A method to determine the next best view with an optimal grasp [45] based on occlusion information will be considered in the future work.

For the case where stems are viewed as obstacles which are not avoided, we are currently working on developing a vision-guided motion planning method, using a deep reinforcement learning algorithm. This approach allows the robot to reach the desired fruits and avoid obstacles using current visual information. The aforementioned next-best-view-based approach will be combined with the motion planning. Furthermore, we aim to explore the estimation of the locations of crops in three dimensions and predict the occlusion order between objects [46].

As our current robot is inappropriate for use outside, we have not tested our harvesting robot in the field. Our next step is to test more types of crops in the field using a UR5 robot and, in addition, consider the autonomous reconstruction of unknown scenes [47].

Supplementary Materials: The following are available online at http://www.mdpi.com/1424-8220/20/1/93/s1. Video: Autonomous fruit and vegetable harvester with a low-cost gripper using a 3D sensor.

Author Contributions: Conceptualization, T.Z.; methodology, T.Z., Z.H. and W.Y.; software, Z.H. and W.Y.; validation, T.Z., Z.H. and W.Y.; formal analysis, T.Z.; investigation, T.Z.; resources, J.L.; data curation, X.T.; writing-original draft preparation, T.Z.; writing-review and editing, H.H.; visualization, T.Z.; supervision, T.Z., H.H.; project administration, H.H.; funding acquisition, H.H. All authors have read and agreed to the published version of the manuscript.

Funding: This work was funded in parts by NSFC (61761146002, 61861130365), Guangdong Higher Education Innovation Key Program (2018KZDXM058), Guangdong Science and Technology Program (2015A030312015), Shenzhen Innovation Program (KQJSCX20170727101233642), LHTD (20170003), and the National Engineering Laboratory for Big Data System Computing Technology.

Conflicts of Interest: The authors declare no conflict of interest.

References

1. Kapach, K.; Barnea, E.; Mairon, R.; Edan, Y.; Ben-Shahar, O. Computer vision for fruit harvesting robots-state of the art and challenges ahead. *Int. J. Comput. Vis. Robot.* **2012**, *3*, 4–34. [CrossRef]
2. Bac, C.W.; Hemming, J.; Tuijl, B.A.J.V.; Barth, R.; Wais, E.; Henten, E.J.V. Performance evaluation of a harvesting robot for sweet pepper. *J. Field Robot.* **2017**, *34*, 6. [CrossRef]
3. Bac, C.W.; van Henten, E.J.; Hemming, J.; Edan, Y. Harvesting robots for high-value crops: State-of-the-art review and challenges ahead. *J. Field Robot.* **2014**, *31*, 888–911. [CrossRef]
4. Kamilaris, A.; Prenafeta-Boldu, F.X. Deep learning in agriculture: A survey. *Comput. Electron. Agric.* **2018**, *147*, 70–90. [CrossRef]
5. Arad, B.; Kurtser, P.; Barnea, E.; Harel, B.; Edan, Y.; Ben-Shahar, O. Controlled lighting and illumination-independent target detection for real-time cost-efficient applications the case study of sweet pepper robotic harvesting. *Sensors* **2019**, *19*, 1390. [CrossRef] [PubMed]
6. Sa, I.; Lehnert, C.; English, A.; McCool, C.; Dayoub, F.; Upcroft, B.; Perez, T. Peduncle detection of sweet pepper for autonomous crop harvesting—Combined color and 3-d information. *IEEE Robot. Autom. Lett.* **2017**, *2*, 765–772. [CrossRef]
7. Luo, L.; Tang, Y.; Lu, Q.; Chen, X.; Zhang, P.; Zou, X. A vision methodology for harvesting robot to detect cutting points on peduncles of double overlapping grape clusters in a vineyard. *Comput. Ind.* **2018**, *99*, 130–139. [CrossRef]
8. Eizicovits, D.; van Tuijl, B.; Berman, S.; Edan, Y. Integration of perception capabilities in gripper design using graspability maps. *Biosyst. Eng.* **2016**, *146*, 98–113. [CrossRef]
9. Eizicovits, D.; Berman, S. Efficient sensory-grounded grasp pose quality mapping for gripper design and online grasp planning. *Robot. Auton. Syst.* **2014**, *62*, 1208–1219. [CrossRef]
10. He, K.; Gkioxari, G.; Dolla'r, P.; Girshick, R. Mask r-cnn. *IEEE Trans. Pattern Anal. Mach. Intell.* **2017**, 2961–2969.

11. van Henten, E.J.; Slot, D.A.V.; Cwj, H.; van Willigenburg, L.G. Optimal manipulator design for a cucumber harvesting robot. *Comput. Electron. Agric.* **2009**, *65*, 247–257. [CrossRef]

12. Chiu, Y.C.; Yang, P.Y.; Chen, S. Development of the end-effector of a picking robot for greenhouse-grown tomatoes. *Appl. Eng. Agric.* **2013**, *29*, 1001–1009.

13. Lehnert, C.; English, A.; McCool, C.S.; Tow, A.M.W.; Perez, T. Autonomous sweet pepper harvesting for protected cropping systems. *IEEE Robot. Autom. Lett.* **2017**, *2*, 872–879. [CrossRef]

14. Yamamoto, S.; Hayashi, S.; Yoshida, H.; Kobayashi, K. Development of a stationary robotic strawberry harvester with picking mechanism that approaches target fruit from below (part 2)—Construction of the machine's optical system. *J. Jpn. Soc. Agric. Mach.* **2010**, *72*, 133–142.

15. Hayashi, S.; Shigematsu, K.; Yamamoto, S.; Kobayashi, K.; Kohno, Y.; Kamata, J.; Kurita, M. "Evaluation of a strawberry-harvesting robot in a field test. *Biosyst. Eng.* **2010**, *105*, 160–171. [CrossRef]

16. Hemming, J.; Bac, C.; van Tuijl, B.; Barth, R.; Bontsema, J.; Pekkeriet, E.; van Henten, E. A robot for harvesting sweet-pepper in greenhouses. In Proceedings of the International Conference of Agricultural Engineering, Zurich, Switzerland, 6–10 July 2014.

17. Hayashi, S.; Takahashi, K.; Yamamoto, S.; Saito, S.; Komeda, T. Gentle handling of strawberries using a suction device. *Biosyst. Eng.* **2011**, *109*, 348–356. [CrossRef]

18. Dimeas, F.; Sako, D.V. Design and fuzzy control of a robotic gripper for efficient strawberry harvesting. *Robotica* **2015**, *33*, 1085–1098. [CrossRef]

19. Bontsema, J.; Hemming, J.; Pekkeriet, E. Crops: High tech agricultural robots. In Proceedings of the International Conference of Agricultural Engineering, Zurich, Switzerland, 6–10 July 2014.

20. Baeten, J.; Donn'e, K.; Boedrij, S.; Beckers, W.; Claesen, E. Autonomous fruit picking machine: A robotic apple harvester. *Springer Tracts Adv. Robot.* **2007**, *42*, 531–539.

21. Haibin, Y.; Cheng, K.; Junfeng, L.; Guilin, Y. Modeling of grasping force for a soft robotic gripper with variable stiffness. *Mech. Mach. Theory* **2018**, *128*, 254–274. [CrossRef]

22. Mantriota, G. Theoretical model of the grasp with vacuum gripper. *Mech. Mach. Theory* **2007**, *42*, 2–17. [CrossRef]

23. Han, K.-S.; Kim, S.-C.; Lee, Y.B.; Kim, S.C.; Im, D.H.; Choi, H.K.; Hwang, H. Strawberry harvesting robot for bench-type cultivation. *J. Biosyst. Eng.* **2012**, *37*, 65–74. [CrossRef]

24. Tian, G.; Zhou, J.; Gu, B. Slipping detection and control in gripping fruits and vegetables for agricultural robot. *Int. J. Agric. Biol. Eng.* **2018**, *11*, 45–51. [CrossRef]

25. Zhang, T.; Zhang, W.; Gupta, M.M. An underactuated self-reconfigurable robot and the reconfiguration evolution. *Mech. Mach. Theory* **2018**, *124*, 248–258. [CrossRef]

26. Zhang, T.; Zhang, W.; Gupta, M. A novel docking system for modular self-reconfigurable robots. *Robotics* **2017**, *6*, 25. [CrossRef]

27. Bac, C.; Hemming, J.; van Henten, E. Robust pixel-based classification of obstacles for robotic harvesting of sweet-pepper. *Comput. Electron. Agric.* **2013**, *96*, 148–162. [CrossRef]

28. McCool, C.; Sa, I.; Dayoub, F.; Lehnert, C.; Perez, T.; Upcroft, B. Visual detection of occluded crop: For automated harvesting. In Proceedings of the 2016 IEEE International Conference on Robotics and Automation (ICRA), Stockholm, Sweden, 16–21 May 2017; pp. 2506–2512.

29. Blasco, J.; Aleixos, N.; Molto, E. Machine vision system for automatic quality grading of fruit. *Biosyst. Eng.* **2003**, *85*, 415–423. [CrossRef]

30. Ruiz, L.A.; Molt'o, E.; Juste, F.; Pla, F.; Valiente, R. Location and characterization of the stem–calyx area on oranges by computer vision. *J. Agric. Eng. Res.* **1996**, *64*, 165–172. [CrossRef]

31. Zemmour, E.; Kurtser, P.; Edan, Y. Dynamic thresholding algorithm for robotic apple detection. In Proceedings of the 2017 IEEE International Conference on Autonomous Robot Systems and Competitions (ICARSC), Coimbra, Portugal, 26–28 April 2017; pp. 240–246.

32. Zemmour, E.; Kurtser, P.; Edan, Y. Automatic parameter tuning for adaptive thresholding in fruit detection. *Sensors* **2019**, *19*, 2130. [CrossRef]

33. Ren, S.; He, K.; Girshick, R.; Sun, J. Faster r-cnn: Towards real-time object detection with region proposal networks. In *Advances in Neural Information Processing Systems*; MIT Press: Cambridge, UK, 2015; pp. 91–99.

34. Sa, I.; Ge, Z.; Dayoub, F.; Upcroft, B.; Perez, T.; McCool, C. Deep-fruits: A fruit detection system using deep neural networks. *Sensors* **2016**, *16*, 1222. [CrossRef]

35. Zhang, T.; You, W.; Huang, Z.; Lin, J.; Huang, H. A Universal Harvesting Gripper. Utility Patent. Chinese Patent CN209435819, 27 September 2019. Available online: http://www.soopat.com/Patent/201821836172 (accessed on 27 September 2019).

36. Zhang, T.; You, W.; Huang, Z.; Lin, J.; Huang, H. A Harvesting Gripper for Fruits and Vegetables. Design Patent, Chinese Patent CN305102083, 9 April 2019. Available online: http://www.soopat.com/Patent/201830631197 (accessed on 9 April 2019).

37. Donahue, J.; Jia, Y.; Vinyals, O.; Hoffman, J.; Zhang, N.; Tzeng, E.; Darrell, T. Decaf: A deep convolutional activation feature for generic visual recognition. In Proceedings of the International Conference on Machine Learning, Berkeley, CA, USA, 21-26 June 2014; pp. 647–655.

38. Baur, J.; Schu¨tz, C.; Pfaff, J.; Buschmann, T.; Ulbrich, H. Path planning for a fruit picking manipulator. In Proceedings of the International Conference of Agricultural Engineering, Garching, Germany, 6–10 June 2014.

39. Katyal, K.D.; Staley, E.W.; Johannes, M.S.; Wang, I.-J.; Reiter, A.; Burlina, P. In-hand robotic manipulation via deep reinforcement learning. In Proceedings of the 30th Conference on Neural Information Processing Systems, Barcelona, Spain, 5–8 December 2016.

40. Krizhevsky, A.; Sutskever, I.; Hinton, G.E. Imagenet classification with deep convolutional neural networks. In *Advances in Neural Information Processing Systems*; MIT Press: Cambridge, UK; Lake Tahoe, NV, USA, 2012; pp. 1097–1105.

41. Quigley, M.; Conley, K.; Gerkey, B.; Faust, J.; Foote, T.; Leibs, J.; Wheeler, R.; Ng, A.Y. Ros: An open-source robot operating system. In Proceedings of the ICRA Workshop on Open Source Software, Kobe, Japan, 12–17 May 2009; pp. 1–6.

42. Davis, J.; Goadrich, M. The relationship between precision-recall and roc curves. In Proceedings of the 23rd International Conference on Machine Learning, Pittsburgh, PA, USA, 25–29 June 2006; pp. 233–240.

43. Tseng, H.; Chang, P.; Andrew, G.; Jurafsky, D.; Manning, C. A conditional random field word segmenter for sighan bakeoff 2005. In Proceedings of the fourth SIGHAN Workshop on Chinese language Processing, Jeju Island, Korea, 14–15 October 2005; pp. 168–171.

44. Wu, C.; Zeng, R.; Pan, J.; Wang, C.C.; Liu, Y.-J. Plant phenotyping by deep-learning-based planner for multi-robots. *IEEE Robot. Autom. Lett.* **2019**, *4*, 3113–3120. [CrossRef]

45. Barth, R.; Hemming, J.; van Henten, E.J. Angle estimation between plant parts for grasp optimisation in harvest robots. *Biosyst. Eng.* **2019**, *183*, 26–46. [CrossRef]

46. Lin, D.; Chen, G.; Cohen-Or, D.; Heng, P.-A.; Huang, H. Cascaded feature network for semantic segmentation of rgb-d images. In Proceedings of the IEEE International Conference on Computer Vision, Honolulu, HI, USA, 21–26 July 2017; pp. 1311–1319.

47. Xu, K.; Zheng, L.; Yan, Z.; Yan, G.; Zhang, E.; Niessner, M.; Deussen, O.; Cohen-Or, D.; Huang, H. Autonomous reconstruction of unknown indoor scenes guided by time-varying tensor fields. *ACM Trans. Graph.* **2017**, *36*, 1–15. [CrossRef]

Article

Development of Willow Tree Yield-Mapping Technology

Maxime Leclerc [1], Viacheslav Adamchuk [1,*], Jaesung Park [1] and Xavier Lachapelle-T. [2]

[1]   Department of Bioresource Engineering, Faculty of Agricultural and Environmental Sciences, McGill University, Ste-Anne-de-Bellevue, QC H9X 3V9, Canada; maxime.leclerc3@mail.mcgill.ca (M.L.); jaesung.park@mail.mcgill.ca (J.P.)

[2]   Ramea Phytotechnologies, Saint-Roch-de-l'Achigan, QC J0K 3H0, Canada; xlachapelle@ramea.co

*   Correspondence: viacheslav.adamchuk@mcgill.ca

Received: 26 February 2020; Accepted: 3 May 2020; Published: 6 May 2020

Abstract: With today's environmental challenges, developing sustainable energy sources is crucial. From this perspective, woody biomass has been, and continues to be, a significant research interest. The goal of this research was to develop new technology for mapping willow tree yield grown in a short-rotation forestry (SRF) system. The system gathered the physical characteristics of willow trees on-the-go, while the trees were being harvested. Features assessed include the number of trees harvested and their diameter. To complete this task, a machine-vision system featuring an RGB-D stereovision camera was built. The system tagged these data with the corresponding geographical coordinates using a Global Navigation Satellite System (GNSS) receiver. The proposed yield-mapping system showed promising detection results considering the complex background and variable light conditions encountered in the outdoors. Of the 40 randomly selected and manually observed trees in a row, 36 were successfully detected, yielding a 90% detection rate. The correctly detected tree rate of all trees within the scenes was actually 71.8% since the system tended to be sensitive to branches, thus, falsely detecting them as trees. Manual validation of the diameter estimation function showed a poor coefficient of determination and a root mean square error (RMSE) of 10.7 mm.

Keywords: precision agriculture; yield estimation; machine vision; willow tree

---

1. Introduction

Biomass has garnered much interest as an alternative sustainable energy source through combustion and transformation. Biomass comprises natural materials from living, or recently dead, plants, trees, or animals which are recycled as fuel in industrial production. More specifically, forest biomass consists of all parts of the tree, such as the trunk, bark, branches, needles, leaves, and roots [1]. At the basic level, photosynthesis allows trees to convert light energy, carbon dioxide, and water into biomass and oxygen. Since trees fix carbon while growing, the production of woody biomass demonstrates a neutral carbon balance when cultivated in low-quality areas such as marginal and abandoned fields. To maximize carbon fixation, research has focused on tree species that yield high amounts of rapid biomass production. Suitable species for biomass production include Willow (*Salix* spp.), Poplar (*Populus* spp.) and Eucalyptus (*Eucalyptus* spp.). Forest biomass can be processed as biofuel in a solid, liquid, and gaseous phase, which can then be burnt to generate useful energy, usually in the form of electrical energy [2].

As biomass production can positively impact sustainability in the energy industry, Precision Agriculture (PA) has had a positive impact on sustainability in the agricultural industry. In the case of crop production, this technology aims at implementing site-specific management (SSM) solutions for agricultural inputs [3]. By doing so, farmers can optimize their operations, reducing the effect of these inputs on the environment. PA technologies rely heavily on the Global Navigation Satellite System

(GNSS), which can be used to locate agriculture vehicles in real-time in a field environment. One of the earliest and most common applications of GNSS-based systems is yield monitoring and mapping. Yield monitoring systems are designed to estimate crop yield and/or yield quality in real-time during harvesting [4]. The system is then able to link yield estimates to geographical coordinates to produce a yield map. The map is updated in real-time on a display that can be viewed by the operator. After harvesting, it can be retrieved for further analysis to ascertain its effectiveness in terms of performance, impact of management practices, locating low- and high-yielding zones, and so on.

Although yield maps are essential tools for today's agriculture, not all crops can be monitored for yield. Currently on the market, yield-monitoring systems are available for grain (e.g., corn, soybean, wheat), cotton, and sugarcane. Consequently, farmers growing other field crops (e.g., specialty crops, vegetables, fruits) must still rely on prior techniques to determine crop yield, where spatial resolution is far inferior to yield monitoring systems.

Methods have been proposed to estimate the yield of specialty crops based on optical feature detection [5–8]. The acquisition of image data using vision sensors, such as conventional cameras, can be analyzed in real-time or post-harvest by computer-vision (CV) algorithms to generate yield maps. The main advantage of optical feature detection is its flexibility since multiple types of feature can be extracted from an image such as color, texture, and shape. Yield-monitoring systems can be built with the capability of extracting distinctive features of a crop to use the output to produce a yield map.

A machine vision (MV)-based yield-monitoring system was developed for vegetable crops [6]. Vegetable segmentation was performed through the watershed algorithm and color data was used for classification. The system was designed to detect crop flow at the individual level, classify them according to shape and size, and count them. Testing was done with shallot onions, but the algorithm was constructed in a way to make it transferable to charlotte onions, Chinese radish, carrots, and lettuce. Results showed a coefficient of determination (R^2) of 0.46 for the overall accuracy of the system. A single RGB camera was used and occlusion, as well as variable light conditions, were the limiting factors for this research.

An image segmentation framework for fruit detection and yield estimation in apple orchards had been investigated [7]. A spherical video camera capable of capturing a full 360° panoramic view was mounted on a remotely controlled testing platform. The framework included contextual information concerning relevant metadata to evaluate state-of-the-art convolutional neural network (CNN) and multiscale multilayered perceptions to perform pixel-wise fruit segmentation. Once the binary image was obtained, watershed segmentation followed by a circle Hough transform algorithm to detect and count the apples. Results showed that the combination of CNN pixel-wise classification and watershed segmentation produced a higher coefficient of determination (R^2) at 0.83 compared to post-harvest fruit counts obtained from grading and counting machines.

An automated apple yield monitor using a stereo rig was demonstrated [8]. The system was mounted on an autonomous orchard vehicle fitted with a controlled artificial light source for nighttime data acquisition. Images from both sides of each tree row were captured. The algorithm started by hue and saturation segmentation, followed by the detection of local maxima in the grayscale intensity profile (i.e., specular reflection) as the feature indicating apples. Once detected, the intensity profile over four lines (i.e., going through the local maxima) of 21 pixels each was checked for roundness to remove false positives. Apples were then registered and counted as crop yield estimation. Errors of −3.2% and 1.2% in yield were reported for red and green apples, respectively.

A method for robust detection of trees using color images in a forest environment was published [9]. The proposed approach was characterised by a seed point generation algorithm followed by a multiple segmentation algorithm. Output segments were refined by morphological operations before being evaluated by a quality function based on the segment's specific property (i.e., area of convex hull, area, orientation, eccentricity, solidity, and height). For each seed point, the segment with the best quality score was selected and all selected segments were fused in one frame which represented the final output of the tree detection algorithm. For that research, 30 images containing 197 trees were collected in a forest environment. The tree detection rate was reported as 86.8% with a corresponding precision of 81.4%.

The goal of this research was to develop a new yield mapping system for willow trees based on MV technology. The system was designed and built in partnership with Ramea Phytotechnologies, Inc. (Ramea), based in Saint-Roch-de-l'Achigan, Quebec. The system was designed to gather the physical characteristics of willow trees and tag them to their geographical coordinates on-the-go, while the trees are being harvested. Features assessed include the number of trees harvested and their diameter. Furthermore, the system has potential as a fast and non-destructive yield estimation tool which could be used to assess carbon dioxide (CO_2) sequestration of willow trees grown in a short-rotation forestry (SRF) system over time.

Since methods for growing willows in SRF vary greatly depending on countries and regions [2], the developed willow tree yield mapping technology must be malleable enough to make this research relevant to industry partners and to producers around the world. Optical sensors have the capacity of sensing more than one feature such as color, intensity, geometry and texture [10]. Considering that the number of stems and morphological characteristics of the willow trees were assessed, and not bulk mass, the use of an optical, multi-sensing, platform was desirable. Moreover, for systems to be adopted in the agricultural industry, they must be affordable, rugged, and easy to install and manage. Thus, willow tree yield mapping based on MV was the preferred and selected method to accomplish the goals of this project.

2. System Components

In terms of fruit detection and localization, several studies have been carried out to assess the performance of different vision sensor technologies. In doing so, researchers reported on the main issues involved in using camera sensors in an outdoor environment. Variations in light conditions, as well as variations of color, shape, and size of the target, are common problems that researchers face [10]. Moreover, complex plant canopy structures can create occlusion of targets, which makes detection even more challenging [11]. For this research, cameras using monochrome imagers were preferred to diminish the effect of variations in light conditions that are prominent in outdoor scenes.

To improve target segmentation in the complex background environment, a camera featuring 3D vision was also favored to better extract the willow's feature from the scenes. By using two imagers, it is possible to extract depth information from a pair of images. This technology is called stereovision. It aims to mimic the mammalian binocular vision system by collecting a stereo pair of images of a scene from two cameras with a known geometrical relationship. Using a matching algorithm, the system matches key-points of one stereo-pair image to the other. It then estimates the disparity between matching key-points and uses this to compute depth according to the system's intrinsic and extrinsic parameters [12].

2.1. Sensors

To acquire raw data, the RealSense D435 RGB-D camera from Intel (Intel Corporation, Santa Clara, California) and a Global Positioning System (GPS) receiver with Wide Area Augmentation System (WAAS) differential correction capability model 19x-HVS from Garmin (Garmin Ltd., Olathe, Kansas) were used. The RealSense D435 (Table 1) features infrared active stereovision technology to create depth maps using an embedded infrared projector. Its large Field Of View (FOV) makes it ideal for an outdoor setting with large targets at close range. The camera is fitted with one red green blue (RGB) imager and one stereo depth module containing two monochrome imagers. Images from the stereo depth module are processed by an embedded processor (Vision Processor D4) to generate depth maps. The RealSense can output four data streams at once in real-time: RGB stream, two monochrome imager streams, and the depth map stream. The on-board Vision Processor D4 makes this camera suitable for an embedded solution since it does not require the host computer to perform the heavy computation required by the stereo-matching algorithm.

Table 1. RealSense D435 camera specifications.

Depth Technology	Active IR Stereo
Depth Stream FOV (Horizontal × Vertical × Diagonal)	87° ± 3° × 58° ± 1° × 95° ± 3°
Depth Stream Resolution	Up to 1280 × 720
Depth Stream Frame Rate	Up to 90 fps
Minimum Depth Distance (Min-z)	0.1 m
Maximum Depth Range	Approximately * 10 m
RGB Sensor FOV (Horizontal × Vertical × Diagonal)	69.4° × 42.5° × 77° ± 3°
Connector	USB-C 3.1 Gen 1

* Varies depending on calibration, scene, and lighting condition.

For this research, a resolution of 1280 × 720 was used to obtain the best pixel-to-millimeter conversion factor possible. The infrared projector was turned off, since it is designed to improve performance in low texture scenes such as a dimly lit white wall [13,14].

2.2. Sensors Position and Mount

Early in the research, the front end of the tractor was considered to be the best location for the camera. The region of interest (ROI) for the willow trees was set between 3.05 m and 3.66 m measured from the soil surface. To achieve this with the RealSense FOV, the camera was mounted at 1.78 m from the ground surface and given a 28.5° tilt angle measured from the ground plane. The horizontal axis of the camera FOV was parallel to the row and the perpendicular distance between the camera and the row was 1.22 m.

The camera and GNSS mount (Figure 1) were designed as a multi-purpose assembly. In addition to providing mounting points for both sensors, it served as a guard against branches and sunlight. Branches can be expected to be leaning outside of the row if they have been hit while harvesting the adjacent row. Contact between a branch and the stereovision camera could result in loss of frame quality and physical damage to hardware (i.e., camera and data transfer cable). As the camera is tilted, direct sunlight can overexpose the optical sensors resulting in a significant decline of the frame's information. The mount was designed to protect the camera when the sun is shining directly above. The sensor mount is also designed to allow easy fine-tuning of the camera position and angle within the field. Extra mounting points were also integrated to secure the assembly to the tractor in the event that future vibration became excessive.

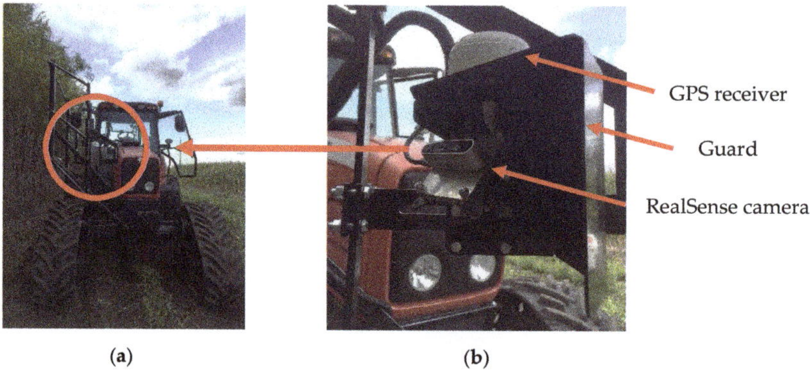

Figure 1. (a) Overall view of final camera position; (b) close view of final camera and Global Positioning System (GPS) mount.

3. Willow Tree Algorithm

The multiple steps that had to be taken in order to count the number of willow trees and to estimate their diameter for each frame are described below and summarized in Figure 2. From this point, when the term "monochrome imager" is used, it implies the leftmost imager of the RealSense D435 sensor (looking at the front view). Since the depth map is computed from this frame's perspective of a scene, they are naturally aligned from one another and additional computation is avoided. The frame produced by the monochrome imager will be referred to as the "grayscale image".

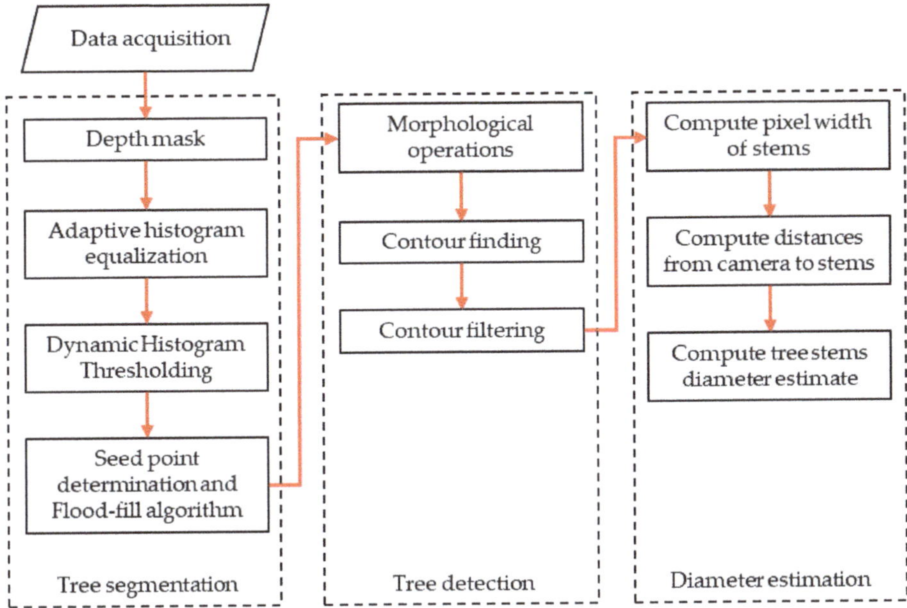

Figure 2. Willow tree algorithm flowchart.

Python was chosen as the programming language for the algorithm and was heavily dependent on NumPy and OpenCV libraries. As the RealSense camera is marketed to developers, a low-level software development kit (SDK) was freely provided by Intel through GitHub (GitHub, Inc., San Francisco, CA, USA) for the RealSense.

The Willow Tree algorithm runs using multiple parameters inputted in open-sourced and user-defined functions. It is relevant to mention that no formal procedures took place to optimize those values. For steps where parameters were involved, system performance and processing time were important criteria with special attention for overfitting since the system was not validated on an extensive data set.

As this system is the first of its kind to the knowledge of the researchers (i.e., MV-based yield monitoring system for willow trees grown in SRF), it was of interest to see if classical CV operations and algorithms such as masks, histograms, adaptive histogram equalization and contours properties could achieve tree detection and diameter estimation with good results. More complex methods such as machine learning and deep learning were not examined for the development of this algorithm since potential problems with classical methods were not known for this application [15]. Furthermore, for the system to be adopted, the algorithm must be able to run on an embedded system, with finite computing power, and affordable for farmers while being designed for a harsh environment.

3.1. Segmentation

To perform the segmentation, several steps were carried out using the depth map and the grayscale image to yield a segmented binary frame. The first was the computation of a depth mask, followed by the application of adaptive histogram equalization on the grayscale image. Then, both outputs were used as inputs in a dynamic histogram thresholding function where seed points were determined and input in the connected components algorithm flood-fill. The flood-fill algorithm output was the product of the segmentation procedure.

3.1.1. Depth Mask

The computation of a binary mask using the depth map was a necessary prior step to the dynamic histogram thresholding. The goal was to create a rough mask of the row being harvested to be input in the histogram calculation. This ensured that most of the background pixels were not considered in the histogram. A thresholding plane was first initialized to set the depth value for every pixel to zero if it was greater than the threshold plane at a specific location. Since the camera was tilted at 28.5° from the soil surface, the threshold plane was tilted by the same amount to account for a greater thresholding distance for the top pixel row compared to the bottom pixel row of the depth map. The top was computed as 2.33 m and the bottom was set as a constant at 1.23 m, which corresponds to the center of the row relative to the camera position. Using the vertical FOV of the monochrome imager, the height of a pixel in meters was computed based on the perpendicular distance from the camera to the threshold plane. Next, simple trigonometry was applied to find the length in meters of the hypotenuse formed between the camera center (fix), the bottom pixel row (fix) and the currently accessed row. This method was iteratively repeated until the top row was reached. This approach was found to be more computationally efficient than rotating the depth map itself.

Before inputting the depth map in the thresholding function, a median filter was applied to remove some of the noise with a 5×5 pixel square kernel. This type of filter works by replacing the kernel's center value by the median of all pixels under itself [16]. It was effective at removing the within region noise while keeping the tree stems sharper than a standard Gaussian filter which was important to promote accurate diameter estimation. The equation below shows the final thresholding operation.

$$Depth\ mask_{i,j} = \begin{cases} 1 & \text{if Depth median}_{i,j} < \text{Threshold plane}_{i,j} \\ 0 & \text{otherwise} \end{cases} \tag{1}$$

3.1.2. Adaptive Histogram Equalization

Adaptive histogram equalization was used to increase the contrast in the grayscale image from the monochrome imager. The goal was to increase the intensity of the tree stem pixels relative to the background and non-tree foreground objects. The OpenCV implementation of the contrast-limited adaptive histogram equalization (CLAHE) algorithm was used. This algorithm was first introduced by Pizer, et al. [17] for post-processing of still medical images. It starts by dividing the input image into regions. There are three classes of regions: the corner regions (one in each corner), the border regions (adjacent to the image border), and the inner regions. Then, the CLAHE algorithm calculates the histogram of each individual region independently. From the function input parameters, a clip limit is calculated and applied to the histograms. Those histograms are then redistributed so as not to exceed the clip limit. Lastly, the cumulative distribution functions of the resultant contrast-limited histograms are determined and a linear combination of the four nearest regions are taken for grayscale mapping. The major drawback of this first implementation was its expensive computation. However, Reza [18] proposed another approach that made the CLAHE algorithm possible for real-time applications. CLAHE, with a clip limit of 40 and 8×8 pixel subregions, was applied as a pre-processing step before being input in the dynamic histogram thresholding function. The resulting frame is shown in Figure 3. CLAHE was used over non-adaptive histogram equalization methods as it yielded a frame with a greater contrast between tree objects and the background; thus, making trees easier to segment using the dynamic histogram thresholding methods explained in the next section.

(a)   (b)

Figure 3. (**a**) Original grayscale image from the infrared imager; (**b**) grayscale image after applying the contrast-limited adaptive histogram equalization (CLAHE) algorithm.

3.1.3. Dynamic Histogram Thresholding

A user-defined dynamic histogram thresholding function was implemented to find a list of pixel intensities that were most likely representing a willow tree in the CLAHE equalized frame. While developing the Willow Tree algorithm, it was noticed that when applying the depth mask to a histogram function of the corresponding grayscale image, the histogram region representing the tree pixel count spiked significantly compared to other regions.

The function was designed to take five arguments (i.e., image, mask, peak range, low limit, count difference) and output a Python list of the intensity values (0 to 255 scale) that best represent the tree objects. The first step was to compute the histogram of the complete input image with one bin for every pixel intensity value and then to normalize the obtained histogram object from 0 to 255. Those two operations use functions already available in the OpenCV library. Next, the location of the maximum value was found, and the peak range parameter was applied where the central element was the maximum value of the normalized histogram object (Equation (2)). Figure 4 graphically represents the location of the parameters for the histogram of Figure 3b).

$$\textit{Intensity range boundaries} = \textit{maximum pixel count} \pm \frac{\textit{peak range}}{2} \qquad (2)$$

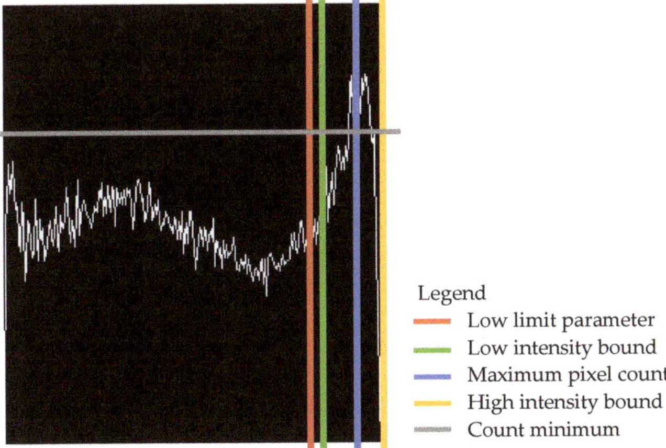

Figure 4. Histogram curve with thresholding parameters and boundaries displayed.

In this case, the function was run with a peak range value of 50. A low and high clipping value equal to the low limit parameter and the total number of bins were applied to restrict the intensity range in one area of the histograms as well as handling run time errors in Python. The following step was to check if the normalized pixel count values found in the intensity range respect the pixel count difference relative to the maximum normalized pixel count found earlier. To perform this test, the difference in Equation (3) was computed. If the difference was greater than zero, the intensity value was added to the final list of pixel intensity values which best represent willow trees in the current frame.

$$Difference_i = intensity\ range_i - (maximum\ pixel\ count - count\ difference) \tag{3}$$

$$Pixel\ intensity\ list_i = \{intensity\ range_i\ if\ difference_i > 0 \tag{4}$$

3.1.4. Seed Point Determination and Flood-Fill Algorithm

In this final step of the segmentation procedure, the intensity values found in the previous step are used to find seed point coordinates to be used in the flood-fill algorithm. The output is a binary image where the white pixels represent the segmented areas that are most likely to be willow trees using depth and intensity information from the scene. The first step was to create a frame, called background removed, from which the seed points could be found. The following equation describes this operation.

$$Background\ removed_{i,j} = \begin{cases} input\ image_{i,j} & if\ Depth\ mask_{i,j} \neq 0 \\ 0 & otherwise \end{cases} \tag{5}$$

In this case, the input image is the same as with the previous step, which is the CLAHE equalized grayscale image. Next, for every intensity value in the pixel intensity list, the background removed image was scanned and the seed point image array is given a value of 1 at the location where a match is found.

$$Seed\ points_{i,j} = \begin{cases} 1 & if\ Backgroud\ removed_{i,j} = Pixel\ intensity\ list_k \\ 0 & otherwise \end{cases} \tag{6}$$

Using the argwhere() function in the NumPy library, a list of all (x,y) coordinates of the marked pixels in the seed point array was created. Finally, the coordinates were iteratively used in the OpenCV implementation of the flood-fill algorithm. Two other image arrays were given to the function; the CLAHE equalized infrared image and an empty mask. The grayscale image was used to compute the connected components of a seed point and for which they had their value set to 1 for the corresponding pixel in the mask. For two pixels to be connected, they must agree with the following statement:

$$Mask_{i,j} = \begin{cases} 1 & if\ input_{x,y} - lowdiff \leq input_{i,j} \leq input_{x,y} + updiff \\ 0 & otherwise \end{cases} \tag{7}$$

where, $input_{i,j}$ is the pixel intensity of the currently observed pixel and $input_{x,y}$ is the pixel intensity of the input image at the seed point coordinate. The parameters lowdiff and updiff are the maximal lower and higher brightness difference, respectively, between the currently observed pixel and its corresponding seed point. In this case, it is a fixed range implementation since the condition always refers to the original seed points [16]. For this research, a value of 10 and 30 for lowdiff and updiff, respectively, were used.

More points were added to the mask as the function scan the list of seed point coordinates. The running time was limited by adding an if statement to check if a value of 1 was already assigned in the mask at the coordinate of the next seed point to be analyzed. This condition reduced the amount of time flood-fill would be run. For example, Figure 5a contains 11,300 points, but only 748 were used to produce the segmentation result in Figure 5b. After the iteration of all seed points, the mask array was reported as the final segmented binary image of the current scene

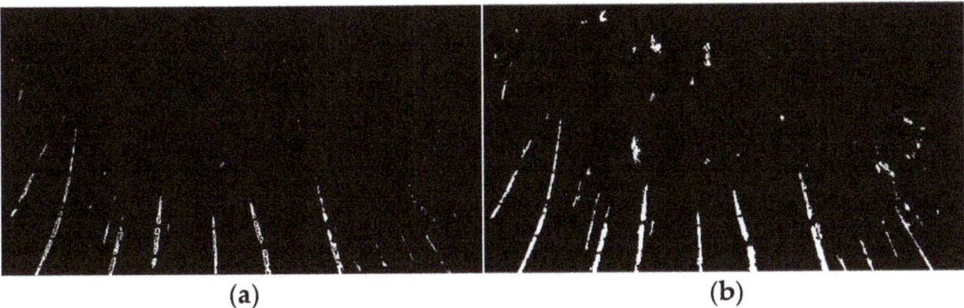

(a)   (b)

Figure 5. (**a**) Resultant seed point frame from the background removed image; (**b**) segmentation output binary frame.

3.2. Tree Detection

In the detection procedure of the willow tree algorithm, a morphological closing operation was applied to the segmented binary image before using the OpenCV contour function to find all close contours that could represent willow trees. A user-defined function was designed to filter out unwanted contours and retain only tree contours under the same assumption. Contours that were still present after the filtering operation were considered by the algorithm as true detected willow trees.

The morphological closing operation was undertaken to reduce noise that might have emerged in the flood-fill algorithm. It is commonly used after connected-components algorithms to remove elements resulting purely from noise and to connect nearby regions [16]. The closing operation is simply two morphological operations that are applied one after the other. The first is dilatation, where the pixel at the center of the kernel is replaced by the local maximum of all pixels covered by the kernel. For the binary image, the dilatation operation has the effect of "growing" each filled region. The second operation is erosion. It is simply the inverse of dilation, where the pixel at the center of the kernel is replaced by the local minimum of all pixels covered by the kernel. The effect of erosion on a binary image is to "reduce" the size of each region [16]. For this research, the kernel size used was a solid 2X15 kernel, applied with only one iteration. The kernel was set as an upward rectangle since it was the shape the target objects were expected to be. A morphological close was applied using the OpenCV implementation.

Once the noise had been removed, the binary image was input to the contour-finding function provided in the OpenCV library (i.e., *find* contours() function). It outputs a list of contour objects found in the binary frame. At the basic level, contour objects are a list of points that represent a curve within the image [16]. Contour object makes it possible to retrieve features about filled regions such as area, center, and bounding box. Those features were exploited to filter the raw contour list and find parameters for the true tree contours.

Contour filtering used two criteria to remove non-tree contours. The first was the area to remove small filled regions that were most likely unwanted foreground objects or background residuals. The area of each contour was computed using a built-in function in OpenCV. A fixed threshold area of 600 was applied for this experiment. Next, it was assumed that if a contour is a tree, the contour must be present on the bottom pixel row of the frame, so it does not "appear" somewhere in the frame. To test this assumption, each contour was individually drawn in an empty image array. If all pixel values of the bottom row were equal to zero, the observed contour did not meet the requirement and was then removed from the contour list. Contours that were left in the list were considered as true detected willow tree regions. The willow tree algorithm then moves on to assign an identification (ID) number and to find the diameter of all detected trees within the frame. Figure 6b shows the final output.

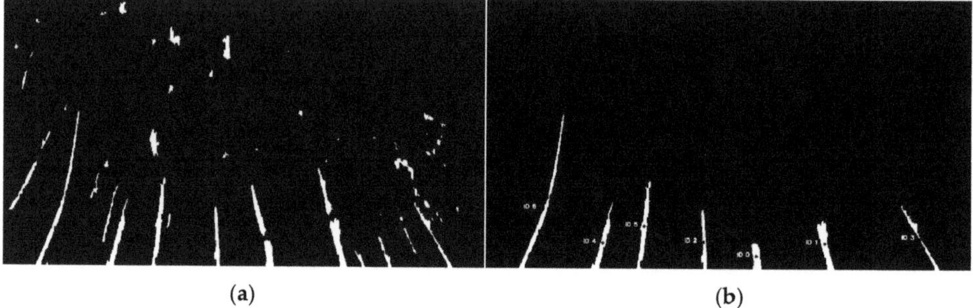

(a)   (b)

Figure 6. (**a**) Segmentation result after morphological closing was applied; (**b**) Detection procedure output frame.

3.3. Diameter Estimation

Once the willow trees were detected, the algorithm proceeded by computing the diameter of the trees. As a first step, the pixel width of all remaining contours at the bottom row of the image array was found. Moreover, all trees were at a different distance from the camera, so the depth at this location was also found. Lastly, those features were fed into a function to find the pixel to millimeter ratio and the diameter in millimeters for all detected trees.

To find the pixel width and depth of a detected tree, its corresponding contour was drawn in an empty image array. Then, the bottom pixel row of the drawn contour and the depth median frame were extracted. A new array, including both aligned rows, was created. The array was then trimmed on the column axis, where zeros were present in the contour row. The function traversed the row starting from both ends until it reached a non-zero element. From there, two possible cases were handled; in the first case, the contour row is solid (meaning no zeros within the trimmed array) and for the second case some zeros are present.

Finding the pixel width and depth for the first case was straightforward. Once the zero columns were trimmed, the length of the array was saved as the pixel width and the depth was computed as the median of all depth values at the coordinates corresponding to the non-trimmed pixels. For the second case, the array was split in two at the location of the first zero encounters while traversing the contour row. Then, both sections were trimmed and traversed again for zeros. If no further zeros were found, the function recoded the pixel width and depth for every section. The section showing a greater pixel width was assumed to belong to the true tree. The consequence of handling all contours as in the first case would be to significantly overestimate the tree's diameter.

As depth values did not always exist at the non-trimmed pixel coordinates, the zeros within the depth row were removed, so they did not affect the median result. In the particular case where all depth values were not available at the non-trimmed pixel coordinate, the median of the depth for all other contours in a specific frame was assigned. This was to be expected since depth values were not always available for all pixels due to low confidence within the RealSense stereo-matching algorithm.

To compute the estimate of the tree's diameter, a function takes the argument found in the previous step (i.e., pixel width, depth) along with the depth scale and the FOV of the monochrome imager, which are a fixed constant acquired through the camera's SDK. The depth scale was first used to convert the depth value of an observed tree from the camera scale to millimeters. Then, the half-width of the frame in millimeters for the converted distance was found. Using this value, the pixel-to-millimeter ratio was found for the observed tree. The final step was to multiply the pixel width with the obtained conversion factor. The steps described above to find any $tree_i$ diameter are shown by the following equation,

$$Distance_i\ (mm) = Depth\ value_i * Depth\ scale * 1000 \tag{8}$$

$$Half\ frame\ width_i\ (mm) = \tan\!\left(\tfrac{FOV_x}{2} * \tfrac{\pi}{180}\right) * Distance_i \tag{9}$$

$$Ratio_i \left(\frac{mm}{pixel_x} \right) = \frac{Half\ frame\ width_i}{Frame\ size_x / 2} \tag{10}$$

$$Diameter_i\ (mm) = Pixel\ width_i * Ratio_i \tag{11}$$

where *Depth scale* is the camera's extrinsic parameter to convert the depth from the camera's 3D coordinate system to meters; *Distance* is the distance between the tree and the camera in millimeters; FOV_x is the camera's FOV over the x-axis; *Half frame width* is the equivalent length of 640 pixels on the x-axis (i.e., half the normal resolution of 1280 pixels) in millimeters for a particular scene at $Distance_i$; *Frame size$_x$* is the normal resolution on the x-axis in pixels (i.e., 1280 pixels); $Ratio_i$ is the pixel to millimeter conversion factor for $tree_i$ over the x-axis; and *Diameter* is the output diameter value for $tree_i$ in millimeters.

3.4. Field Trials

All field trials were conducted in Ramea fields located in Saint-Roch-de-l'Achigan. Preliminary field trials were run in the first half of August 2019. The goal was to test the camera position as discussed and test for any hardware issues on site. In the final field trials, five classes of trees were defined, and eight trees were randomly selected per class. A total of 40 trees were observed from the same row on the same day. The classes were defined with the Ramea team. They represent their internal classification system for the manufacturing of green fences and noise barriers. Table 2 shows the morphological characteristics of all classes.

Table 2. Willow tree classification classes.

Class Number	Diameter	Straightness	Marker
1	28.57 mm < d < 34.92 mm	Curvature < 3.05 m	White
2	28.57 mm < d < 34.92 mm	Curvature > 3.05 m	Green
3	34.92 mm < d	Curvature < 3.05 m	Blue
4	34.92 mm < d	3.05 m < Curvature < 3.66 m	Yellow
5	34.92 mm < d	Curvature > 3.66 m	Red

Two ranges for diameter were looked at: greater than 28.57 mm but smaller than 34.92 mm and strictly greater than 34.92 mm. Trees having a diameter less than 28.57 mm were not targeted for diameter characterization by Ramea, since they would be converted to mulch instead of being kept as rods for manufacturing green fences and noise barriers. It is essential to mention that diameter measurement for classification was taken at 152.4 mm from the ground surface. Diameter measurements for data validation were taken at 1.88 m from the soil surface, which was the real-world height of pixels located on the bottom row of the analyzed frame using a caliper with sub-millimeter accuracy. The stem diameter was measured on the plane parallel to the row line and perpendicular to the ground plane. Straightness was a subjective measure of how straight the stem was relative to itself for the 3.05 m to 3.66 m height section measured 152 mm from the ground surface. Table 2 describes where an excessive curvature is located relative to the height range of interest, if any.

Each randomly chosen tree was identified using colored electrical tape as a marker. The electrical tape was applied at 1.5 m above the soil surface. A caliper was used to measure the stem diameter at 1.78 m and 0.15 m from the ground surface. As stems are not perfectly round in nature, the diameter was measured on a plane parallel to the tree row, which was coplanar to the plane used by the camera to estimate diameter. To ensure repeatability, a custom-built measuring stick was prepared to indicate the height where every measurement should be taken. This tool reduced the chance for errors as well as making data collection faster. The tool was used to adjust the tilt of the RealSense camera as well. Once it was placed straight up at the center of the row, the camera was tilted upward until the corresponding mark (3.81 m) on the stick could be seen in the monochrome imager output frame.

Data was then recorded by driving the tractor in front of the row as if harvesting were taking place. In order not to compromise the tree segmentation procedure, the colored electrical tape was placed below the RealSense camera FOV. To capture the markers, another camera, model KeyMission, featuring a wide FOV color sensor from Nikon (Nikon Corporation, Tokyo, Japan), was used. The latest model was able to capture the color markers in addition to the ROI of the RealSense. However, this technique required manual work to match the marked trees with the correct tree ID to perform data validation. As the Willow Tree algorithm was not ready for live testing during the harvesting season, all four streams of the RealSense were saved as one .bag file along with one text file containing raw GPS strings and one .mp4 file from the KeyMission for later analysis. Final field trials were conducted in late September 2019. The tractor was a Massey Ferguson (AGCO, Duluth, Georgia) model 6485 equipped with a track system form SoucyTrack (SoucyTrack, Quebec, Canada). The operator aligned the tractor center line with the center of the next parallel willow tree row allowing the camera to be at roughly 1.22 m from the center of the row as designed. The data acquisition was carried out at a speed of 2.3 km/h with the engine crank-shaft speed set at 2200 rotation per minutes (RPM). Those parameters are typical for the harvesting operation at Ramea. Figure 7 shows the complete hardware installed during the final field trials carried out on a sunny afternoon with little cloud cover.

Figure 7. (a) Final field trials; (b) sensing components layout.

3.5. Data Analysis

To assess the performance of the tree-detection procedure, 10 labeled images out of the 37 images containing the 40 observed trees were randomly selected and manually validated. Additionally, stratified sampling of frames with respect to the number of detected trees was undertaken. Two categories were created, frames containing 7 or less detected trees and frames containing 10 or more detected trees. For each category, 3 frames were randomly selected, so a total of 6 frames were selected as stratified samples. Thus, the complete set of images for which manual validation has been performed was 16 out of 37 images from the same pass in the field (i.e., 10 random samples and 6 stratified samples). This was done to better capture the distribution of the number of detected trees across frames for statistical analysis. Three variables were examined in each frame: number of correctly detected trees, number of falsely detected trees, and number of undetected trees. Since it can be ambiguous to distinguish trees and branches visually, clear criteria had been specified to count a tree as a tree and not a branch (and vice-versa). Criteria are defined in Table 3.

Table 3. Tree detection validation criteria.

Criteria Number	Criteria
1	Trees must be visible in the grayscale's bottom pixel row as one complete solid stem.
2	The stem must not be clustered in the same contour to be considered as correctly detected trees.
3	Stems must be at least 10 pixels wide at the haft height of the frame.

The performance of the diameter estimation procedure was assessed using the 40 observed trees in Ramea fields. They corresponded to eight randomly selected trees per class for the five classes observed in the same row and on the same day. Since two ranges of diameters are examined within all classes (i.e., 28.57 mm < d < 34.92 mm, 34.92 mm < d), classes with the same range were grouped together. Thus, results from Class 1 and Class 2 are shown together as well as Class 3, Class 4, and Class 5. Root mean square error (RMSE) was computed according to class group as well as for all data points together. A linear regression with intercept fix at the origin between the observed and estimated diameter was undertaken without regard to class. Moreover, the two highest residuals (i.e., above and below regression lines) were removed from the dataset for further analysis. Hence, they were not included in the linear regression.

4. Results and Discussion

4.1. Hardware Reliability

During the field trials, no issues related to hardware occurred. Mechanical and electrical connections were sturdy, and the sensor mount did not show excessive vibration. No loss of power, connection fault with sensors or loss of data took place. Considering the relatively high machine vibration due to the installed track system, the system hardware proved to be rugged and reliable.

4.2. Tree Detection Results

Manual validation results of the tree detection procedure are shown in Figure 8. From the 16 images, a total of 135 trees were detected where 71.8% were correctly detected and 28.2% were falsely detected. Furthermore, 19 trees were not detected, which represents 16.4% of the total detected count. However, from the falsely detected group, 89.7% were branch objects and the remaining were clusters of trees for which the algorithm filled in the same contour. To support those results, it was also found that 90% of the 40 observed trees were correctly detected. This means that the algorithm was more sensitive than expected to branch objects. From the manually validated images, no non-tree or non-branch objects had been detected, which indicated that the segmentation procedure was able to properly remove the background and none-tree like objects in the foreground. On the horizontal axis of Figure 8, ID 4 to ID 37 are the 10 randomly sampled frames out the original 40 frames dataset. To the right of ID 37, ID 26 to ID 0 represents the 6 stratified sampled frames.

Undetected trees were mostly due to fragmented filled regions, which in fact, represented a single tree. Since the contour filtering function deletes regions smaller than a certain threshold before confirming its presence on the bottom pixel row, if only a small region was observed, the entire contour was deleted. Thus, all other contours representing the same tree were deleted as well and the tree was not detected. The Willow Tree algorithm has no features to regroup fragmented contours of the same object.

Looking at the distribution of the undetected trees within the sample of 40 observed trees, one tree per class was undetected, excluding Class 5, which equates to a 90% detection rate. Table 4 summarizes those findings.

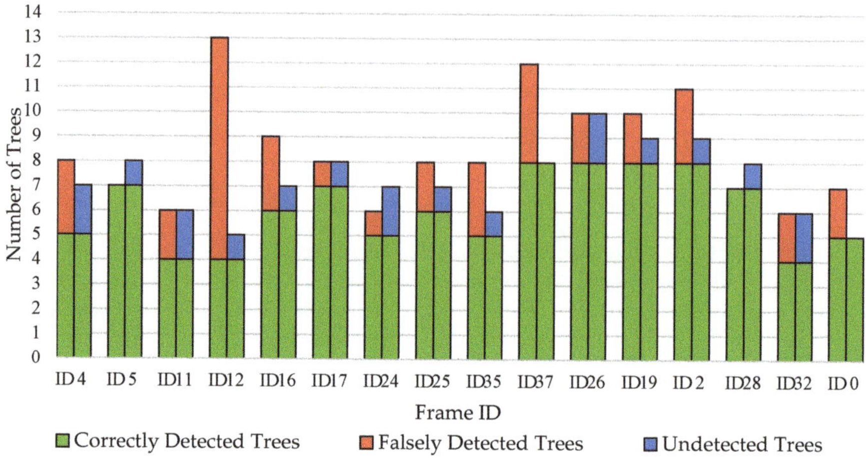

Figure 8. Manual validation compared to algorithm results.

Table 4. Results of tree detection of the 40 observed trees.

Number of Trees	Tree Class				
	1	**2**	**3**	**4**	**5**
Observed	8	8	8	8	8
Correctly detected	7	7	7	7	8
Undetected	1	1	1	1	0

To examine detection precision, a linear regression between the true number of trees and the estimated number of trees was undertaken with the same 10 randomly sampled and the 6 stratified sampled frames (Figure 9). As can be seen, the algorithm tends to overestimate the count of true trees in frames. Knowing that the system has a high amount of falsely detected trees due to branches, this behavior was to be expected. ID 12 had the highest residual from the fitted regression line and so, it was not included in the statistical analysis but still shown as an outlier in Figure 9. The resultant linear regression features an R^2 of 0.40 and a RMSE of 1.37 trees.

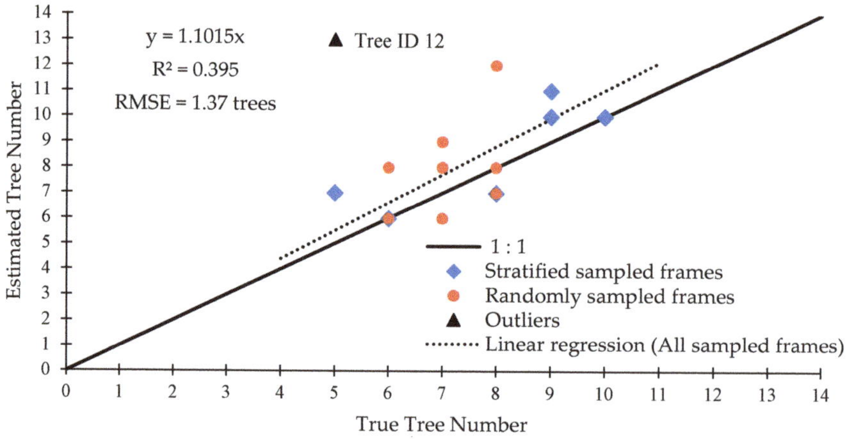

Figure 9. True tree count vs. estimated tree count per frame.

High falsely detected tree rates were mainly due to branches being detected as trees. The willow tree algorithm was sensitive to branches that were the closest to the camera since they could be picked up by depth maps; hence, their pixel intensity value was included in the determination of seed point coordinates to be input in the flood-fill algorithm. In the detection procedure, some of the filled regions representing branches were not filtered out since they were large enough and present on the bottom of the pixel row. Thus, improvement in contour filtering is needed to handle such cases without diminishing the number of correctly detected trees. One solution would be the implementation of a machine-learning technique to improve tree classification. Contour features could be fed into a classifier, such as support vector machine (SVM), which would be able to distinguish between contours representing trees and branches. SVM classifier has demonstrated a high accuracy rate compared to other machine-learning algorithms for fruit detection in outdoor scenes [10]. However, to create a robust machine-learning model, training data would need to be acquired at a different time of the day (e.g., morning, afternoon), in different weather conditions (e.g., sunny, cloudy), and at a different time of the year (i.e., spring, summer, fall). This, of courses, requires more time and resources but have the potential to improve results.

4.3. Diameter Estimation Results

Due to detection performance, it was possible to analyze 36 of 40 data points. The slope of the regression was found to be 0.78, which indicated that the function tends to underestimate tree diameter. The RMSE for the full dataset was found to be 10.7 mm, which represents 37.5% and 30.7% of the low and high diameter range boundaries, respectively. RMSE for group Classes 1–2 and Classes 3–4–5 was 10.6 mm and 10.8 mm, respectively. The overall coefficient of determination R^2 was found to be poor at 0.099 as shown in Figure 10.

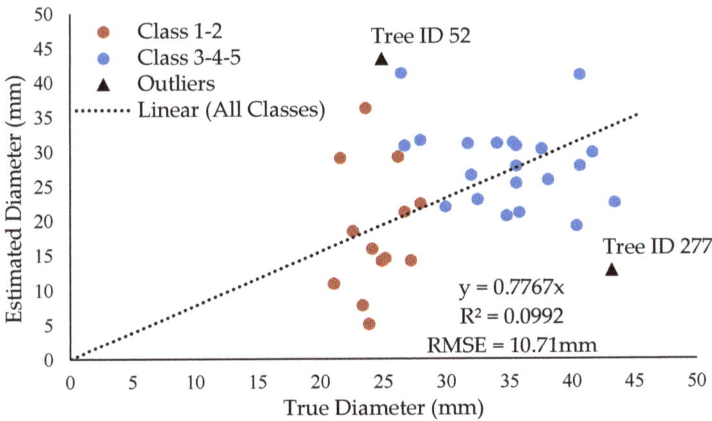

Figure 10. Measured tree diameter compared to estimated tree diameter.

Tree ID 277 and Tree ID 52 were the furthest outliers from the fitted regression line for the lowest and highest diameter estimation, respectively. Analyzing the situation at these points, provided insight into the system behavior. For Tree ID 277, the system output was 12.8 mm in diameter, where the measured diameter was 43.2 mm. After the diagnosis, a possible cause was found to be the poor quality of CLAHE equalized grayscale images. CLAHE equalized frames are used as input in the dynamic histograms thresholding function to compute the histograms and find seed point coordinates. In this case, some parts of the trees within the CLAHE frame were darker. Since the flood parameters in the flood-fill algorithm are taken as constant, the function was not able to connect the darker components to the final segmented binary image. Other contours of the same scene appeared skeletal as well. Figure 11 displays the CLAHE frame and the filtered contours frame.

(a)   (b)

Figure 11. (**a**) CLAHE equalized frame; (**b**) segmentation procedure output with labeled detected trees.

For Tree ID 52, the system output a diameter of 43.4 mm whereas the measured diameter was only 24.9 mm. Again, a diagnostic of the processed frames was conducted. In contrast to the previous case, here, the CLAHE equalized was too bright in the background, leading to the non-tree object being connected to a pixel pertaining to true tree objects as the clear distinction in pixel intensity was more subtle. This effect caused the filled region to be larger and hence, overestimated the diameter.

Even if the use of color images was avoided to increase the robustness of the algorithm in varying light conditions, the effect was not completely prevented with grayscale images. Since the RealSense was tilted upward to improve depth map quality, it was also prone to over saturation when beams of light found their way through the tree canopy and into the camera's optical sensors. To address this problem, the low and high difference parameters used by the flood-fill algorithm could adapt dynamically to current frames. It is possible that intensity location of the peak count value in the dynamic histogram thresholding function could be used as an indicator.

4.4. Row Transect

Because data were acquired for a single willow tree row, spatial interpolation of willow tree count would not have been appropriate. Instead, a transect graph (Figure 12) was produced to show the variability of yield (i.e., number of trees) along the row. Since the operator must stop harvesting to unload before moving on to the next load, data collection must stop once for every cycle. To produce a yield map for an entire field, all data from single loads need to be merged and analyzed together. A spatial interpolation using the inverse distance weighting method could then be used.

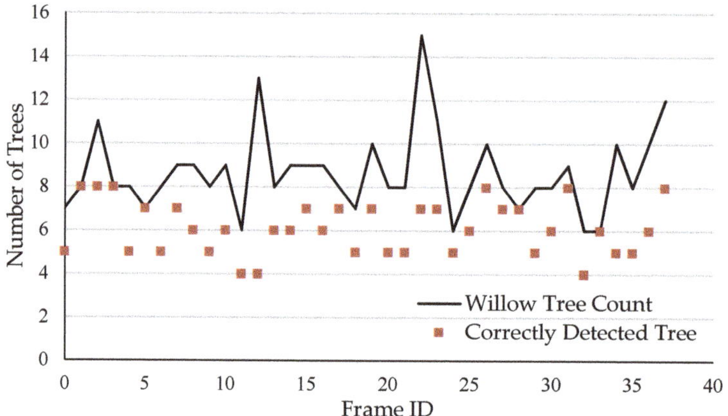

Figure 12. Willow tree yield transect.

5. Conclusions

A willow tree yield-mapping technology was developed based on the number and diameter of tree stems harvested. An RGB-D camera using stereovision technology was implemented to collect grayscale images and compute the depth maps of willow tree row scenes grown in a SRF system, prior to harvesting. The system also has potential as a fast and non-destructive yield estimation tool which could be used to assess carbon dioxide (CO_2) sequestration of willow trees grown in SRF over time. Data from a GNSS receiver were integrated to tag image data to their corresponding geographical coordinates. A CV algorithm was developed to detect willow trees in images and estimate their diameter. To achieve this, the algorithm relied on equalized grayscale images using the CLAHE algorithm and depth maps information. The RGB stream of the camera was not used. The system was designed and built in partnership with Ramea Phytotechnologies, Inc.

The system was able to correctly detect 71.8% of tree stems and estimate their diameter with an RMSE of 10.7 mm. To achieve this, the willow tree algorithm relied on equalized grayscale images using the CLAHE algorithm and depth map information. Detection errors were primarily due to the detection of branches as tree objects which represented 89.7% of the number of falsely detected trees. Diameter estimation errors were mostly due to the low contrast between pixels pertaining to tree objects and backgrounds despite the implementation of the CLAHE algorithm. To increase system performance, the contour filtering procedure needs to be strengthened and the system must be able to better handle cases of oversaturation in scenes. This system is the first of its kind and provides a promising first step for further development of a more robust and commercially viable product.

Author Contributions: Conceptualization, X.L.-T., and M.L.; methodology, M.L. and V.A.; software, M.L. and J.P.; validation, M.L.; formal analysis, M.L.; writing—original draft preparation, M.L.; review and editing, V.A.; supervision and project administration, V.A.; funding acquisition, M.L., X.L.-T., and V.A. All authors have read and agreed to the published version of the manuscript.

Funding: Research was funded by the Natural Sciences and Engineering Research Council of Canada (NSERC), grant number NSERC EGP 536515-18.

Acknowledgments: The authors would like to thank Ramea Phytotechnologies, Inc for their time, input, and field-based advice throughout this research.

References

1. Canadian Forest Services-Policy Notes. *Is Forest Bioenergy Good for the Environment?* Natural Resources Canada, Canadian Forest Service: Ottawa, ON, Canada, 2010; ISBN 978-1-100-17372-6.
2. Labrecque, M.; Lajeunesse, S.L. *Guide de Production de Saules en Culture Intensive sur Courtes Rotations*; Agri-Réseau: Québec, QC, Canada, 2017.
3. Gebbers, R.; Adamchuk, V.I. Precision Agriculture and Food Security. *Science* **2010**, *327*, 828. [CrossRef] [PubMed]
4. Fulton, J.; Hawkins, E.; Taylor, R.; Franzen, A. Yield monitoring and mapping. *Precis. Agric. Basics* **2018**, 63–78.
5. Ghatrehsamani, S.; Ampatzidis, Y. A Review: Vision Systems Application on Yield Mapping. In Proceedings of the 2019 ASABE Annual International Meeting, St. Joseph, MI, USA, 7–10 July 2019; p. 1.
6. Boatswain Jacques, A.; Adamchuk, V.I.; Cloutier, G.; Clark, J.J.; Leclerc, M. A Machine Vision Yield Monitor for Vegetable Crops. In Proceedings of the 2017 ASABE Annual International Meeting, Spokane, WA, USA, 16–19 July 2017; p. 1.
7. Bargoti, S.; Underwood, J.P. Image segmentation for fruit detection and yield estimation in apple orchards. *J. Field Robot.* **2017**, *34*, 1039–1060. [CrossRef]
8. Wang, Q.; Nuske, S.; Bergerman, M.; Singh, S. Automated Crop Yield Estimation for Apple Orchards. In *Experimental Robotics: The 13th International Symposium on Experimental Robotics*; Desai, J.P., Dudek, G., Khatib, O., Kumar, V., Eds.; Springer International Publishing: Berlin/Heidelberg, Germany, 2013; pp. 745–758.

9. Hellström, T.; Ostovar, A.; Hellström, T.; Ostovar, A. Detection of trees based on quality guided image segmentation. In Proceedings of the Second International RHEA Conference, Madrid, Spain, 21–23 May 2014; pp. 21–23.

10. Gongal, A.; Amatya, S.; Karkee, M.; Zhang, Q.; Lewis, K. Sensors and systems for fruit detection and localization: A review. *Comput. Electron. Agric.* **2015**, *116*, 8–19. [CrossRef]

11. Manoj, K.; Qin, Z. Mechanization and Automation Technologies in Specialty Crop Production. *Resour. Mag.* **2012**, *19*, 16–17.

12. Cyganek, B.; Siebert, J.P. *An Introduction to 3D Computer Vision Techniques and Algorithms*; John Wiley & Sons: Chischester, UK, 2011.

13. Grunnet-Jepsen, A.; Sweetser, J.N.; Woodfill, J. *Best-Known-Methods for Tuning Intel® RealSense™ D400 Depth Cameras for Best Performance*; Intel Corporation: Satan Clara, CA, USA, 2018; Volume 1.

14. Grunnet-Jepsen, A.; Sweetser, J.N.; Winer, P.; Takagi, A.; Woodfill, J. *Projectors for Intel® RealSense™ Depth Cameras D4xx*; Intel Support, Interl Corporation: Santa Clara, CA, USA, 2018.

15. Ku, J.; Harakeh, A.; Waslander, S.L. In Defense of Classical Image Processing: Fast Depth Completion on the CPU. In Proceedings of the 2018 15th Conference on Computer and Robot Vision (CRV), Toronto, ON, Canada, 8–10 May 2018; pp. 16–22.

16. Kaehler, A.; Bradski, G. *Learning OpenCV 3: Computer Vision in C++ with the OpenCV Library*; O'Reilly Media, Inc.: Sebastopol, CA, USA, 2016.

17. Pizer, S.M.; Amburn, E.P.; Austin, J.D.; Cromartie, R.; Geselowitz, A.; Greer, T.; ter Haar Romeny, B.; Zimmerman, J.B.; Zuiderveld, K. Adaptive histogram equalization and its variations. *Comput. Vis. Graph. Image Process.* **1987**, *39*, 355–368. [CrossRef]

18. Reza, A.M. Realization of the Contrast Limited Adaptive Histogram Equalization (CLAHE) for Real-Time Image Enhancement. *J. VLSI Signal Process. Syst. Signal Image Video Technol.* **2004**, *38*, 35–44. [CrossRef]

Article

Deep Learning and Machine Vision Approaches for Posture Detection of Individual Pigs

Abozar Nasirahmadi [1,*], **Barbara Sturm** [1], **Sandra Edwards** [2], **Knut-Håkan Jeppsson** [3], **Anne-Charlotte Olsson** [3], **Simone Müller** [4] **and Oliver Hensel** [1]

1   Department of Agricultural and Biosystems Engineering, University of Kassel, 37213 Witzenhausen, Germany
2   School of Natural and Environmental Sciences, Newcastle University, Newcastle upon Tyne NE1 7RU, UK
3   Department of Biosystems and Technology, Swedish University of Agricultural Sciences, 23053 Alnarp, Sweden
4   Department Animal Husbandry, Thuringian State Institute for Agriculture and Rural Development, 07743 Jena, Germany
*   Correspondence: abozar.nasirahmadi@uni-kassel.de

Received: 22 July 2019; Accepted: 28 August 2019; Published: 29 August 2019

Abstract: Posture detection targeted towards providing assessments for the monitoring of health and welfare of pigs has been of great interest to researchers from different disciplines. Existing studies applying machine vision techniques are mostly based on methods using three-dimensional imaging systems, or two-dimensional systems with the limitation of monitoring under controlled conditions. Thus, the main goal of this study was to determine whether a two-dimensional imaging system, along with deep learning approaches, could be utilized to detect the standing and lying (belly and side) postures of pigs under commercial farm conditions. Three deep learning-based detector methods, including faster regions with convolutional neural network features (Faster R-CNN), single shot multibox detector (SSD) and region-based fully convolutional network (R-FCN), combined with Inception V2, Residual Network (ResNet) and Inception ResNet V2 feature extractions of RGB images were proposed. Data from different commercial farms were used for training and validation of the proposed models. The experimental results demonstrated that the R-FCN ResNet101 method was able to detect lying and standing postures with higher average precision (AP) of 0.93, 0.95 and 0.92 for standing, lying on side and lying on belly postures, respectively and mean average precision (mAP) of more than 0.93.

Keywords: convolutional neural networks; livestock; lying posture; standing posture

---

1. Introduction

Computer vision techniques, either three-dimensional (3D) or two-dimensional (2D), have been widely used in animal monitoring processes and play an essential role in assessment of animal behaviours. They offer benefits for the monitoring of farm animals due to the wide range of applications, cost and efficiency [1]. Examples of employing machine vision techniques in monitoring of animal health and behaviour have been reviewed [1–3]. In monitoring of pigs, interpretation of animal behaviours, particularly those relating to health and welfare as well as barn climate conditions, is strongly related to their locomotion and posture patterns [3,4]. In order to determine lying and standing postures in pigs, 3D cameras have been widely used, due to their possibility of offering different colours in each pixel of an image based on the distance between the object and the depth sensor. One such effort was reported by [5], in which the monitoring of standing pigs was addressed by using the Kinect sensor. In this research, initially noise from depth images was removed by applying a spatiotemporal interpolation technique, then a background subtraction method was applied to detect standing pigs. Lao et al. [6] recognized lying, sitting and standing behaviours of sows based on 3D images. In another study, a Kinect sensor was also employed to localize standing pigs in depth images for automatic detection of aggressive events [7]. 2D images have also been used for locomotion and lying posture detection, which were mainly based on pixel movement [8] or features of ellipses fitted to the animals [9,10].

In most of the studies, due to different light sources and resolution (quality) of captured images, problems of a high level of noise and the generation of a great deal of data cause challenges for machine vision-based monitoring of livestock. It has been reported that machine learning techniques have the possibility to tackle these problems [11,12]. The most popular machine learning techniques employed for analysis of optical data for monitoring of pigs are included models (i.e., linear discriminant analysis (LDA), artificial neural networks (ANN), support vector machine (SVM)). However, in recent years, deep learning approaches, a fast-growing field of machine learning, have been used in image classification, object recognition, localization and object detection [13]. Deep learning is similar to ANN, but with deeper architecture and the ability for massive learning capabilities which leads to higher performance [14]. Deep learning has recently been applied in 2D and 3D pig-based machine vision studies. An example is the separation of touching pigs in 3D images using a low-cost Kinect camera, based on the convolutional neural network (CNN), with high accuracy of around 92% [15].

Deep learning has also been used for detection and recognition of pigs' or sows' behaviours in different imaging systems. A detection system based on the Kinect v2 sensor, the faster regions with convolutional neural network features (Faster R-CNN) technique in combination with region proposal network (RPN) and Zeiler and Fergus Net (ZFnet), were developed by [16] to identify different postures of sows (i.e., standing and sitting behaviours, and sternal, ventral and lateral recumbency). Yang et al. [17] used deep learning for automatic recognition of sows' nursing behaviours in 2D images, with an accuracy of 97.6%, sensitivity of 90.9% and specificity of 99.2%. Faster R-CNN and ZFnet were also employed to recognize individual feeding behaviours of pigs by [18], where each pig in the barn was labelled with a letter. Their proposed method was able to recognise pigs' feeding behaviours with an accuracy of 99.6% during the study. Segmentation of sow images from background in top view 2D images was also addressed by means of Fully Convolutional Network (FCN), built in Visual Geometry Group Network with 16 layers (VGG16), in another study by [19]. Their results showed the possibility of employing deep learning approaches for segmentation of the animal from different background conditions.

Due to the fast growth of pig production around the world, the issue of real-time monitoring of the animals, particularly in large scale farms, becomes more crucial [1]. Monitoring of an individual pig's posture during the whole lifetime in large scale farms is almost impossible for farmers or researchers, due to the labour- and time-intensive nature of the task. Utilizing state-of-the-art machine learning, along with machine vision techniques, for monitoring of groups of pigs in different farming conditions could offer the advantage of early problem detection, delivering better performance and lower costs. Thus, the main objective of this study was to develop a machine vision and a deep learning-based method to detect standing and lying postures of individual pigs in different farming conditions, including a variety of commercial farm systems.

2. Material and Methods

2.1. Housing and Data Collection

The data sets for this study were derived from three different commercial farms in Germany and Sweden. In Germany, two farms for weaning and fattening of commercial hybrids of Pietrain× (Large White × Landrace) were selected. A set of four pens in a room in each farm were chosen; these had fully slatted floors with central concrete panels and plastic panels on both sides (weaning farm), and fully slatted concrete floors (fattening farm). In a Swedish fattening farm, two rooms with two pens were selected, each having part-slatted concrete flooring with some litter on the solid concrete in the lying area. Pigs were of commercial Hampshire × (Landrace × Large White) breeding.

The images used in this study were recorded by top view cameras (VIVOTEK IB836BA-HF3, and Hikvision DS-2CD2142FWD-I). The cameras were connected via cables to servers and video images from the cameras were recorded simultaneously and stored on hard disks. Recording in different farming conditions allowed for capture of images with pigs of different skin colour and age, housed under various floor type and illumination conditions, which facilitated the development of a robust deep learning model. Examples of images used for development of the detection model are illustrated in Figure 1.

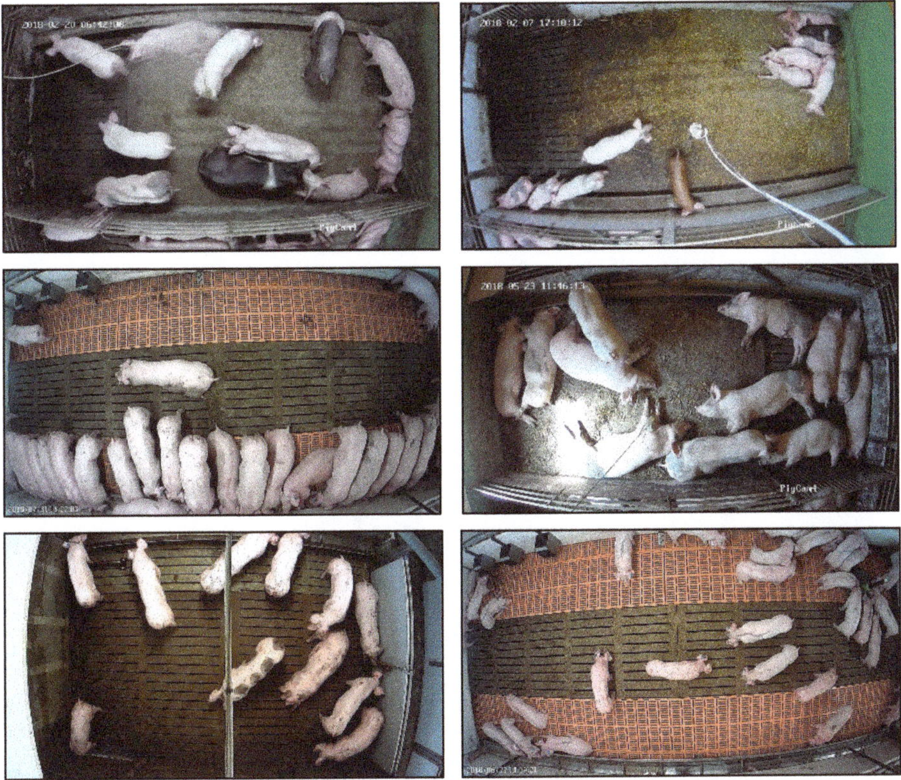

Figure 1. Example of images used for development of the detection algorithms in this study.

2.2. Proposed Methodology

According to the findings of Nasirahmadi et al. [12], when in a side (lateral) lying posture pigs lie in a fully recumbent position with limbs extended, and when in a belly (sternal) lying posture the limbs are folded under the body. Due to the lack of available benchmark datasets for pigs' lying and standing postures, three classes of standing, lying on belly and lying on side were defined in this work. Examples of these postures are represented in Figure 2. Most of the possible scenarios of individual pig standing and lying postures within the group have been considered. The proposed detection methods were developed and implemented in python 3.6 and OpenCV 3.4 using TensorFlow [20]. Experiments and the deep learning training were conducted on a Windows 10 with a NVIDIA GeForce GTX 1060 GPU with 6GB of memory.

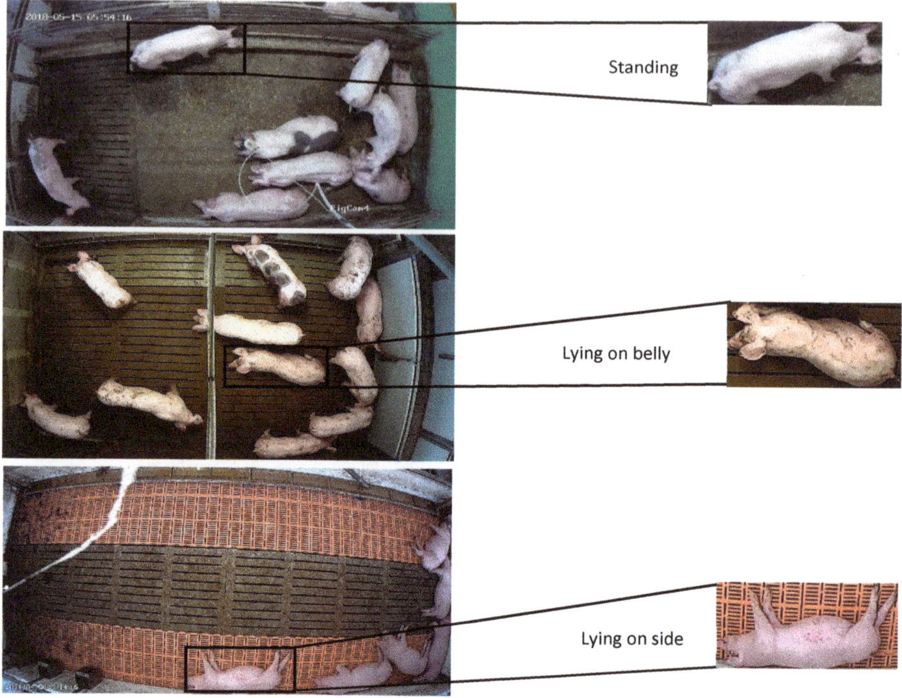

Figure 2. Examples of three posture classes used in this study in different farming conditions.

In order to establish the lying and standing postures, images from the three commercial farms over a period of one year were collected from surveillance videos. A total of 4900 images (4500 for training and validation, and 400 for testing) from this dataset, which incorporated various farm, animal age, and animal and background colour conditions, were selected to provide the training (80%: 3600) and validation (20%: 900) datasets. The images were selected at 1–2 week intervals from the beginning to the end of every batch, and different samples of images were collected randomly in one day of these selected weeks. Furthermore, another 400 images were taken randomly from the image data set as a testing set. The testing data were independent of the training and validation sets, and were used for an evaluation of the detection phase. All images were first resized to 1280 × 720 (width × height), then annotated using the graphical image annotation tool LabelImg, created by [21]. The annotations included the identification of standing and lying postures and were saved as XML files in the PASCAL VOC format. Since each image included different pigs (7–32 in number) based on farm conditions and time of recording, the total number of labelled postures was 52,785. The information on each posture class is summarized in Table 1.

Table 1. Details of image data sets used for posture detection.

Posture Classes	Training Process			Testing Process
	Number of Individual Postures			
	Training	Validation	Total	Test
Standing	11,632	4372	16,004	1839
Lying on side	11,435	4085	15,520	1489
Lying on belly	15,781	5480	21,261	1659
Total samples	**38,848**	**13,937**	**52,785**	**4987**

In this study, various combinations of deep learning-based models (i.e., Faster R-CNN (Inception V2, ResNet50, ResNet101 and Inception-ResNet V2), R-FCN (ResNet101) and SSD (Inception V2)) were developed with the aim of finding the best posture detection technique in 2D images. Two architectures of the ResNet model (i.e., ResNet50 and ResNet101) were applied. ResNet is a deep residual network introduced by [22] and has been widely used in object detection and localization tasks. Both ResNet50 and ResNet101 are based on repeating of four residual blocks. These feature extractors are made of three convolution layers of 1×1, 3×3 and 1×1. They also have an additional connection joining the input of the first 1×1 convolution layer to the output of the last convolution 1×1 layer. Additionally, residual learning in the ResNet model resolves the training by fusing filtered features with original features. ResNet50 and 101 are very deep networks and contain 50 and 101 layers. ResNet applies skip connections and batch normalization and provides short connection between layers [23]. Furthermore, ResNets allow direct signal propagation from the first to the end layer of the network without the issue of degradation. Inception V2 GoogLeNet [24] with layer-by-layer structure was also implemented as a feature extractor of the input pig posture images. The feature extractors which map input images to feature maps, characterize computational cost as well as performance of networks. Inception V2 GoogLeNet is based on the repetition of building blocks and extracts features concatenated as output of the module [24]. It is made of a pooling layer and different convolutional layers. The convolution layers are of different sizes of 1×1, 3×3, 5×5 and help to extract features from the same feature map on different scales and improve performance. Inception V2, by factorization of filters n×n to a combination of $1 \times n$ and $n \times 1$ convolutions, has led to better performance and reduction in representational bottlenecks. This has resulted in the widening of the filter banks instead of deepening, to remove the representational bottleneck without increase in computational cost and number of parameters [25]. Finally, Inception-ResNet V2, which was proposed in 2017 [26], was utilized as a feature extractor in this work. Inception-ResNet V2 is one of the state-of-the-art approaches in image classification and is based on combination of residual connections and Inception architectures. Inception-ResNet V2 is an Inception style network which uses residual connections instead of filter concatenation [26]. Further details on all parameters of the proposed feature extractors can be found in [22,24,26].

2.2.1. Faster R-CNN

The Faster R-CNN method [27] was used to monitor standing, side and belly lying postures of pigs. Figure 3 shows the architecture and different steps of the Faster R-CNN utilized in posture detection. Various feature extraction methods (Inception V2, ResNet50, ResNet101 and Inception-ResNet V2) were applied and the regional proposal network (RPN), which is a fully convolutional network for generating object proposals, was used to produce a proposed region for each image and generate feature maps in the final layer by predicting the classes [25]. The feature maps were fed into the region of interest (RoI) pooling to extract regions from the last feature maps. Each RoI (region of pigs) was then determined to a confidence score. Finally, the feature vector from the RoI was fed into several fully connected (FC) layers of a softmax classifier to obtain the final assurance scores and a regression bounding box to localize the coordinates of the detected objects.

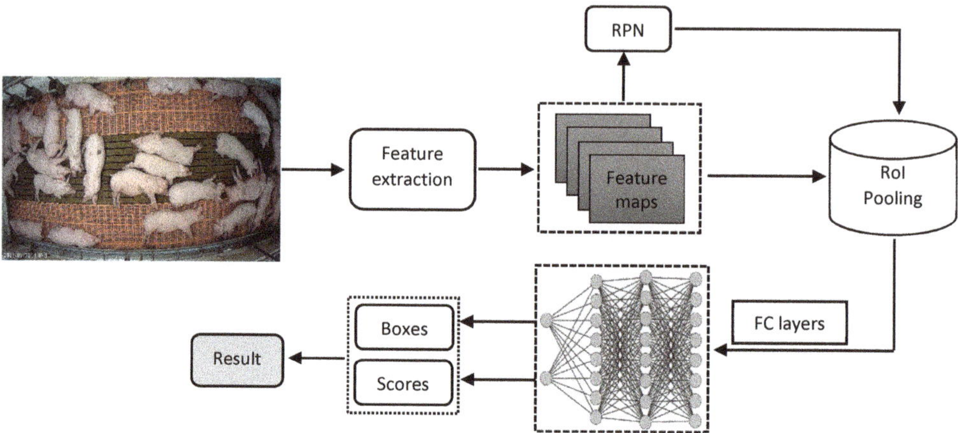

Figure 3. Schematic diagram of the faster regions with convolutional neural network (Faster R-CNN) used in this study.

2.2.2. R-FCN

R-FCN [28] is an improvement of the Faster R-CNN and is proposed as an object detection technique using CNNs. The structure of the proposed R-FCN applied in this study is shown in Figure 4. The R-FCN consists of a feature extractor (ResNet101 in this study) and several FC layers behind an RoI pooling layer. The RoI pooling layer uses position sensitive maps to address the issue of translation invariance [29]. In the R-FCN, the computation is shared across the whole image by creating a deeper FCN without increasing the speed overhead. Region proposal and classification are done by use of RPN, followed by the position sensitive maps and RoI pooling, respectively. For training of the R-FCN model in this study, the same hyperparameters as the faster R-CNN were used.

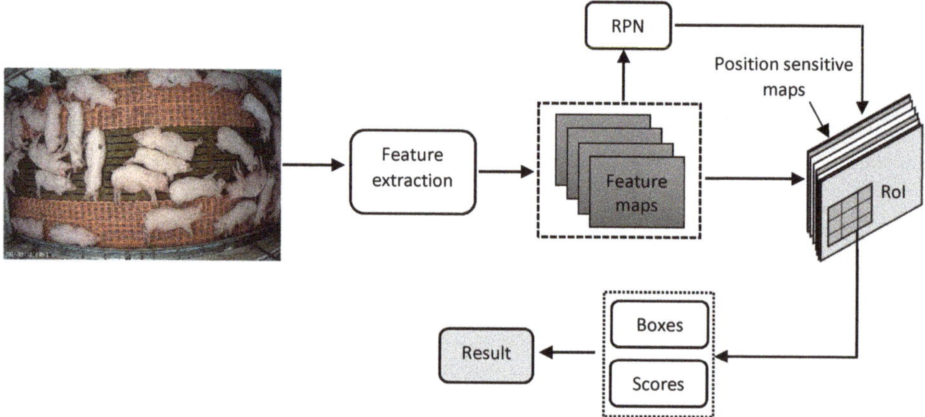

Figure 4. Schematic diagram of the region-based fully convolutional network (R-FCN) used in this study.

2.2.3. SSD

The SSD was proposed by Liu et al. [30] and is based on a bounding box regression principle. In the SSD, the input images are first divided into small kernels on different feature maps to predict anchor boxes of each kernel (Figure 5). In comparison with the Faster R-CNN and R-FCN techniques,

SSD uses a single feed forward CNN to produce a series of bounding boxes and scores for the presence of the objects in each box [29]. Finally, the classification scores of each region are computed based on the score obtained in the previous section. The SSD model, instead of applying proposal generation, encloses the whole process into a single network, which leads to less computational time.

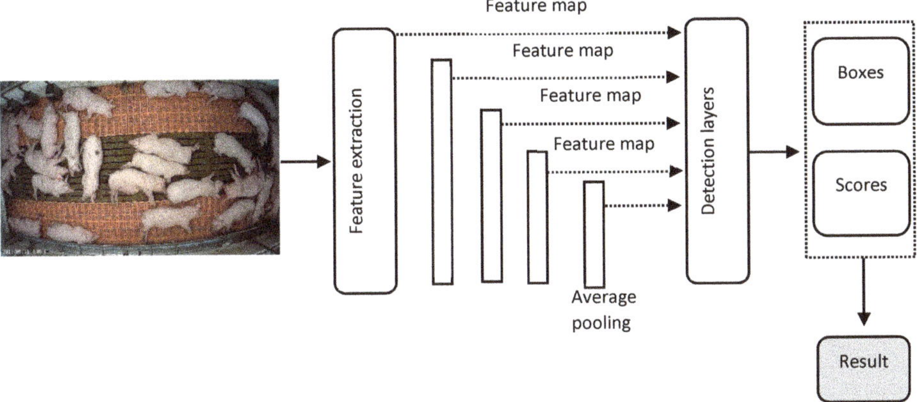

Figure 5. Schematic diagram of the single shot multibox detector (SSD) used in this study.

The initial learning rate is a hyperparameter and shows adjustment of weight of the networks.

Too high learning rates may cause poor converge or a network unable to converge, and a low initial learning rate will result in slow convergence [31]. So, in this study different initial learning rates of 0.03, 0.003, 0.0003 were selected to train the Faster R-CNN, R-FCN and SSD networks. The momentum algorithm accumulates the weighted average of the previous training gradient and continues to move in that direction. The past weighted average updates the current gradient direction, and the momentum coefficient specifies to what extent the update needs to increase or decrease [32]. In this study, training of all networks was conducted with a momentum 0.9. The right number of iterations is important in training of the models, as too small numbers will result in poor performance while a high number of iterations may cause a weakening of the generalization ability of the trained model [32]. So, according to the number of samples [32] the iteration number, which shows the number of weight updates during the training process, was chosen to be 70,000 for all networks in this study. Furthermore, for data augmentation a random horizontal flip method for all detectors was used.

3. Experimental Results and Discussion

The big challenges in monitoring of pigs' behaviour, particularly in large-scale farms, are cost, time and labour-intensity of the visual investigation. The importance of automatic monitoring of group and individual behaviours of pigs has led to the development of image processing techniques with the ability to monitor a large number of pigs. However, due to the variability in farm and animal conditions, especially in commercial farms, the development of a robust monitoring technique with the ability to detect individual animal behaviours has been examined in this study.

The performance obtained for each posture class of the models in the test data set with each initial learning rate is illustrated in Table 2. In order to evaluate the detection capability of the proposed models, the widely used average precision (AP) of each class and mean average precision (mAP) were calculated. High values for the AP and mAP show the acceptable performance of the R-FCN ResNet101, Faster R-CNN ResNet 101 and Faster R-CNN Inception V2 detection approaches for scoring of standing, lying on belly and lying on side postures among group-housed pigs when compared to the other models. Additionally, the learning rate of 0.003 gave the best results in the mentioned detection models. The evaluation metrics illustrate a trend for improvement when the learning rate decreases from 0.03 to 0.003, however these values declined at the learning rate of 0.0003. This finding is in line with the previous finding that the detection performance changes with the same trend in learning rates in various scenarios [33].

Table 2. Performance of the detection phase in test data set in various learning rates.

Classes	Learning Rate	Feature Extractor	AP			
			Standing	Lying on Side	Lying on Belly	mAP
Faster R-CNN	0.03	Inception V2	0.82	0.87	0.88	0.86
Faster R-CNN	0.03	ResNet50	0.80	0.85	0.83	0.83
Faster R-CNN	0.03	ResNet101	0.87	0.86	0.81	0.85
Faster R-CNN	0.03	Inception-ResNet V2	0.79	0.83	0.77	0.80
R-FCN	0.03	ResNet101	0.88	0.88	0.87	0.88
SSD	0.03	Inception V2	0.69	0.70	0.68	0.69
Faster R-CNN	0.003	Inception V2	0.90	0.93	0.91	0.91
Faster R-CNN	0.003	ResNet50	0.85	0.92	0.89	0.88
Faster R-CNN	0.003	ResNet101	0.93	0.92	0.89	0.91
Faster R-CNN	0.003	Inception-ResNet V2	0.86	0.89	0.84	0.86
R-FCN	0.003	ResNet101	0.93	0.95	0.92	0.93
SSD	0.003	Inception V2	0.76	0.79	0.74	0.76
Faster R-CNN	0.0003	Inception V2	0.85	0.90	0.89	0.87
Faster R-CNN	0.0003	ResNet50	0.85	0.86	0.87	0.86
Faster R-CNN	0.0003	ResNet101	0.87	0.89	0.87	0.88
Faster R-CNN	0.0003	Inception-ResNet V2	0.80	0.85	0.79	0.81
R-FCN	0.0003	ResNet101	0.90	0.90	0.88	0.89
SSD	0.0003	Inception V2	0.75	0.80	0.72	0.76

The training and validation loss of the models at a learning rate of 0.003 are presented in Figure 6. As can be seen, there are more fluctuations in the validation curves than in the training ones for each model, which could be due to the lower size of data set used in the validation phase [34]. The graphs show the decrease in the loss of training and validation steps of the developed models, and the values converged at a low value when the training process was finished, which is an indicator of a sufficiently trained process. After training the model, 400 images (not used for training and validation) were used as a test set (including 4987 lying and standing postures) to evaluate the correctness of the posture detection approaches. The test data set contained images from the three farms of the study which were selected from different times of the day where animals had lying, feeding, drinking and locomotion activities in the barn.

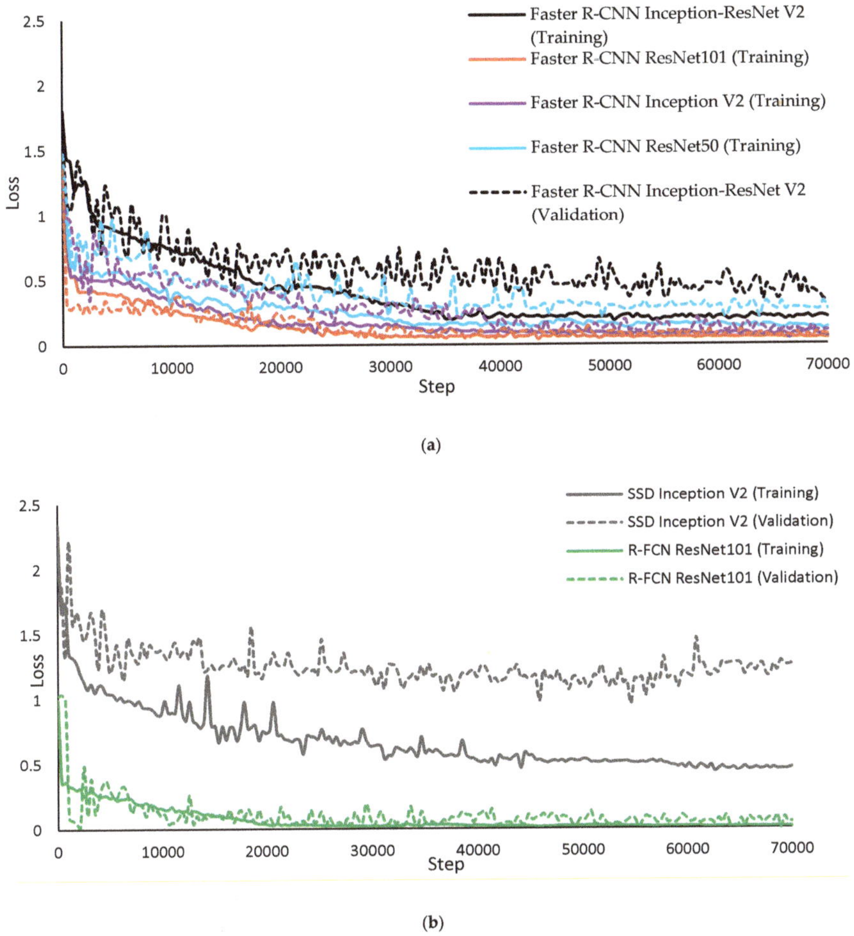

Figure 6. Training and validation loss during the (**a**) Faster R-CNN, (**b**) R-FCN and SSD training processes.

Figure 7 illustrates examples of detected postures by Faster R-CNN, R-FCN and SSD models in the different farming conditions in the learning rate of 0.003. As can be seen from the figure, some of the developed algorithms have the ability to detect different lying and standing postures under all tested commercial farming conditions. In some cases, there were multiple postures in an image, so the possibility of detecting various postures in an image is also shown in Figure 7.

It can be observed from Table 2 that one of the major misdetections was with respect to standing and belly lying postures. Due to the similarity between these postures in top view images (Figure 8), some incorrect detection was observed. However, in lying on side the animals have a different top view image (Figure 2) compared to the other postures, and the number of incorrect detections was lower.

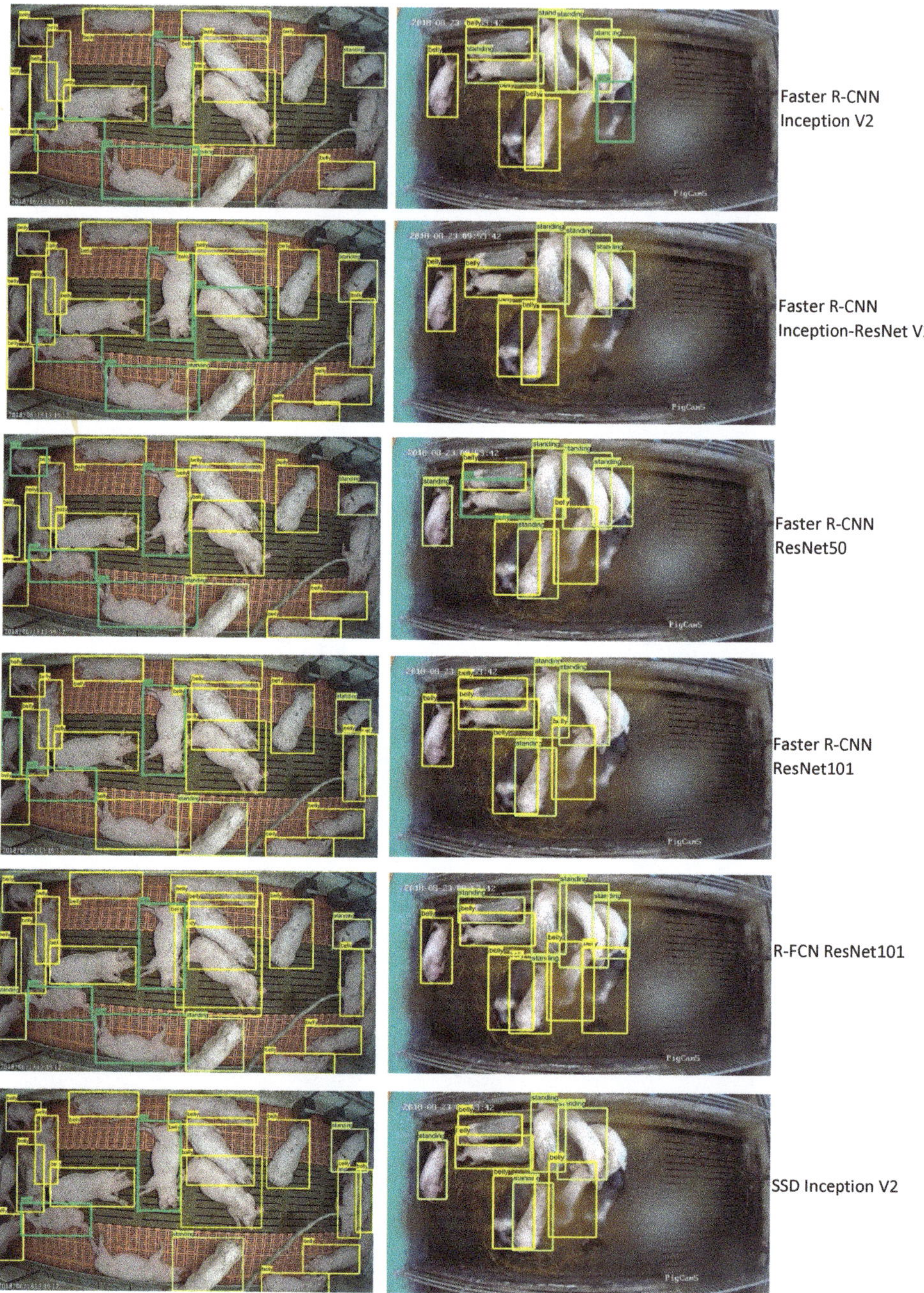

Figure 7. Examples of detected standing (light green rectangle), lying on belly (yellow rectangle) and side posture (green rectangle) of six different models in various farming conditions.

The confusion matrix of the selected R-FCN ResNet101 model in the test dataset is shown in Table 3. As can be seen in the table, the number of misdetections between standing and belly lying postures are higher than the other postures.

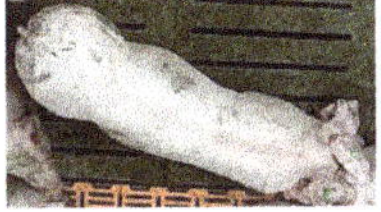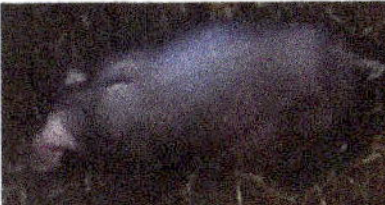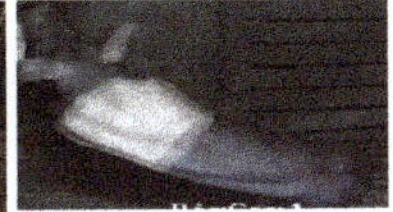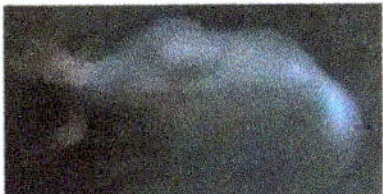

Figure 8. Sample of images of standing posture which are similar to belly lying posture in top view.

Table 3. Confusion matrix of the proposed R-FCN ResNet101 in the test data set at learning rate of 0.003.

		Predicted Class		
		Standing	Lying on side	Lying on belly
Actual class	Standing	1672	25	89
	Lying on side	18	1382	31
	Lying on belly	71	16	1523

The proposed R-FCN ResNet101 technique for detection of standing and lying postures in pigs under commercial farming conditions by 2D cameras provided a high level of detection performance (e.g., mAP of 0.93), which is illustrated in Figure 7. In a sow posture study, Faster R-CNN along with RPN and ZF-net models were able to detect standing, sitting and lying behaviours with a mAP of 0.87 using Kinect v2 depth images [16]. In another study, monitoring of lying and nursing behaviour of sows using a deep learning approach (FCN and VGG16) resulted in achieved accuracy, sensitivity and specificity of 96.4%, 92.0% and 98.5%, respectively [17]. However, the study reported by [17] was carried out to monitor one sow per image, which has less probability of detection mistake compared to our study with number of pigs varying from 7–32 per image. The use of Faster R-CNN as a detector and ZF-net as a feature extractor showed high values of precision and recall (99.6% and 86.93%, respectively) in a study by [18]. However, compared to our study, they had just four animals in each image and each animal's body was labelled as A, B, C, and D for tracking their behaviours, which would help to improve detection performance. There were some misdetections of the test data set for each posture in our study. This was mainly due to the similarity between standing and belly lying postures in top view images (Figure 8, as previously discussed) and to the problem of covering of the camera lens with fly dirt as time progressed, reducing the visibility in images [10]. Furthermore, the shape change of an animal's body caused by the fisheye lens effect at the edge of the pen (particularly in the weaning farm), and the existence of a water pipe in the middle of the pen (at the fattening farm) which caused some invisible area in the image, gave some misdetection in the models. Low resolution of the image data impacts on the performance of machine vision techniques, and the model is not able to extract enough features to perform accurate detection. As shown in this study, misdetection was more likely for pigs with black skin colour as they have a greater similarity with the background (pen floor) colour.

Furthermore, in this study testing time of the state-of-the-art detection models was calculated. The speed of each proposed method was measured based on the frames per second (FPS) [35]. The results show a FPS of 6.4 for the Faster R-CNN Inception V2, 3.6 for the Faster R-CNN ResNet50, 3.8 for the Faster R-CNN ResNet101, 2.3 for the Faster R-CNN Inception-ResNet V2, 5 for the R-FCN ResNet101, and 8.8 for the proposed SSD Inception V2. According to Table 2, R-FCN ResNet101 showed the highest performance as well as an acceptable FPS (5) for the pig posture detection. It was also found that the Faster R-CNN Inception V2 had high enough performance for lying-on-side posture detection along with a FPS of 6.4. Since the lying posture can be used as a parameter of pig comfort assessment in various thermal conditions, the Faster R-CNN Inception V2 model can be beneficial in terms of higher processing speed. The SSD model had the highest rate of FPS (8.8), however the lowest performance was obtained in this model, which was not sufficiently reliable in pig lying and standing detection.

The proposed R-FCN model could be used to monitor pig postures in the group to control climate and barn conditions continuously. In this context, three days of recorded video data were fed into the model in order to score the number of the lying and standing postures. Results of the scoring phase are shown in Figure 9. Data from early morning (5:50) until evening (16:00), which includes different animals' activities, were utilised and the number of the postures continuously saved in excel files.

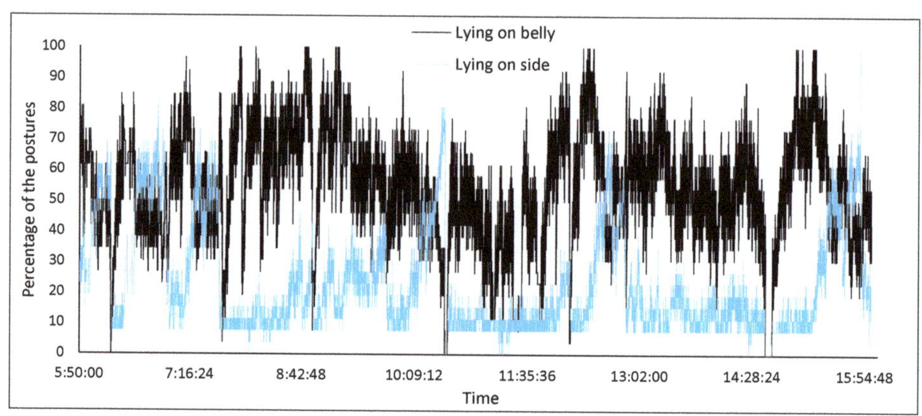

(a)

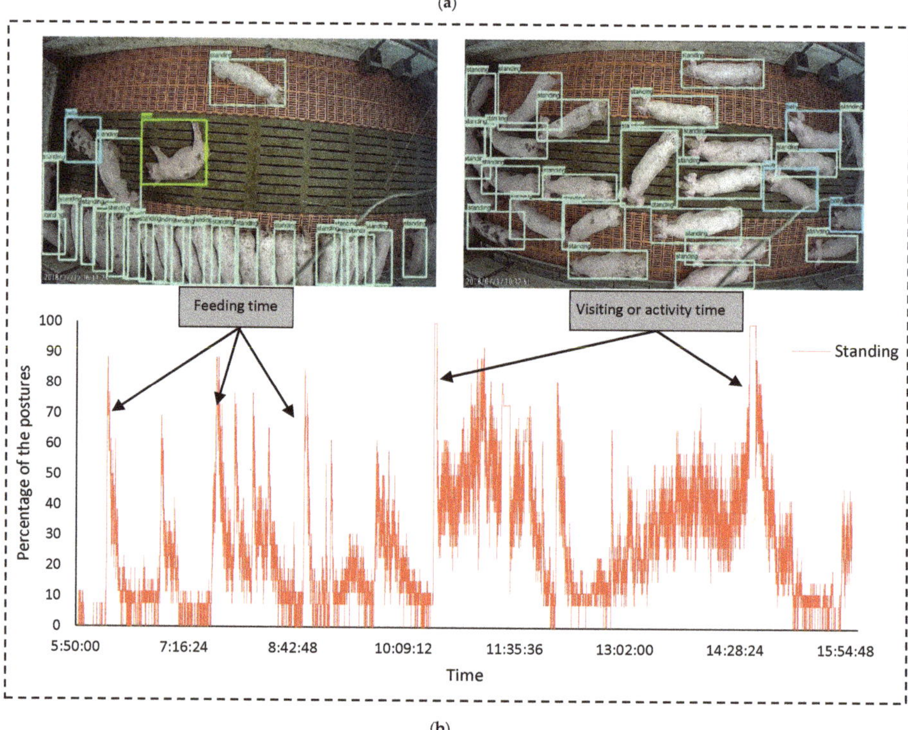

(b)

Figure 9. (a) Results of scoring (in percentage) of the lying and standing postures across the day. (b) Standing posture (light blue rectangle), lying on belly (blue rectangle) and side posture (green rectangle).

As shown in the scoring diagrams, the algorithm can continuously monitor the activity and postures of group-housed pigs in the barn. In the graph of standing postures, activity peaks were apparent at feeding and activity times or when the animals were checked by the farmer. The lying and standing activity patterns of the animals, which were scored using the proposed automated method, offer the advantage of early detection of health and welfare issues in pig farms [3]. For example, changes in lying and standing activities levels could be used for the detection of lameness and tail biting events [36].

In a previous study, Nasirahmadi et al. [12] showed that image processing and a linear SVM model was able to score different lying postures (sternal and lateral) in commercial farming conditions. However, the performance of the scoring technique was highly dependent on the output of the image processing method, which led to some wrong scoring in the pigs lying postures. In this study, due to the use of different deep learning approaches, high enough precision for practical use was obtained. The monitoring approach described in this study could be a valuable tool to detect changes in the number of pigs in the standing, lying on side or belly postures to improve welfare and health surveillance. However, this new method needs to be adapted and evaluated with a wider range of farming conditions (both commercial and research) in future, which may require a greater number of images for training of the model or use of other feature extraction methods.

4. Conclusions

In pig husbandry, it is important to monitor the animals' behaviour continuously between birth and slaughter. However, this can only be achieved using robust machine learning and machine vision techniques capable of processing image data from varied farming conditions. In this study, two commercial farms (weaning and fattening) in Germany and one commercial fattening farm in Sweden were used to record 2D video data for a year and provide an image database. In order to have various imaging conditions, pens with different floor type and colour, various numbers of pigs per pen (from 7 to 32) and varying ages and colours of pig were selected. The techniques proposed in this study were based on using image data from the surveillance system and Faster R-CNN, R-FCN and SSD methods. These models were trained using 80% (3600) of the image data and validated by 20% (900) of the 2D images, with a total of around 52,785 standing, lying on belly and lying on side postures in the images. The trained model was then evaluated using 400 new images. Results of the testing phase showed a high level of mAP and good processing speed for the different postures in the R-FCN ResNet101 model. This technique performed well in the monitoring of various lying and standing postures of pigs in image data acquired from commercial farms, which has been the limitation of most 2D and 3D conventional machine vision techniques. The proposed model has the flexibility and good robustness toward different pig farm lighting conditions, age of the animals and skin colour, which can be an important step toward developing real-time, continuous computer-based monitoring of livestock farms.

Author Contributions: Conceptualization, A.N., B.S. and S.E.; methodology, A.N., B.S., S.E. and S.M; software, A.N; validation, A.N., B.S. and S.M.; formal analysis, A.N., B.S. and S.M. investigation, A.N., B.S., S.E. and S.M.; resources, H.J., A.-C.O. and S.M.; data curation, A.N.; writing—original draft preparation, A.N.; writing—review and editing, B.S., K.-H.J., A.-C.O., S.E., S.M. and O.H.; project administration, B.S. and A.N.; funding acquisition, B.S., S.E., A.N. and O.H., please turn to the CRediT taxonomy for the term explanation. Authorship must be limited to those who have contributed substantially to the work reported.

Funding: This research was funded by the European Union's Horizon 2020 research and innovation programme, grant number "No 696231". The work was financially supported by the German Federal Ministry of Food and Agriculture (BMEL) through the Federal Office for Agriculture and Food (BLE), grant number "2817ERA08D" and The Swedish Research Council Formas, grant number "Dnr 2017-00152".

Acknowledgments: We thank the funding organizations of the SusAn ERA-Net project PigSys.

Conflicts of Interest: The authors declare no conflict of interest.

References

1. Nasirahmadi, A.; Edwards, S.A.; Sturm, B. Implementation of machine vision for detecting behaviour of cattle and pigs. *Livestock Sci.* **2017**, *202*, 25–38. [CrossRef]
2. Frost, A.R.; Tillett, R.D.; Welch, S.K. The development and evaluation of image analysis procedures for guiding a livestock monitoring sensor placement robot. *Comput. Electron. Agric.* **2000**, *28*, 229–242. [CrossRef]

3. Matthews, S.G.; Miller, A.L.; Clapp, J.; Plötz, T.; Kyriazakis, I. Early detection of health and welfare compromises through automated detection of behavioural changes in pigs. *Vet. J.* **2016**, *217*, 43–51. [CrossRef] [PubMed]

4. Olsen, A.W.; Dybkjær, L.; Simonsen, H.B. Behaviour of growing pigs kept in pens with outdoor runs: II. Temperature regulatory behaviour, comfort behaviour and dunging preferences. *Livestock Prod. Sci.* **2001**, *69*, 265–278.

5. Kim, J.; Chung, Y.; Choi, Y.; Sa, J.; Kim, H.; Chung, Y.; Park, D.; Kim, H. Depth-Based Detection of Standing-Pigs in Moving Noise Environments. *Sensors* **2017**, *17*, 2757. [CrossRef] [PubMed]

6. Lao, F.; Brown-Brandl, T.; Stinn, J.P.; Liu, K.; Teng, G.; Xin, H. Automatic recognition of lactating sow behaviors through depth image processing. *Comput. Electron. Agric.* **2016**, *125*, 56–62. [CrossRef]

7. Lee, J.; Jin, L.; Park, D.; Chung, Y. Automatic recognition of aggressive behavior in pigs using a kinect depth sensor. *Sensors* **2016**, *16*, 631. [CrossRef]

8. Ott, S.; Moons, C.P.H.; Kashiha, M.A.; Bahr, C.; Tuyttens, F.A.M.; Berckmans, D.; Niewold, T.A. Automated video analysis of pig activity at pen level highly correlates to human observations of behavioural activities. *Livestock Sci.* **2014**, *160*, 132–137. [CrossRef]

9. Kashiha, M.A.; Bahr, C.; Ott, S.; Moons, C.P.; Niewold, T.A.; Tuyttens, F.; Berckmans, D. Automatic monitoring of pig locomotion using image analysis. *Livestock Sci.* **2014**, *159*, 141–148. [CrossRef]

10. Nasirahmadi, A.; Richter, U.; Hensel, O.; Edwards, S.; Sturm, B. Using machine vision for investigation of changes in pig group lying patterns. *Comput. Electron. Agric.* **2015**, *119*, 184–190. [CrossRef]

11. Rieke, N.; Tombari, F.; Navab, N. Computer Vision and Machine Learning for Surgical Instrument Tracking: Focus: Random Forest-Based Microsurgical Tool Tracking. *Comput. Vis. Assist. Healthc.* **2018**, 105–126. [CrossRef]

12. Nasirahmadi, A.; Sturm, B.; Olsson, A.C.; Jeppsson, K.H.; Müller, S.; Edwards, S.; Hensel, O. Automatic scoring of lateral and sternal lying posture in grouped pigs using image processing and Support Vector Machine. *Comput. Electron. Agric.* **2019**, *156*, 475–481. [CrossRef]

13. Cao, C.; Liu, F.; Tan, H.; Song, D.; Shu, W.; Li, W.; Zhou, Y.; Bo, X.; Xie, Z. Deep learning and its applications in biomedicine. *Genet. Proteomics Bioinf.* **2018**, *16*, 17–32. [CrossRef] [PubMed]

14. Kamilaris, A.; Prenafeta-Boldú, F.X. Deep learning in agriculture: A survey. *Comput. Electron. Agric.* **2018**, *147*, 70–90. [CrossRef]

15. Ju, M.; Choi, Y.; Seo, J.; Sa, J.; Lee, S.; Chung, Y.; Park, D. A Kinect-Based Segmentation of Touching-Pigs for Real-Time Monitoring. *Sensors* **2018**, *18*, 1746. [CrossRef]

16. Zheng, C.; Zhu, X.; Yang, X.; Wang, L.; Tu, S.; Xue, Y. Automatic recognition of lactating sow postures from depth images by deep learning detector. *Comput. Electron. Agric.* **2018**, *147*, 51–63. [CrossRef]

17. Yang, A.; Huang, H.; Zhu, X.; Yang, X.; Chen, P.; Li, S.; Xue, Y. Automatic recognition of sow nursing behaviour using deep learning-based segmentation and spatial and temporal features. *Biosyst. Eng.* **2018**, *175*, 133–145. [CrossRef]

18. Yang, Q.; Xiao, D.; Lin, S. Feeding behavior recognition for group-housed pigs with the Faster R-CNN. *Comput. Electron. Agric.* **2018**, *155*, 453–460. [CrossRef]

19. Yang, A.; Huang, H.; Zheng, C.; Zhu, X.; Yang, X.; Chen, P.; Xue, Y. High-accuracy image segmentation for lactating sows using a fully convolutional network. *Biosyst. Eng.* **2018**, *176*, 36–47. [CrossRef]

20. Abadi, M.; Agarwal, A.; Barham, P.; Brevdo, E.; Chen, Z.; Citro, C.; Corrado, G.S.; Davis, A.; Dean, J.; Devin, M.; et al. TensorFlow: Large-scale machine learning on heterogeneous systems. *arXiv* **2016**, arXiv:1603.04467.

21. Tzutalin, LabelImg. Git Code. Available online: https://github.com/tzutalin/labelImg,2015 (accessed on 1 February 2018).

22. He, K.; Zhang, X.; Ren, S.; Sun, J. Deep residual learning for image recognition. In Proceedings of the IEEE Conference, Las Vegas, NV, USA, 26 June–1 July 2016.

23. da Silva, L.A.; Bressan, P.O.; Gonçalves, D.N.; Freitas, D.M.; Machado, B.B.; Gonçalves, W.N. Estimating soybean leaf defoliation using convolutional neural networks and synthetic images. *Comput. Electron. Agric.* **2019**, *156*, 360–368. [CrossRef]

24. Szegedy, C.; Liu, W.; Jia, Y.; Sermanet, P.; Reed, S.; Anguelov, D.; Erhan, D.; Vanhoucke, V.; Rabinovich, A. Going deeper with convolutions. In Proceedings of the IEEE Conference on Computer Vision and Pattern Recognition, Boston, MA, USA, 7–12 June 2015.

25. Arcos-García, Á.; Álvarez-García, J.A.; Soria-Morillo, L.M. Evaluation of Deep Neural Networks for traffic sign detection systems. *Neurocomputing* **2018**, *316*, 332–344. [CrossRef]

26. Szegedy, C.; Ioffe, S.; Vanhoucke, V.; Alemi, A.A. Inception-v4, inception-resnet and the impact of residual connections on learning. In Proceedings of the Thirty-First AAAI Conference on Artificial Intelligence, San Francisco, CA, USA, 4–10 February 2017.

27. Ren, S.; He, K.; Girshick, R.; Sun, J. Faster R-CNN: towards real-time object detection with region proposal networks. *Adv. in Neural inf. Process. Syst.* **2015**, *39*, 1137–1149. [CrossRef] [PubMed]

28. Dai, J.; Li, Y.; He, K.; Sun, J. R-FCN: Object Detection via Region-based Fully Convolutional Networks. In Proceedings of the Neural Information Processing Systems (NIPS 2016), Barcelona, Spain, 5 December–10 December 2016.

29. Fuentes, A.; Yoon, S.; Kim, S.; Park, D. A robust deep-learning-based detector for real-time tomato plant diseases and pests recognition. *Sensors* **2017**, *17*, 2022. [CrossRef] [PubMed]

30. Liu, W.; Anguelov, D.; Erhan, D.; Szegedy, C.; Reed, S.; Fu, C.Y.; Berg, A.C. Ssd: Single shot multibox detector. In Proceedings of the European Conference on Computer Vision, Amsterdam, The Netherlands, 8–16 October 2016.

31. Wei, X.; Yang, Z.; Liu, Y.; Wei, D.; Jia, L.; Li, Y. Railway track fastener defect detection based on image processing and deep learning techniques: A comparative study. *Eng. Appl. Artif. Intel.* **2019**, *80*, 66–81. [CrossRef]

32. Zou, Z.; Zhao, X.; Zhao, P.; Qi, F.; Wang, N. CNN-based statistics and location estimation of missing components in routine inspection of historic buildings. *J. Cult. Herit.* **2019**, *38*, 221–230. [CrossRef]

33. Tang, T.A.; Mhamdi, L.; McLernon, D.; Zaidi, S.A.R.; Ghogho, M. Deep learning approach for network intrusion detection in software defined networking. In Proceedings of the International Conference on Wireless Networks and Mobile Communications (WINCOM), Fez, Morocco, 26–29 October 2016.

34. Andersen, R.S.; Peimankar, A.; Puthusserypady, S. A deep learning approach for real-time detection of atrial fibrillation. *Expert Syst. Appl.* **2019**, *115*, 465–473. [CrossRef]

35. Shen, Z.Y.; Han, S.Y.; Fu, L.C.; Hsiao, P.Y.; Lau, Y.C.; Chang, S.J. Deep convolution neural network with scene-centric and object-centric information for object detection. *Image Vis. Comput.* **2019**, *85*, 14–25. [CrossRef]

36. Nalon, E.; Conte, S.; Maes, D.; Tuyttens, F.A.M.; Devillers, N. Assessment of lameness and claw lesions in sows. *Livestock Sci.* **2013**, *156*, 10–23. [CrossRef]

Article

Body Dimension Measurements of Qinchuan Cattle with Transfer Learning from LiDAR Sensing

Lvwen Huang [1,2,*], Han Guo [1], Qinqin Rao [1], Zixia Hou [1], Shuqin Li [1,2,*], Shicheng Qiu [1], Xinyun Fan [3] and Hongyan Wang [4]

[1] College of Information Engineering, Northwest A&F University, Yangling, Xianyang 712100, China; loraine@nwafu.edu.cn (H.G.); R-QinQ@nwafu.edu.cn (Q.R.); houzixia@nwafu.edu.cn (Z.H.); 1342524215@nwafu.edu.cn (S.Q.)

[2] Key Laboratory of Agricultural Internet of Things, Ministry of Agriculture and Rural Affairs, Yangling, Xianyang 712100, China

[3] College of Computer Science, Wuhan University, Wuhan 430072, China; fxy.rebecca@163.com

[4] Western E-commerce Co., Ltd., Yinchuan 750004, China; nxwhy01@126.com

* Correspondence: huanglvwen@nwafu.edu.cn (L.H.); lsq_cie@nwafu.edu.cn (S.L.); Tel.: +86-137-0922-3117 (L.H.); +86-137-5997-2183 (S.L.)

Received: 7 October 2019; Accepted: 18 November 2019; Published: 19 November 2019

Abstract: For the time-consuming and stressful body measuring task of Qinchuan cattle and farmers, the demand for the automatic measurement of body dimensions has become more and more urgent. It is necessary to explore automatic measurements with deep learning to improve breeding efficiency and promote the development of industry. In this paper, a novel approach to measuring the body dimensions of live Qinchuan cattle with on transfer learning is proposed. Deep learning of the Kd-network was trained with classical three-dimensional (3D) point cloud datasets (PCD) of the ShapeNet datasets. After a series of processes of PCD sensed by the light detection and ranging (LiDAR) sensor, the cattle silhouettes could be extracted, which after augmentation could be applied as an input layer to the Kd-network. With the output of a convolutional layer of the trained deep model, the output layer of the deep model could be applied to pre-train the full connection network. The TrAdaBoost algorithm was employed to transfer the pre-trained convolutional layer and full connection of the deep model. To classify and recognize the PCD of the cattle silhouette, the average accuracy rate after training with transfer learning could reach up to 93.6%. On the basis of silhouette extraction, the candidate region of the feature surface shape could be extracted with mean curvature and Gaussian curvature. After the computation of the FPFH (fast point feature histogram) of the surface shape, the center of the feature surface could be recognized and the body dimensions of the cattle could finally be calculated. The experimental results showed that the comprehensive error of body dimensions was close to 2%, which could provide a feasible approach to the non-contact observations of the bodies of large physique livestock without any human intervention.

Keywords: transfer learning; deep learning; body dimensions; point cloud; Kd-network; feature recognition; FFPH; non-contact measurement

---

1. Introduction

The availability of three-dimensional (3D) sensing models for large animals is becoming more and more significant in many different agricultural applications [1–3]. For the healthy cultivation and genetic breeding of large animals, periodic measurement of the animals' body dimensions is necessary for breeders and researchers to master the growing complications related to pregnancy, laming, and animal diseases [4–7]. Large-scale measuring at a hold frame by skilled inspectors (traditional ways consist of a meter stick and metric tapes [8]) is ubiquitous, which has costs in terms of heavy manual labor, the animal's stress response, and low efficiency due to long fatigue or differences in the individual experience among inspectors [9]. Very often, large animals cannot be effectively modeled on the spot by means of a series of classical 3D modeling methods due to their geometrical complexity or texture at growing periods, and 3D imaging or range scanners have been widely used to acquire the shape of an animal [10]. With the great achievements and breakthroughs of deep convolutional neural networks (DCNNs) in light detection and ranging (LiDAR) data classification [11–13], deep models could be reconstructed for the body measuring of large animals. This paper follows the state-of-the-art DCNN models and classical PCD processing methods, focusing particularly on the possibility of measuring the body dimensions of Qinchuan cattle, one of five most excellent beef breeds in China.

Qinchuan cattle are historic, famous, and have excellent performance in meat production. For an adult cow, the average height at the withers is about 132 cm and its average body weight reaches 420 kg, whilst for an adult bull, the average height at the withers is about 148 cm and its average body weight can reach above 820 kg. With the growing demand for meat from the vast population, the number of Qinchuan cattle is on the rise. To master the periodic feedback of growth and nutrition status in mass production, it is necessary to be able to automatically measure them in a feasible and effective way.

1.1. Livestock Body Measuring with LiDAR

With the development of 3D information technology, LiDAR (light detection and ranging) sensor scanning and structural light measurements have become important methods to quickly obtain 3D spatial data. They can be used to quickly and accurately obtain the coordinate data of the measured object surface and realize the non-contact measurement of the object. Compared with traditional measurement methods, 3D scanning measurement technology has many advantages such as fast speed, good real-time performance, and high precision [14]. According to different research objectives, relevant studies on the measurement requirements of 3D data in animal husbandry can be roughly divided into three categories: (i) to improve the breeding quality and precision feeding and to prevent diseases such as claudication; (ii) to analyze the body condition of specific parts; and (iii) to study the acquisition methods of body weight, body dimensions, etc. To generate a 3D cattle body model based on depth image, the point cloud data of cattle can be collected by a Kinect sensor to obtain 3D information of the cattle body to quantify the evaluation of cattle body condition [15]. By using the advantages of 3D depth image, researchers combined a thermal imaging camera and proposed a system for measuring the body shape and temperature of black cattle to complete the periodic quality assessment during the growth of cattle [16]. Some researchers have used a depth camera to obtain the body dimension data of pigs to estimate their weight, and found that the average absolute error of the nonlinear model measured by non-contact depth camera was 40% lower than that of the same nonlinear model measured by hand [17]. A dual web-camera high-resolution system was developed to obtain the 3D position of homologous points in a scene, which can be used to estimate the size and weight of live goats [18].

Due to the wide applications of livestock body measurement with 3D data, many processing and analysis methods have been proposed for different purposes. In order to measure pig body dimensions, the random sample consensus (RANSAC) algorithm can be used to remove the background point cloud and the foreground point cloud can be extracted with Euclidean clustering [19]. Additionally, some researchers have applied spatiotemporal interpolation techniques to remove moving noises in the depth images and then detect the standing-pigs with the undefined depth values around them [20]. Azzaaro et al. [21] used the linear and polynomial kernel principal component analysis to reconstruct shapes of cows with a linear combination of the basic shapes constructed from the example database and model validation showed that the polynomial model performed better than other state-of-the-art methods in estimating body condition score (BCS). These successful studies show that it is feasible to use 3D depth sensors to measure cattle body dimensions.

1.2. Applications of Deep Learning

In recent years, with sustainable development in the field of artificial intelligence and the continuous improvement of deep learning methods, image classification and recognition technology are developing toward a more intelligent direction. At present, image feature extraction is mainly divided into two categories: manual feature and mechanized feature extraction. Most representative and classical manual features are scale invariant feature transform (SIFT) [22], and histogram of oriented gradients (HOG) [23], and so on. These features are widely used for image classification of small datasets; however, in the case of large datasets, it is difficult to extract appropriate features from images [24]. Therefore, deep learning is commonly considered to extract high-level features of images to resolve this problem and reduce the impact of low-level features of manual feature extraction on image classification performance. At present, there have been many in-depth studies on two-dimensional (2D) image processing [25], and good results have been achieved in classification and recognition and other tasks. However, 3D data processing is still in its infancy. With the advent of 3D sensors such as IFM O3D303, Microsoft Kinect, Google Tango, and so on, 3D data have grown rapidly, and the recognition or classification based on 2D images has some limitations on special occasions. Image processing is developing from 2D to 3D, and the construction of deep learning networks for 3D image will be the hotspot of future research [26].

For 3D data sensed by LiDAR, deep learning techniques have been developed and applied for different occasions. To extract and classify tree species from mobile LiDAR data, deep learning techniques have been used to generate feature abstractions of the waveform representations of trees, which contribute to the improvement of classification accuracies of tree species [27]. A building detection approach based on deep learning utilizing the fusion of LiDAR data and orthophotos has been presented, and comparison experiments show that the proposed model outperforms the support vector machine (SVM) models in working area [28]. A way to segment individual maize from terrestrial LiDAR data has been proposed by combining deep leaning and regional growth algorithms, and it shows the possibility of solving the individual maize segmentation problem from LiDAR data through deep leaning [29].

The development of a neural network of 3D data has experienced three stages of PointNet [30], PointNet++ [31], and the Kd-network [32]. PointNet mainly resolved the problem of the disorder of PCD. The point cloud feature was abstracted point by point, and then the global feature vector was obtained by using symmetric function. PointNet++ has a hierarchical structure based on PointNet, divides more child point cloud in each level, and uses PointNet to extract the features of each point cloud. PointNet in local point cloud recognition has obvious advantages in accuracy, which is up to 95%. The wide applications of convolutional neural networks (CNNs) in 2D images provide the theoretical basis and technical support for the classification of PCD. A team designed and implemented a visualization technology to optimize the CNN model and deeply understand the functions and operations of the intermediate layers, which is the classic study of CNN visualization [33]. The Kd-network can perform fast computation through a 3D indexing structure, where the parameter sharing mechanism is applied and the representations from leaf nodes to the roots are calculated. However, this method needs to sample point clouds and to construct Kd-trees for every iteration and can employ multiple Kd-trees to represent a single object [34].

3D sensing for the non-contact measurement of body conditions has a good effect and effectively reduces the possibility of injuries to animal and inspectors [35]. With the development of deep learning models proposed for PCD, the evolving applications of PCD with deep learning could improve the accuracy of object classification and measurement, and similarly, the application of transfer learning could refine a deep model to an object. Transfer learning is an emerging research direction of machine learning, and unlike deep learning, it is mostly applied to cases of insufficient training data. The prior knowledge acquired by deep learning, namely the training results, can be applied to the relevant recognition fields. Most 3D data are not enough or perfect to satisfy all demands of different occasions and there also exists a lack of sufficient PCD of large livestock to train and verify deep models.

Therefore, it is necessary to resolve the problem of the different distributions of the training and test dataset by means of transfer learning [36]. To improve the generalization ability of the deep model, the features and related examples could be transferred from a massive dataset to a trace dataset [37]. The deep model framework has a practical impact on the actual operation performance and classification accuracy. In this paper, the framework of PyTorch was employed after performance comparisons of the mainstream deep learning frameworks, whose application is more concise and flexible than the TensorFlow framework of PCD.

1.3. Main Purposes

There are two key problems in realizing the automatic measurement of body dimensions of live Qinchuan cattle through transfer learning including how to preprocess the original PCD and how to transfer the features of PCD to obtain the target cattle body and how to automatically recognize the feature points of the body dimensions. For these problems, the original contributions of this paper can be summarized as follows:

- A new processing fusion for the 3D PCD of cattle is proposed. The original cattle PCD sensed by the LiDAR sensor was filtered by conditional, statistical, and voxel filtering, and then segmented by methods of Euclidean and RANSAC clustering. After the normalization of PCD and orientation correction of body shape, the fast point feature histogram (FPFH) was extracted to retrieve the body silhouettes and local surfaces.
- A 3D classification framework of the target cattle body based on transfer learning is presented. The PyTorch framework of the Kd-network was trained by the ShapeNet PCD dataset. The prior knowledge, the case-based transfer learning of the TrAdaBoost algorithm retrained by the collected cattle silhouettes, was applied to transfer the 3D silhouette of the point cloud and to classify the target cattle body point cloud. The PCD of the cattle body was normalized to extract the candidate surfaces of the feature points, and with extraction of FPFH, the feature points of the cattle body dimensions could be recognized.

The rest of this paper is organized as follows: Section 2 describes the proposed methods in detail. In Section 3, the automatic extraction experimentation of the feature points of body dimensions was carried out on the collected PCD of several groups of live Qinchuan cattle. Section 4 describes the experimental results. Section 5 discusses the performance of the cattle body dimension measurements. The last section summarizes the considerations, conclusions, and future works.

2. Materials and Methods

An overview of the proposed methodology is shown in Figure 1, which combines the Conda software package management system under the PyTorch deep learning programming framework to develop the deep learning network in the environment of Python 3.6 and CUDA 9.1. The LiDAR sensor of IFM O3D303 (IFM Inc., Essen, Germany) was used to acquire the PCD of cattle body shape. Visual Studio 2015 was combined with the software development suite of IFM and Point Cloud Library (PCL), the C++ language was adopted to acquire and preprocess the collected PCD, and the acquired point cloud dataset of the cattle body shape was combined with the obtained deep learning network information for transfer learning. Finally, the recognition of the feature points of the cattle body dimensions was realized, and normalization of the cattle body point cloud and selection of the feature points were carried out to obtain the measurement data of cattle body dimensions.

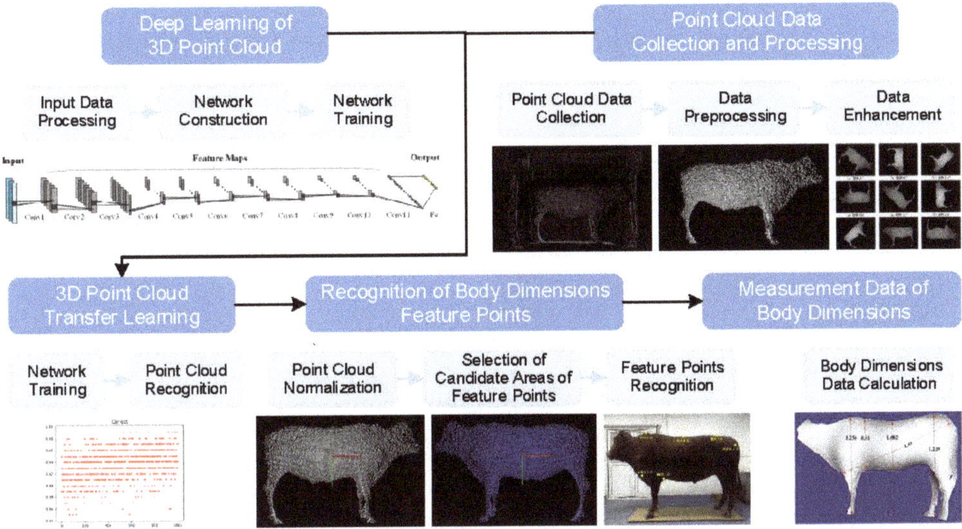

Figure 1. Flowchart of body dimensions measurement of Qinchuan cattle with transfer learning from LiDAR sensor.

2.1. D Point Cloud Deep Learning Network

The construction of a neural network is influenced by the defects of PCD collection and the characteristics of the point cloud itself. Three problems will be encountered in the acquisition process: data missing, noise, and rotation variability [38]. In addition, the characteristics of 3D PCD also affect the construction of a neural network including unstructured data, invariance of arrangement, and change in point cloud number [39].

To resolve the problems in the acquisition process and the point cloud's own characteristics, in this paper, the Kd-tree network [40] was first used to make the 3D PCD have a fixed representation method and arrangement order. Then, the CNN was used to construct the deep learning network structure to realize the deep learning of the 3D point cloud.

The size of the input PCD in this paper was 2^{11}, and there were 16 categories. Therefore, 11 convolution layers were set, and each layer corresponded to each layer in the tree. The deep learning network framework was composed of 11 convolution layers and one full connection layer. The specific model hierarchy is shown in Figure 2.

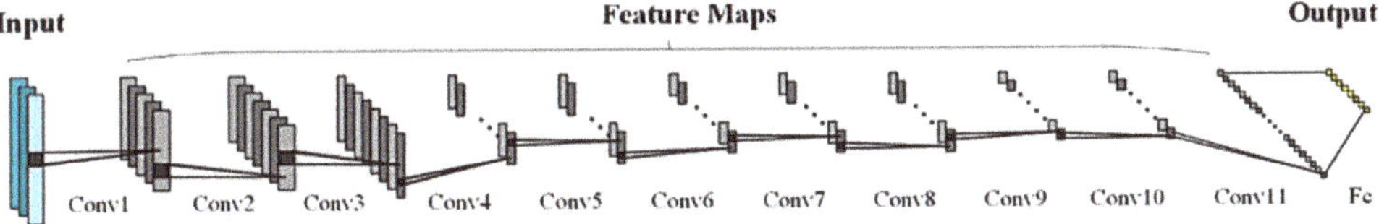

Figure 2. Network model hierarchy.

Similar to the CNN, the network could share the weights in the same level, and used a bottom-up approach and the process was layered. In some sense, the representation of spatial positions on a certain layer was obtained through linear and nonlinear operations in the representation of multiple surrounding positions on the previous layer. In the same layer of the Kd-tree, the receptive field of any two nodes did not overlap [41].

Since the data of deep learning are not usually linearly distributed, the activation function needs to be used to add nonlinear factors [42]. Common activation functions include the Sigmoid function, Tanh function, and ReLU function. The ReLU function is the modified linear unit function, which is the mainstream neuron activation function in deep learning in recent years. Its mathematical expression is shown in Equation (1). When $x > 0$, the function gradient is identically equal to 1. Therefore, in the process of back propagation, the parameters of the first few layers of the network can be updated quickly to alleviate the problem of the disappearing gradient. Compared with the other two functions, the ReLU function is linear and unsaturated, which can significantly accelerate the convergence rate of the neural network [43].

$$f(x) = max(0, x) \tag{1}$$

where x is the input neuron.

As there is no body shape dataset for animals to train the network, the 3D contour shape dataset of PCD in ShapeNet was selected. The training data contained 16 classes, a total of 15,990 samples of PCD files, which included planes (2421), bags (68), hats (49), cars (1641), chairs (3371), headsets (62), guitars (708), knives (352), lights (1391), laptops (400), motorcycles (181), cups (165), guns (247), rockets (59), skateboards (136), and tables (4739). (data source: [44]).

Different training epochs have different influences on deep models. Smaller training epochs lead to under fitting and very large training errors. When the training epoch is increased to a certain number, the accuracy can basically remain unchanged or change a little bit with more and more computing facilities. If continuing to increase the epochs, it could waste more training time and more facilities [45]. The training epoch of 100, 500, 1000, 2000, and 3000, respectively, was selected to train the deep model, and the training results are shown in Figure 3, where the abscissa in each figure is the training epoch, and the ordinate is the corresponding accuracy rate. Table 1 is a comparison of the average accuracy rate with each training epoch. From Figure 3 and Table 1, as the training epoch increased, so did the accuracy rate. If the training epoch increases to or exceeds 1000, the changing tendency turns slowly. Considering the training consumption-time, learning rate, and computing loss, the training epoch was set as 1000 for this Kd-network model.

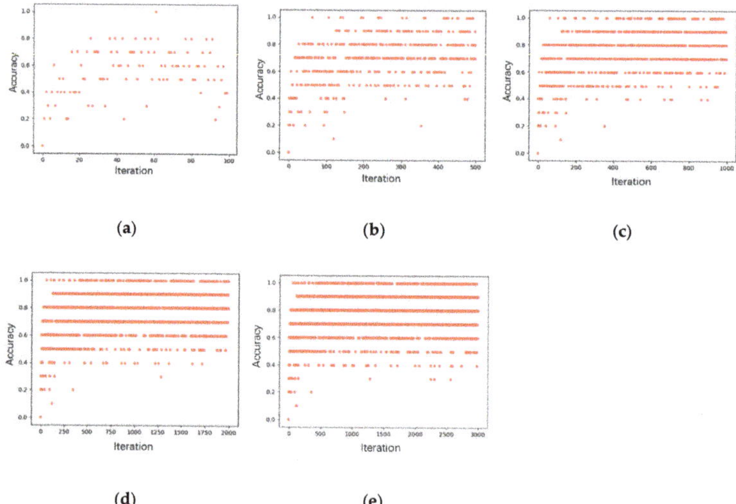

Figure 3. Accuracy rates with different training epochs. (**a**) Accuracy rates with 100 training epochs; (**b**) accuracy rates with 500 training epochs; (**c**) accuracy rates with 1000 training epochs; (**d**) accuracy rates with 2000 training epochs; (**e**) accuracy rates with 3000 training epochs.

Table 1. Comparison of the accuracy rate with different training times.

Training Epochs	Average Accuracy Rate
100	55.2%
500	69.8%
1000	76.1%
2000	76.8%
3000	77.3%

The learning rate reflects the speed of gradient descent (SGD) in training, whose size can affect the convergence errors. If the learning rate is set too large, the error could be difficult to converge. If the learning rate is set too small, the consuming time could increase to cause the local optimization [46]. The learning rate of 0.001, 0.003, 0.005, 0.007, and 0.009, respectively, was set to train the deep network. The training results are shown in Figure 4, where the abscissa is the training epoch and the ordinate is the corresponding accuracy. It can also be seen from Table 2, that when the learning rate was less than 0.003, the average accuracy rate increased with the increase in the learning rate; when the learning rate was over 0.003, the average accuracy rate decreased with the increase in the learning rate. Considering the global SGD, computing facilities, accuracy, and number of samples, the learning rate can be set as 0.003 to achieve better training results. By selecting an appropriate training epoch and learning rate as above-mentioned, the comprehensive accuracy rate of 3D PCD could reach up to 89.6%, where the deep model can effectively extract, recognize, and classify 3D PCD.

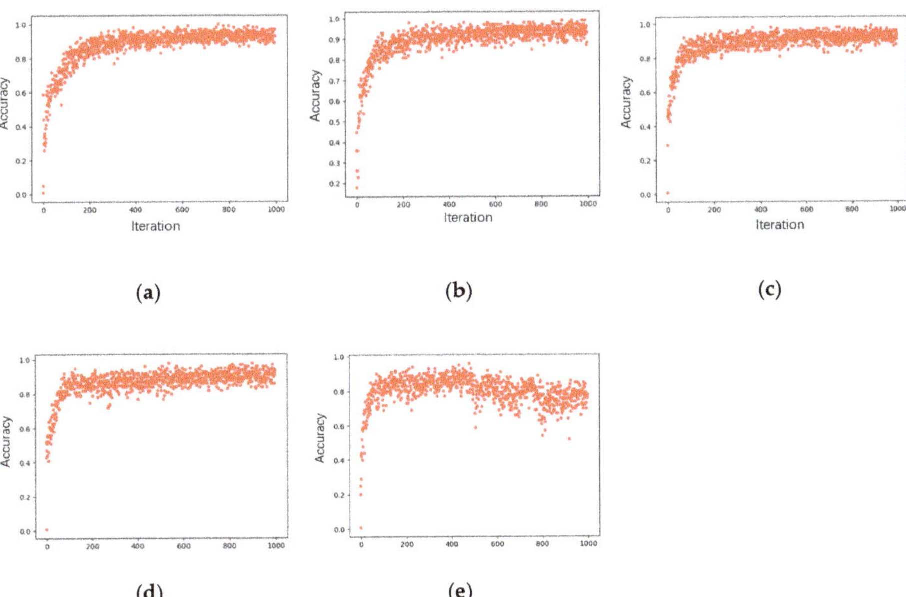

Figure 4. Accuracy rates with different learning rates. (**a**) Accuracy rates with a learning rate of 0.001; (**b**) accuracy rates with a learning rate of 0.003; (**c**) accuracy rates with a learning rate of 0.005; (**d**) accuracy rates with a learning rate of 0.007; (**e**) accuracy rates with a learning rate of 0.009.

Table 2. Comparison of accuracy with different learning rates.

Learning Rate	Average Accuracy Rate
0.001	87.4%
0.003	89.6%
0.005	88.5%
0.007	86.7%
0.009	79.9%

2.2. Cattle Body Point Cloud Recognition Based on Transfer Learning

2.2.1. Data Acquisition and Preprocessing

The LiDAR sensor of IFM O3D303, based on the principle of Time of Flight (ToF), was employed to acquire the PCD of cattle shape. By sending continuous light pulses to the target, the sensor receives the light returned by the target, and calculates the flight time of the detected light pulses to obtain the distance of the target [47]. As the industrial depth sensor has relatively low resolutions, the captured PCD is sparse, therefore, every five frames, files could be stored to meet the body shape computation of enough density. To capture more concise and more precise PCD, the live cattle to be measured could be guided to be calm, or stationary or walk slowly in front of the LiDAR sensor by farmers or workers, and the cattle could be led to pass the measuring passage one by one, and not to trot or jump greatly. The original results of the 3D PCD are illustrated in Figure 5, where the 3D PCD after transformation clearly showed the basic contour of the target cattle with background information.

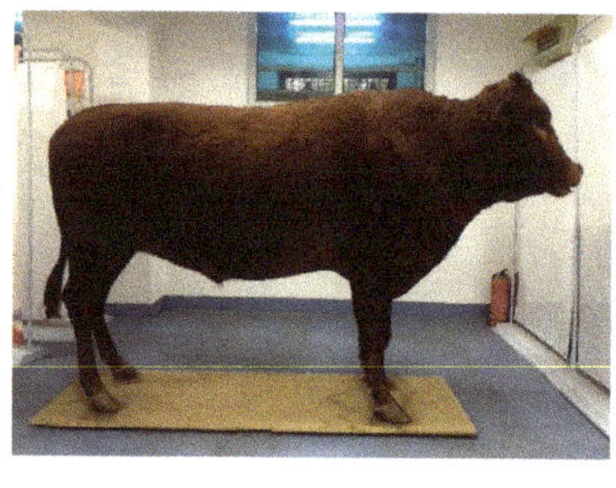

(a)

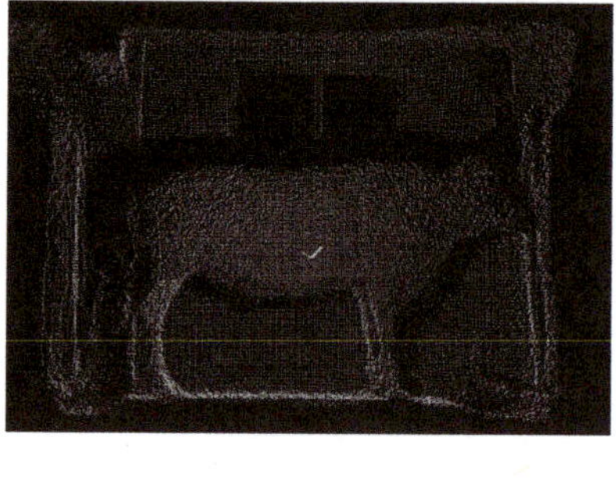

(b)

Figure 5. 3D PCD (Point Cloud Data) acquisition for Qinchuan cattle (real specimen) with the LiDAR sensor where the 3D image shows the basic silhouette of the target cattle. (**a**) Shown in RGB; (**b**) Shown in 3D image.

To obtain better results of segmentation, clustering, and feature extraction, the original PCD obtained needs to be preprocessed by a series of classical point cloud processing methods to cancel the background information such as different occlusions, target surface reflections, equipment errors, and so on [48]. According to the different characteristics of each filter, a preprocessing fusion with different filters was applied to achieve the best filtering results, where the conditional filter [49] was employed to remove the background with different distances, the statistical filter [50] to cancel outliers with different object surfaces, and the voxel filter [51] to compress the PCD of cattle. The fused processing results of multiple filters are shown in Figure 6, where much irrelevant information and outliers could be clearly eliminated. In Figure 6, the regions marked with yellow circles in the left were almost removed, leaving a clear PCD of the target cattle shape and other adherent outliers.

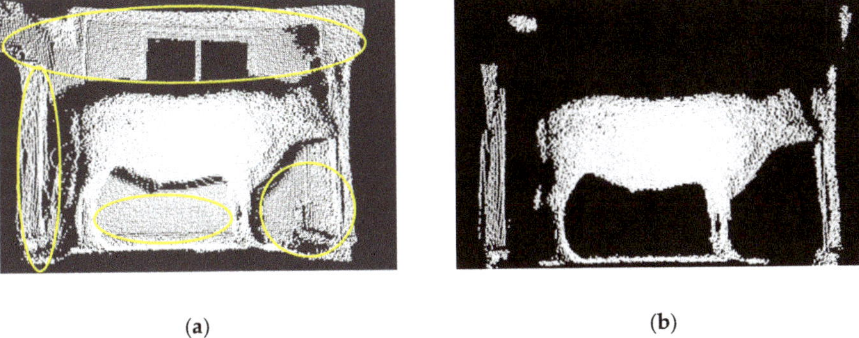

(**a**)   (**b**)

Figure 6. Filtering results with multiple filters. (**a**) Original PCD; (**b**) Results with three filters where most noises and outliers were well removed.

For some adherent outliers after preprocessing, the clustering segmentation [52] was required to obtain the PCD of a single cattle body. Based on Euclidean distance, the clustering segmentation [53] can only divide objects outside the target within a certain distance, and objects close to the target (such as the ground) can be classified into the same clusters, which cannot be separated [54]. To separate the same clusters of different objects, the RANSAC algorithm [55] is needed for plane extraction. The plane model is used to extract the ground PCD close to the cattle body and to segment a single file of cattle PCD. The segmentation results with Euclidean and RANSAC clustering are shown in Figure 7, where only a single cattle body silhouette is well preserved in the point cloud file.

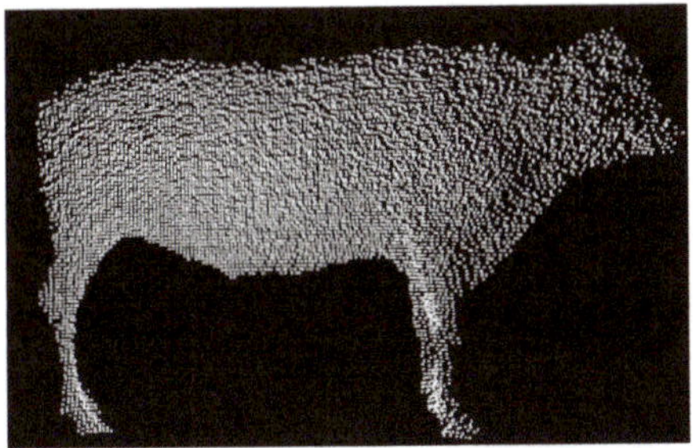

Figure 7. Segmentation with Euclidean clustering and RANSAC, where most of the background adherent to cattle has been canceled.

Since there is no 3D PCD dataset for animals, all data need to be collected manually, so the amount of data obtained is limited [56]. However, a large amount of training data is needed in the training process of neural networks to prevent overfitting. Therefore, the affine transformation [57] was used to enrich the existing PCD of the cattle body silhouette.

By rotating PCD at different angles and by mirroring PCD in horizontal and vertical directions, the PCD of the cattle body can be enriched. As shown in Equation (2), the coordinate $\begin{pmatrix} x & y & z \end{pmatrix}$ of PCD multiples the affine transformation matrix M to get a new coordinate of rotation $\begin{pmatrix} x' & y' & z' \end{pmatrix}$, where the rotation, mirroring, and other transformations can be realized.

$$\begin{pmatrix} x' & y' & z' \end{pmatrix} = M\begin{pmatrix} x & y & z \end{pmatrix} \tag{2}$$

The PCD could be rotated clockwise along the 45°, 90°, 135°, 180°, 225°, 270°, and 315° as well as mirrored through horizontal and vertical transformation. The data augmentation operation expanded the original 251 cattle by nine times to a final dataset of 2510 PCD files of cattle bodies. The transformation results of a single PCD of the cattle body silhouette are shown in Figure 8.

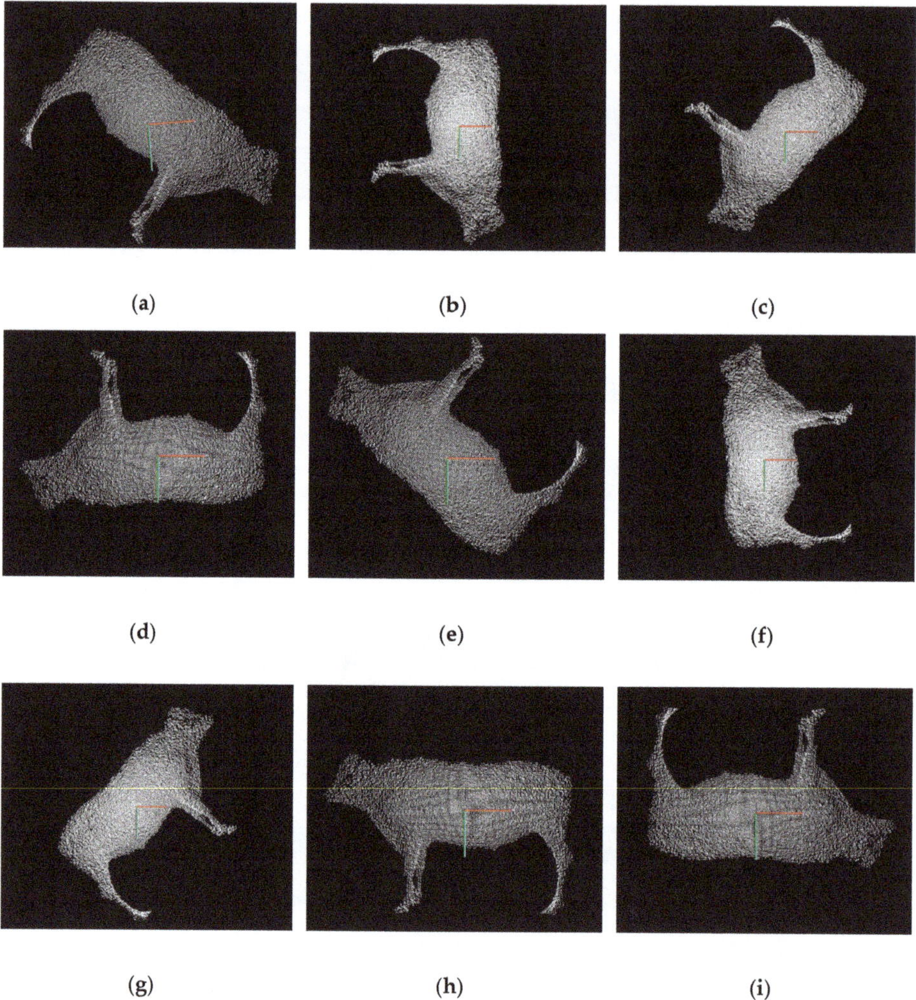

Figure 8. Transformation results of a single point cloud. (**a**) Rotation result with clockwise 45°; (**b**) rotation result with clockwise 90°; (**c**) rotation result with clockwise 135°; (**d**) rotation result with clockwise 180°; (**e**) rotation result with clockwise 225°; (**f**) rotation result with clockwise 270°; (**g**) rotation result with clockwise 315°; (**h**) horizontal mirror result; (**i**) vertical mirror result.

2.2.2. Design of Transfer Learning Network Structure

After the pre-training of the Kd-network, the initial parameters of the 3D deep model were obtained on large-scale ShapeNet network datasets. The spatial feature information of 3D PCD was extracted, and needed to be transferred to recognize the cattle body silhouette. For transfer learning here, the source data and the target data do not need to have the same data distribution [58].

In transfer learning, the PCD dataset of the cattle body was used as input to obtain partial output of the convolution layer that has been pre-trained on the Kd-network, and then use this output to train a fully connected network. Connect the convolution layer with the fully connected layer obtained from the previously pre-trained Kd-network, and then start the training of the transfer learning model. During transferring, there is training and verification in each iteration, and the loss can be back-propagated during training. Meanwhile, the parameters can be optimized, and the training results can be adjusted during verification. Each iteration, the loss value and correct value can be counted, the model with the highest correct value is saved, and then the next iteration of training is conducted. In transfer learning, the weights of all network layers, except the last fully connected layer, are frozen, and only the fully connected layer is modified so that the gradients during back propagation are not calculated, which can effectively avoid the occurrence of overfitting and improve training efficiency [59]. Finally, the original 16 outputs were changed into two outputs, that is, the original 16 classifications were changed into two classifications: cattle body point cloud and other point clouds.

In the training, 2510 enriched PCD of cattle body were applied for input, and the TrAdaboost algorithm [60] was used to transfer the samples. This algorithm can gradually improve the training weight of target samples in the examples of source datasets according to certain weight rules, reduce the weights of non-target samples, and improve the generalization ability of the model.

2.3. Recognition of Feature Points of Live Qinchuan Cattle Body

2.3.1. Normalization of Cattle Body Point Cloud

In the process of collecting cattle data, due to the influence of the position and orientation of collecting equipment, the acquired PCD of the cattle body had a different orientation. To extract uniform features, the normalization method [61] was first used to calculate a unified orientation.

The standard measurement coordinate of cattle body was defined as follows: take the center of the mass of PCD of live Qinchuan cattle silhouettes as the coordinate origin, the body length direction of cattle as the x axis, the body height direction of cattle as the y axis, and the chest width direction of cattle as the z axis. The tail direction of the cattle is the positive direction of the x axis, and the direction pointing vertically to the ground is the positive direction of the y axis. The positive direction of three axes conforms to the right-handed rectangular coordinate system.

Principal component analysis (PCA) [62] was adopted to obtain the coordinate axis of the cattle point cloud and establish a new coordinate system. The processing flow is as follows:

(1) Obtain the center point of the cattle body.

Define the input point set of PCD of the cattle body $P = \{p_i | i = 1, 2, \cdots , n\}$, where n is the number of points in the set. Then, the center point p_m of the cattle body can be calculated by Equation (3).

$$p_m = \frac{1}{n}\Sigma^n_{i=1}p_i \tag{3}$$

(2) Calculate the covariance matrix.

The covariance matrix C_p of the PCD of cattle body can be calculated by Equation (4) with the center point p_m of the cattle body.

$$C_p = \frac{1}{n}\Sigma^n_{i=1}(p_i - p_m)(p_i - p_m)^T \tag{4}$$

(3) PCA coordinate system is established according to feature vectors.

Three non-negative eigenvalues, λ_0, λ_1, and λ_2, can be obtained through the covariance matrix C_p, and Equation (5) is used to calculate the eigenvector e_0, e_1, and e_2, where λ_i is the N-*the* eigenvalue of the covariance matrix C_p and e_i is the corresponding eigenvector of λ_i.

$$C_p e_i = \lambda_i e_i, \; i \in 0, 1, 2 \tag{5}$$

The PCA coordinate system is established by taking the obtained center point p_m of the cattle body as the coordinate origin, and the direction of the obtained characteristic vectors e_0, e_1, and e_2 as the direction of the x axis, y axis, and z axis, respectively.

The PCA coordinate obtained was ambiguous for the orientation of the cattle tail point to the positive direction of the x axis or the negative direction of the x axis. Here, the tail pointed uniformly to the positive direction of the x axis. The uniform orientation of all the PCD of the cattle body needed to be corrected as the highest point of the first half part of the cattle body was higher than that of the latter half part with common sense [63]. Define the point set Q_1 over zero in the point set of cattle P on the x axis after orientation correction, and define the point set Q_2 less than zero on the x axis.

$$Q_1 = \{p_i \in P | x_i > 0\}$$
$$Q_2 = \{p_i \in P | x_i < 0\}$$

The point sets Q_1 and Q_2 were searched to find the maximum points on the y axis direction, which were denoted as q_1 and q_2, respectively. If $q_1 < q_2$, the orientation of the cattle body is in the positive direction of the x axis and the point cloud of the cattle body needs to be corrected. Otherwise, no correction of the point cloud is required. A mirror transformation on point set P is needed and its symmetric data can be obtained by Equation (6) to change the orientation of the cattle body, where $\begin{pmatrix} x & y & z \end{pmatrix}$ represents the coordinates before the mirror, and $\begin{pmatrix} x' & y' & z' \end{pmatrix}$ represents the new coordinates after the mirror. Finally, the tail of the cattle body in the corrected PCD files points to the positive direction of the x axis.

$$\begin{pmatrix} x' & y' & z' \end{pmatrix} = \begin{pmatrix} x & y & z \end{pmatrix} \begin{bmatrix} -1 & 0 & 0 \\ 0 & 1 & 0 \\ 0 & 0 & 1 \end{bmatrix} \tag{6}$$

2.3.2. Extraction of the Candidate Areas of Feature Points

To accelerate the selection efficiency of feature points in the PCD of cattle body, the candidate regions of feature points can be extracted. The curvature is the basic characteristics of the surface shape, reflecting the concave or convex degree of the PCD surface, and has a geometrical invariability of translation, rotation, scaling, and other transformation [64]. The mean curvature H [65] and Gaussian curvature K [66] were applied to extract the feature points in the candidate region of Qinchuan cattle.

It can be known from the differential geometry that there are countless normal planes at points in a surface, and normal curvature k is the curvature of the intersection line between the normal plane and the surface. The normal curvature k describes the curvature degree of the surface in a certain direction, and its maximum k_1 and minimum k_2 are the principal curvature of normal curvature. The normal curvature k in any direction can be calculated by the Euler equation, as shown in Equation (7), where θ is the angle between the osculating plane and the direction of the maximum value k_1.

$$k = k_1 \cos^2 \theta + k_2 \sin^2 \theta \tag{7}$$

The mean curvature of a surface is the mean value of the two-principal curvature, as calculated by Equation (8), which can indicate the concavity and convex of a surface. Gaussian curvature K of a surface is the product of two principal curvatures, as shown in Equation (9), whose positive and negative value can determine the properties of points on the surface. When $K > 0$, the points on the surface are elliptic points. When $K = 0$, the points on the surface are parabolic points. When $K < 0$, the points on the surface are hyperbolic points.

$$H = \frac{1}{2}(k_1 + k_2) \tag{8}$$

$$K = k_1 k_2 \tag{9}$$

The principal curvature [67] of a point on a surface is the representation of the local shape of the point, which is independent of the surface parameters and can be calculated according to the first and second basic forms of the surface. Therefore, the mean curvature and Gaussian curvature can also be obtained by Equations (10) and (11), where I_x, I_y are the first derivatives along the x axis and y axis, and I_{xy}, I_{xx}, and I_{yy} are the corresponding second derivatives.

$$H = \frac{\left(1 + I_x^2\right)I_{yy} - 2I_x I_y I_{xy} + \left(1 + I_y^2\right)I_{xx}}{2\left(1 + I_x^2 + I_y^2\right)^{\frac{3}{2}}} \tag{10}$$

$$K = \frac{I_{xx}I_{yy} - I_{xy}^2}{\left(1 + I_x^2 + I_y^2\right)^2} \tag{11}$$

Mean curvature and Gaussian curvature reflect the concavity and convexity of the surface, and nine kinds of combinations can be obtained according to the differences of positive or negative characteristics, as shown in Table 3. As the sixth kind of combination $H = 0$ and $K > 0$ is contradictory in math, it does not exist, so there are actually eight surface types that can be obtained by the combinations. The surface shapes of the corresponding combinations are shown on the right side of Table 3. Finally, through the calculation of mean curvature and Gaussian curvature, the cattle point cloud was classified according to the concavity and convexity of the feature points on the point cloud surface, and candidate areas of feature points were obtained.

Table 3. Local surface types of points [68].

Combination	Mean Curvature H	Gaussian Curvature K	Surface Type	Surface Shape
1	<0	<0	Saddle valley	
2	<0	=0	Valley	
3	<0	>0	Well	
4	=0	=0	Plane	
5	=0	>0	Does not exist	Does not exist
6	>0	<0	Saddle ridge	
7	>0	=0	Ridge	
8	>0	>0	Peak	

2.3.3. Feature Point Recognition

The measuring data of the body dimensions are the height at the withers, chest depth, back height, waist height, and body length, respectively, as shown in Figure 9. The points that need to be automatically obtained are the upper point of the height at the withers, the lower point of the cattle chest depth, the upper point of the height, the upper point of the waist height, the upper point of the body length, and the lower point of the body length, as shown in Figure 9b. As the height at the withers, chest depth, back height, or waist height is perpendicular to the ground, the lower point is the height at the withers or the lower point of the back height. The lower point of the waist height can be obtained by the y axis coordinate of the ground point as well as the x axis, the z axis coordinate of the corresponding point, and the upper point of the chest depth can be obtained through the x axis, the z

axis coordinate of the lower point of the chest depth, and the highest point in the area regarded as the y axis. Finally, the fast feature point histogram (FPFH) [69] data were used to construct the feature model database of each cattle, and to recognize the feature point of the cattle body.

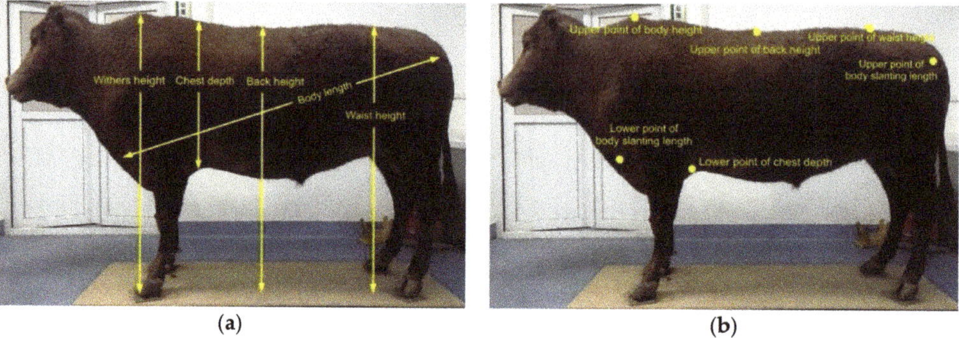

(a)   (b)

Figure 9. Schemes of adult Qinchuan cattle: (**a**) Five body dimensions; (**b**) Positions of feature points to be automatically acquired.

As the recognition process of each feature point is similar, we used the recognition of the upper point of the height at the withers as an example to describe it in detail. First, the FPFH values of the feature points in the candidate areas were calculated one by one according to the saving order, and then were matched with the values in the feature model database. If the difference between the two points is within a certain threshold, the matching is successful, and the point is judged to be the feature point of the height at the withers of the target cattle, and the search and calculation are stopped. Otherwise, match failure occurs and the same searching continues with the saving order until the match succeeds.

3. Results

After the transfer learning model was trained, it took about 25 s to process the dataset and extract the relevant parameters. The non-contact measurements of the body dimensions were performed in Visual Studio 2015 with PCL 1.8 (CPU Intel i7 3.4 G-Eight cores, Gloway DDR4 RAM of 64 GB, Windows 10 of 64-bit, Nvidia GeForce GTX950 of 2G, CUDA 9.1). The running time of all the C++ code was less than five seconds from the acquisition, filtering, and clustering segmentation to reconstruction. Figure 10 shows the result of the transfer learning and training, with the abscissa representing the training epochs and the ordinate representing the corresponding accuracy rate. After training 1000 epochs, the accuracy rate of each training was calculated including all of the training epochs. The average accuracy rate of the deep model was 93.6%, which was greatly improved when compared with the previous average correct rate of deep learning.

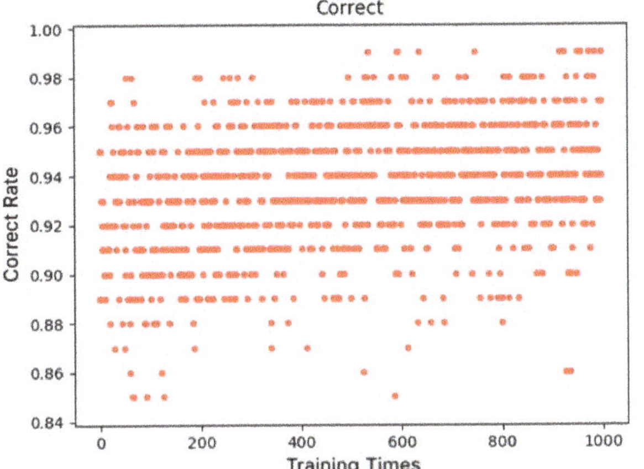

Figure 10. Correct rates of transformation training.

To validate the deep model for measuring the body sizes of live Qinchuan cattle, the 3D PCD of three live Qinchuan cattle were collected at the National Beef Cattle Improvement Center, Ministry of Agriculture and Rural Affairs, Shaanxi Province. The ear tags of the three cattle were Q0392, Q0526, and Q0456, respectively. In the experimentation, six feature points of the body dimensions were extracted from the upper point of the height at the withers, the lower point of the chest depth, the upper point of the back height, the upper point of the waist height, the upper point of the body length, and the lower point of the body length. The visualization results of body dimensions feature points recognition of the three cattle are shown in Figure 11. Figure 12 shows the results calculated by automatically acquiring the feature points of the reconstructed cattle silhouette.

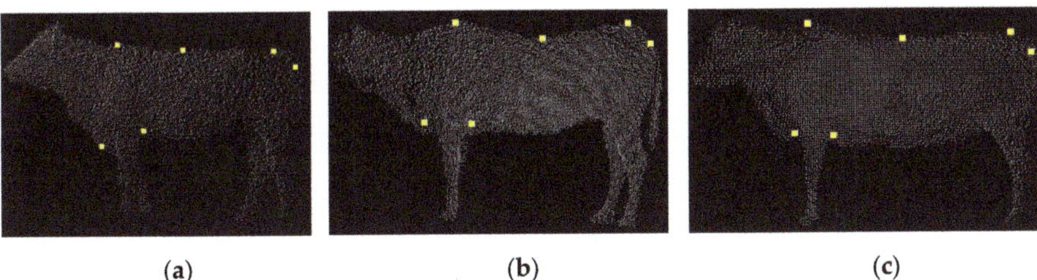

(a)   (b)   (c)

Figure 11. Recognition results of the feature points of three adult Qinchuan cattle: (**a**) Cattle Q0392; (**b**) Cattle Q0526; (**c**) Cattle Q0456.

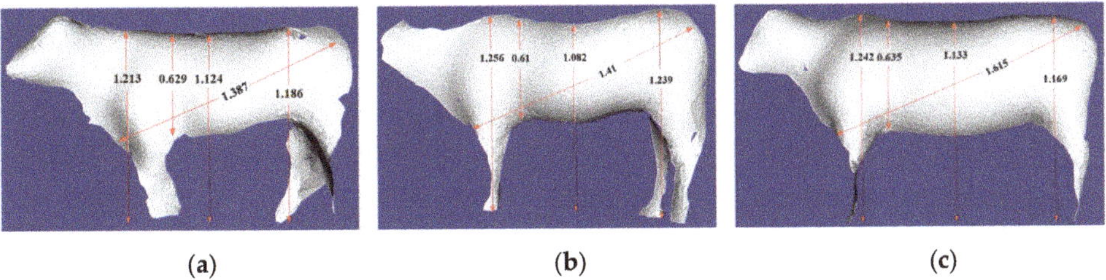

(a)   (b)   (c)

Figure 12. Automatic measurement results of the body dimensions of three live Qinchuan cattle: (**a**) Cattle Q0392; (**b**) Cattle Q0526; (**c**) Cattle Q0456.

4. Discussion

The body dimension data of cattle Q0392, Q0526, and Q0456 extracted by automatic recognition and manual interactive extraction and their errors are shown in Table 4. The maximum error value of cattle Q0392 was body length, which was 2.36%, and caused by the large error during the recognition of feature points of the body length. For cattle Q0526, the chest depth data error value was the largest, which was 1.45%, while the height at the withers error value was the smallest, only 0.08%. For cattle Q0456, the error values of the height at the withers and chest depth was large, while the error of the back height was the smallest, only 0.09%.

Table 4. The data and error values of the body dimensions of the three cattle (unit: m).

Ear Tag of Cattle	Data Extraction Method and Error	Withers Height	Chest Depth	Back Height	Waist Height	Body Length
Q0392	Automatic recognition	1.213	0.629	1.124	1.186	1.387
	Human-machine interaction	1.211	0.630	1.110	1.175	1.355
	Error value	0.17%	0.16%	1.26%	0.94%	2.36%
Q0526	Automatic recognition	1.256	0.610	1.082	1.239	1.410
	Human-machine interaction	1.255	0.619	1.095	1.237	1.414
	Error value	0.08%	1.45%	1.19%	0.16%	0.28%
Q0456	Automatic recognition	1.242	0.635	1.133	1.169	1.615
	Human-machine interaction	1.238	0.637	1.134	1.166	1.612
	Error value	0.32%	0.31%	0.09%	0.26%	0.19%

From Table 4, the overall error of the body dimensions with the proposed deep learning model was within 0.03 m and the average error was about 0.02 m. In the calculated data of the cattle body dimensions, the error rate was basically within 1.5%. A bigger error in the body length of cattle Q0526 occurred through a bigger error of the upper point of the body length. The methodology can effectively achieve the automatic acquisition of feature point data of cattle body dimensions, avoid any manual intervention, simplify the measurement process, and improve the measurement efficiency.

5. Conclusions

This paper proposed an automatic acquisition of body dimensions with deep learning for live Qinchuan cattle. The PyTorch framework of deep learning of 3D point clouds based on a Kd-network was pre-trained by a 3D PCD dataset of ShapeNet. The training epoch and learning rate of the Kd-network deep model were analyzed with an average classification accuracy of 89.6%. After the 3D PCD acquisition of cattle by the LiDAR sensor, preprocessing of the classical filters fusion, and PCD processing of classical segmentation, the PCD of the cattle silhouette was extracted, and after data augmentation, the training dataset of the transfer learning based on the TrAdaBoost algorithm was created. The created dataset, as the target data, was applied to train the transfer learning model with a final average accuracy of 93.6%. After the normalization and orientation correction of the cattle body, the candidate feature regions were extracted with the mean curvature and Gaussian curvature. The FPFH of the candidate regions were extracted and matched. Finally, five linear body dimensions of cattle were calculated with the error rate within 2.36% and measuring errors within 0.03 m, and thus basically meets the measuring demands of real production.

The Qinchuan cattle is one of the five main beef cattle breeds in China, and the most widely used in northwest China. This method is only applicable to short-haired cattle breeds such as Qinchuan cattle and there have been proven large errors or even some big mistakes in our experiments with adult Qinghai yaks. Although some achievements have been made in theory, method and application, there are still many challenges to be resolved. In the near future, we will experiment with other beef breeds or crossbred cattle to validate these methods, which will improve the classification accuracy of the deep model by adjusting the network layers and designing convolution filters of different dimensions.

Author Contributions: L.H. designed the whole research, edited the English language and rewrote this paper. S.L. contributed the whole research. H.G. and Q.R. designed and implemented the measuring system with C++/C# and wrote this manuscript. X.F. designed the fitting model and correction experimentation of the system and wrote the detection of this manuscript. Z.H. and S.Q. assisted gathering all the experimental data during the experiments and manuscript. H.W. contributed the effective experimental solution.

Funding: This research was funded by the Key Research and Development Project in Ningxia Hui Nationality Autonomous Region, grant number 2017BY067.

Acknowledgments: All of the authors thank the research group of Prof. Linsen Zan (Chief Scientist of the National Beef Cattle Improvement Center, Ministry of Agriculture and Rural Affairs, China) for supporting the manual guidance and holding of Qinchuan cattle.

Conflicts of Interest: The authors declare no conflicts of interest.

References

1. Salau, J.; Haas, J.H.; Junge, W.; Bauer, U.; Harms, J.; Bieletzki, S. Feasibility of automated body trait determination using the SR4K time-of-flight camera in cow barns. *SpringerPlus* **2014**, *3*, 225. [CrossRef]

2. Pezzuolo, A.; Guarino, M.; Sartori, L.; Marinello, F. A Feasibility study on the use of a structured light depth-camera for three-dimensional body measurements of dairy cows in free-stall barns. *Sensors* **2018**, *18*, 673. [CrossRef] [PubMed]

3. Guo, H.; Ma, X.; Ma, Q.; Wang, K.; Su, W.; Zhu, D. LSSA_CAU: An interactive 3d point clouds analysis software for body measurement of livestock with similar forms of cows or pigs. *Comput. Electron. Agric.* **2017**, *138*, 60–68. [CrossRef]

4. Pezzuolo, A.; Guarino, M.; Sartori, L.; González, L.A.; Marinello, F. On-barn pig weight estimation based on body measurements by a Kinect v1 depth camera. *Comput. Electron. Agric.* **2018**, *148*, 29–36. [CrossRef]

5. Enevoldsen, C.; Kristensen, T. Estimation of body weight from body size measurements and body condition scores in dairy cows. *J. Dairy Sci.* **1997**, *80*, 1988–1995. [CrossRef]

6. Brandl, N.; Jorgensen, E. Determination of live weight of pigs from dimensions measured using image analysis. *Comput. Electron. Agric.* **1996**, *15*, 57–72. [CrossRef]

7. Wilson, L.L.; Egan, C.L.; Terosky, T.L. Body measurements and body weights of special-fed Holstein veal calves. *J. Dairy Sci.* **1997**, *80*, 3077–3082. [CrossRef]

8. Communod, R.; Guida, S.; Vigo, D.; Beretti, V.; Munari, E.; Colombani, C.; Superchi, P.; Sabbioni, A. Body measures and milk production, milk fat globules granulometry and milk fatty acid content in Cabannina cattle breed. *Ital. J. Anim. Sci.* **2013**, *12*, 107–115. [CrossRef]

9. Huang, L.; Li, S.; Zhu, A.; Fan, X.; Zhang, C.; Wang, H. Non-contact body measurement for qinchuan cattle with LiDAR sensor. *Sensors* **2018**, *18*, 3014. [CrossRef]

10. McPhee, M.J.; Walmsley, B.J.; Skinner, B.; Littler, B.; Siddell, J.P.; Cafe, L.M.; Wilkins, J.F.; Oddy, V.H.; Alempijevic, A. Live animal assessments of rump fat and muscle score in Angus cows and steers using 3-dimensional imaging. *J. Anim. Sci.* **2017**, *95*, 1847–1857. [CrossRef]

11. Rizaldy, A.; Persello, C.; Gevaert, C.; Elberink, S.O.; Vosselman, G. Ground and Multi-Class Classification of Airborne Laser Scanner Point Clouds Using Fully Convolutional Networks. *Remote Sens.* **2018**, *10*, 1723. [CrossRef]

12. He, X.; Wang, A.; Ghamisi, P.; Li, G.; Chen, Y. LiDAR Data Classification Using Spatial Transformation and CNN. *IEEE Geosci. Remote Sens. Lett.* **2018**, *16*, 125–129. [CrossRef]

13. Maltezos, E.; Doulamis, A.; Doulamis, N.; Ioannidis, C. Building Extraction From LiDAR Data Applying Deep Convolutional Neural Networks. *IEEE Geosci. Remote Sens. Lett.* **2018**, *16*, 155–159. [CrossRef]

14. Edson, C.; Wing, M.G. Airborne Light Detection and Ranging (LiDAR) for Individual Tree Stem Location, Height, and Biomass Measurements. *Remote Sens.* **2011**, *3*, 2494–2528. [CrossRef]

15. Maki, N.; Nakamura, S.; Takano, S.; Okada, Y. 3D Model Generation of Cattle Using Multiple Depth-Maps for ICT Agriculture. In Proceedings of the Conference on Complex, Intelligent, and Software Intensive Systems, Matsue, Japan, 4–6 July 2018.

16. Kawasue, K.; Win, K.D.; Yoshida, K.; Tokunaga, T. Black cattle body shape and temperature measurement using thermography and KINECT sensor. *Artif. Life Robot.* **2017**, *22*, 1–7. [CrossRef]

17. Fernandes, A.F.A.; Dorea, J.R.R.; Fitzgerald, R.; Herring, W.; Rosa, G.J.M. A novel automated system to acquire biometric and morphological measurements and predict body weight of pigs via 3D computer vision. *J. Anim. Sci.* **2019**, *97*, 496–508. [CrossRef]

18. Menesatti, P.; Costa, C.; Antonucci, F.; Steri, R.; Pallottino, F.; Catillo, G. A low-cost stereovision system to estimate size and weight of live sheep. *Comput. Electron. Agric.* **2014**, *103*, 33–38. [CrossRef]

19. Wang, K.; Guo, H.; Ma, Q.; Su, W.; Chen, L.; Zhu, D. A portable and automatic Xtion-based measurement system for pig body size. *Comput. Electron. Agric.* **2018**, *148*, 291–298. [CrossRef]

20. Jun, K.; Kim, S.J.; Ji, H.W. Estimating pig weights from images without constraint on posture and illumination. *Comput. Electron. Agric.* **2018**, *153*, 169–176. [CrossRef]

21. Azzaro, G.; Caccamo, M.; Ferguson, J.D.; Battiato, S.; Farinella, G.M.; Guarnera, G.C.; Puglisi, G.; Petriglieri, R.; Licitra, G. Objective estimation of body condition score by modeling cow body shape from digital images. *J. Dairy Sci.* **2011**, *94*, 2126–2137. [CrossRef]

22. Zhou, Z.; Wang, Y.; Wu, Q.M.J.; Yang, C.; Sun, X. Effective and Efficient Global Context Verification for Image Copy Detection. *IEEE Trans. Inf. Forensics Secur.* **2017**, *12*, 48–63. [CrossRef]

23. Omid-Zohoor, A.; Young, C.; Ta, D.; Murmann, B. Toward Always-On Mobile Object Detection: Energy Versus Performance Tradeoffs for Embedded HOG Feature Extraction. *IEEE Trans. Circuits Syst. Video Technol.* **2018**, *28*, 1102–1115. [CrossRef]

24. Zhou, L.; Li, Q.; Huo, G.; Zhou, Y. Image Classification Using Biomimetic Pattern Recognition with Convolutional Neural Networks Features. *Comput. Intell. Neurosci.* **2017**, *2017*, 3792805. [CrossRef] [PubMed]

25. Litjens, G.; Kooi, T.; Bejnordi, B.E.; Setio, A.A.A.; Ciompi, F.; Ghafoorian, M.; van der Laak, J.A.W.M.; van Ginneken, B.; Sanchez, C.I. A survey on deep learning in medical image analysis. *Med. Image Anal.* **2017**, *42*, 60–88. [CrossRef] [PubMed]

26. Wu, Z.; Song, S.; Khosla, A.; Yu, F.; Zhang, L.; Tang, X.; Xiao, J.; Wu, Z.; Song, S.; Khosla, A. 3D ShapeNets: A deep representation for volumetric shapes. In Proceedings of the IEEE Conference on Computer Vision & Pattern Recognition, Boston, MA, USA, 7–12 June 2015.

27. Guan, H.; Yu, Y.; Ji, Z.; Li, J.; Zhang, Q. Deep learning-based tree classification using mobile LiDAR data. *Remote Sens. Lett.* **2015**, *6*, 864–873. [CrossRef]

28. Nahhas, F.H.; Shafri, H.Z.M.; Sameen, M.I.; Pradhan, B.; Mansor, S. Deep Learning Approach for Building Detection Using LiDAR-Orthophoto Fusion. *J. Sens.* **2018**, *7*. [CrossRef]

29. Jin, S.; Su, Y.; Gao, S.; Wu, F.; Hu, T.; Liu, J.; Li, W.; Wang, D.; Chen, S.; Jiang, Y.; et al. Deep Learning: Individual Maize Segmentation From Terrestrial Lidar Data Using Faster R-CNN and Regional Growth Algorithms. *Front. Plant Sci.* **2018**, *22*, 866. [CrossRef]

30. Charles, R.Q.; Hao, S.; Mo, K.; Guibas, L.J. PointNet: Deep Learning on Point Sets for 3D Classification and Segmentation. In Proceedings of the IEEE Conference on Computer Vision & Pattern Recognition, Honolulu, HI, USA, 21–26 July 2017.

31. Qi, C.R.; Li, Y.; Hao, S.; Guibas, L.J. PointNet++: Deep Hierarchical Feature Learning on Point Sets in a Metric Space. In Proceedings of the Neural Information Processing Systems (NIPS), Long Beach, CA, USA, 4–9 December 2017.

32. Klokov, R.; Lempitsky, V. Escape from Cells: Deep Kd-Networks for the Recognition of 3D Point Cloud Models. In Proceedings of the IEEE International Conference on Computer Vision, Venice, Italy, 22–29 October 2017.

33. Zeiler, M.D.; Fergus, R. Visualizing and Understanding Convolutional Networks. In Proceedings of the European Conference on Computer Vision, Sydney, Australia, 1–8 December 2013.

34. Zeng, W.; Gevers, T. 3D ContextNet: K-d Tree Guided Hierarchical Learning of Point Clouds Using Local and Global Contextual Cues. In Proceedings of the European Computer Vision, Munich, Germany, 8–14 September 2018.

35. Marinello, F.; Pezzuolo, A.; Cillis, D.; Gasparini, F.; Sartori, L. Application of Kinect-Sensor for three-dimensional body measurements of cows. In Proceedings of the 7th European Precision Livestock Farming, ECPLF 2015. European Conference on Precision Livestock Farming, Milan, Italy, 15–18 September 2015; pp. 661–669.

36. Pan, S.J.; Yang, Q. A Survey on Transfer Learning. *IEEE Trans. Knowl. Data Eng.* **2010**, *22*, 1345–1359. [CrossRef]

37. Tan, C.; Sun, F.; Tao, K.; Zhang, W.; Chao, Y.; Liu, C. A Survey on Deep Transfer Learning. In Proceedings of the 27th International Conference on Artificial Neural Networks, Rhodes, Greece, 5–7 October 2018.

38. Andujar, D.; Rueda-Ayala, V.; Moreno, H.; Rosell-Polo, J.R.; Escola, A.; Valero, C.; Gerhards, R.; Fernandez-Quintanilla, C.; Dorado, J.; Griepentrog, H. Discriminating crop, weeds and soil surface with a terrestrial LIDAR sensor. *Sensors* **2013**, *13*, 14662–14675. [CrossRef]

39. Wang, Z.; Zhang, L.; Zhang, L.; Li, R.; Zheng, Y.; Zhu, Z. A Deep Neural Network With Spatial Pooling (DNNSP) for 3-D Point Cloud Classification. *IEEE Trans. Geosci. Remote Sens.* **2018**, *56*, 4594–4604. [CrossRef]

40. Silpa-Anan, C.; Hartley, R. Optimised KD-trees for fast image descriptor matching. In *2018 IEEE Conference on Computer Vision and Pattern Recognition*; IEEE Computer Society: Anchorage, AK, USA, 2008; pp. 1–8.

41. Duarte, D.; Nex, F.; Kerle, N.; Vosselman, G. Multi-resolution feature fusion for image classification of building damages with convolutional neural networks. *Remote Sens.* **2018**, *10*, 1636. [CrossRef]

42. Scardapane, S.; Van Vaerenbergh, S.; Totaro, S.; Uncini, A. Kafnets: Kernel-based non-parametric activation functions for neural networks. *Neural Netw.* **2019**, *110*, 19–32. [CrossRef] [PubMed]

43. Eckle, K.; Schmidt-Hieber, J. A comparison of deep networks with ReLU activation function and linear spline-type methods. *Neural Netw.* **2019**, *110*, 232–242. [CrossRef] [PubMed]

44. ShapeNet Datasource. Available online: https://shapenet.cs.stanford.edu/ericyi/shapenetcore_partanno_segmentation_benchmark_v0.zip (accessed on 10 October 2019).

45. Iyer, M.S.; Rhinehart, R.R. A method to determine the required number of neural-network training repetitions. *IEEE Trans. Neural Netw.* **1999**, *10*, 427–432. [CrossRef] [PubMed]

46. Takase, T.; Oyama, S.; Kurihara, M. Effective neural network training with adaptive learning rate based on training loss. *Neural Netw.* **2018**, *101*, 68–78. [CrossRef] [PubMed]

47. Foix, S.; Alenya, G.; Torras, C. Lock-in Time-of-Flight (ToF) Cameras: A Survey. *IEEE Sens. J.* **2011**, *11*, 1917–1926. [CrossRef]

48. Zeybek, M.; Sanlioglu, I. Point cloud filtering on UAV based point cloud. *Measurement* **2019**, *133*, 99–111. [CrossRef]

49. Kushner, H.J.; Budhiraja, A.S. A nonlinear filtering algorithm based on an approximation of the conditional distribution. *IEEE T. Automat. Contr.* **2000**, *45*, 580–585. [CrossRef]

50. Pourmohamad, T.; Lee, H.K.H. The Statistical Filter Approach to Constrained Optimization. *Technometrics* **2019**, 1–10. [CrossRef]

51. Liu, L.; Lim, S. A voxel-based multiscale morphological airborne lidar filtering algorithm for digital elevation models for forest regions. *Measurement* **2018**, *123*, 135–144. [CrossRef]

52. Li, Y.; Li, L.; Li, D.; Yang, F.; Liu, Y. A Density-Based Clustering Method for Urban Scene Mobile Laser Scanning Data Segmentation. *Remote Sens.* **2017**, *9*, 331. [CrossRef]

53. Flores-Sintas, A.; Cadenas, J.M.; Martin, F. Detecting homogeneous groups in clustering using the Euclidean distance. *Fuzzy Set. Syst.* **2001**, *120*, 213–225. [CrossRef]

54. Shaikh, S.A.; Kitagawa, H. Efficient distance-based outlier detection on uncertain datasets of Gaussian distribution. *World Wide Web* **2014**, *17*, 511–538. [CrossRef]

55. Schnabel, R.; Wahl, R.; Klein, R. Efficient RANSAC for point-cloud shape detection. *Comput. Graph. Forum.* **2007**, *26*, 214–226. [CrossRef]

56. Silva, C.; Welfer, D.; Gioda, F.P.; Dornelles, C. Cattle Brand Recognition using Convolutional Neural Network and Support Vector Machines. *IEEE Lat. Am. Trans.* **2017**, *15*, 310–316. [CrossRef]

57. Konovalenko, I.A.; Kokhan, V.V.; Nikolaev, D.P. Optimal affine approximation of image projective transformation. *Sens. Sist.* **2019**, *33*, 7–14.

58. Lu, H.; Fu, X.; Liu, C.; Li, L.; He, Y.; Li, N. Cultivated land information extraction in UAV imagery based on deep convolutional neural network and transfer learning. *J. Mt. Sci.* **2017**, *14*, 731–741. [CrossRef]

59. Wang, L.; Geng, X.; Ma, X.; Zhang, D.; Yang, Q. Ridesharing car detection by transfer learning. *Artif. Intell.* **2019**, *273*, 1–18. [CrossRef]

60. Zhang, Q.; Li, H.; Zhang, Y.; Li, M. Instance Transfer Learning with Multisource Dynamic TrAdaBoost. *Sci. World J.* **2014**. [CrossRef]

61. Guo, H.; Li, Z.; Ma, Q.; Zhu, D.; Su, W.; Wang, K.; Marinello, F. A bilateral symmetry based pose normalization framework applied to livestock body measurement in point clouds. *Comput. Electron. Agric.* **2019**, *160*, 59–70. [CrossRef]

62. Sun, Y.; Li, L.; Zheng, L.; Hu, J.; Li, W.; Jiang, Y.; Yan, C. Image Classification base on PCA of Multi-view Deep Representation. *arXiv* **2019**, arXiv:1903.04814.

63. Kamprasert, N.; Duijvesteijn, N.; Van der Werf, J.H.J. Estimation of genetic parameters for BW and body measurements in Brahman cattle. *Animal* **2019**, *13*, 1576–1582. [CrossRef] [PubMed]

64. Li, J.; Fan, H. Curvature-direction measures for 3D feature detection. *Sci. China Inform. Sci.* **2013**, *9*, 52–60. [CrossRef]

65. Gong, Y. Mean Curvature Is a Good Regularization for Image Processing. *IEEE Trans. Circuits Syst. Video Technol.* **2019**, *29*, 2205–2214. [CrossRef]

66. Meek, D.S.; Walton, D.J. On surface normal and Gaussian curvature approximations given data sampled from a smooth surface. *Comput. Aided Geom. Des.* **2000**, *17*, 521–543. [CrossRef]

67. Tang, Y.; Li, H.; Sun, X.; Morvan, J.; Chen, L. Principal Curvature Measures Estimation and Application to 3D Face Recognition. *J. Math. Imaging Vis.* **2017**, *59*, 211–233. [CrossRef]

68. Gruen, A.; Akca, D. Least squares 3D surface and curve matching. *ISPRS J. Photogramm.* **2005**, *59*, 151–174. [CrossRef]

69. Rusu, R.B.; Blodow, N.; Beetz, M. Fast Point Feature Histograms (FPFH) for 3D Registration. In Proceedings of the IEEE International Conference on Robotics and Automation-ICRA, Kobe, Japan, 12–17 May 2009; pp. 1848–1853.

Article

An Automatic Head Surface Temperature Extraction Method for Top-View Thermal Image with Individual Broiler

Xingguo Xiong [1], Mingzhou Lu [1,*], Weizhong Yang [1], Guanghui Duan [1], Qingyan Yuan [2], Mingxia Shen [1], Tomas Norton [3] and Daniel Berckmans [3,4]

[1]   College of Engineering/Jiangsu Province Engineering Lab for Modern Facility Agriculture Technology & Equipment, Nanjing Agricultural University, Nanjing 210031, China; 32116230@njau.edu.cn (X.X.); ywz@njau.edu.cn (W.Y.); 2018812092@njau.edu.cn (G.D.); mingxia@njau.edu.cn (M.S.)

[2]   Jiangsu Lihua Animal Husbandry CO., LTD., Changzhou 213168, China; yuanqy@lihuamuye.com

[3]   M3-BIORES- Measure, Model & Manage Bioresponses, KU Leuven, Kasteelpark Arenberg 30, B-3001 Leuven, Belgium; tomas.norton@kuleuven.be (T.N.); daniel.berckmans@kuleuven.be (D.B.)

[4]   BioRICS nv, Technologielaan 3, 3001 Leuven, Vlaams-Brabant, Belgium

*   Correspondence: lmz@njau.edu.cn; Tel.: +86-138-1384-1336

Received: 20 September 2019; Accepted: 27 November 2019; Published: 30 November 2019

Abstract: Surface temperature variation in a broiler's head can be used as an indicator of its health status. Surface temperatures in the existing thermograph based animal health assessment studies were mostly obtained manually. 2185 thermal images, each of which had an individual broiler, were captured from 20 broilers. Where 15 broilers served as the experimental group, they were injected with 0.1mL of pasteurella inoculum. The rest, 5 broilers, served as the control group. An algorithm was developed to extract head surface temperature automatically from the top-view broiler thermal image. Adaptive K-means clustering and ellipse fitting were applied to locate the broiler's head region. The maximum temperature inside the head region was extracted as the head surface temperature. The developed algorithm was tested in Matlab® (R2016a) and the testing results indicated that the head region in 92.77% of the broiler thermal images could be located correctly. The maximum error of the extracted head surface temperatures was not greater than 0.1 °C. Different trend features were observed in the smoothed head surface temperature time series of the broilers in experimental and control groups. Head surface temperature extracted by the presented algorithm lays a foundation for the development of an automatic system for febrile broiler identification.

Keywords: broiler surface temperature extraction; thermal image processing; head region locating; adaptive K-means; ellipse fitting

---

1. Introduction

Body temperature is one of the most important indicators of a broiler's health status [1]. When a broiler is infected with bacteria or a virus, the autoimmune system will take effect, resulting in an obvious rise of core body temperature [2]. Therefore, sudden changes in core body temperature can be utilized to identify a sick broiler. Traditional body temperature measurement is achieved by inserting a mercury thermometer into a broiler's rectum [3]. It is laborious and time consuming. At the same time, it could increase the probability of disease spreading between farmers and broilers.

With the development of sensor technology, animal body temperature monitoring based on implantable temperature sensors has been evaluated in the rumen [4,5], vagina [6], and subcutis [7]. However, it is difficult to apply implantable sensors to collect body temperature for animals of small size, such as broilers. A wearable temperature sensor, positioned under a broiler's wing, was utilized by Li, et al. [8] in 2013 to monitor a broiler's under-wing temperature. However, it has the following demerits. Firstly, frequent replacement of the battery is required. Secondly, it is difficult to keep the sensor in a fixed position under the wing. Position shift of the sensor will cause incorrect under-wing temperature. Worse, falling off of the sensor could lead to temperature data loss.

In recent years, thermal imaging technology has been successfully applied to obtain superficial temperature for cows [9], cattle [10], pigs [11], and so on. Based on the obtained superficial temperature, animal health was evaluated [12,13]. In the field of poultry breeding, thermal imaging has also been employed to obtain the surface temperature for broilers. For example, Giloh, Shinder, and Yahav [1] extracted broiler surface temperature from a thermal image. They found that broiler surface temperature is correlated to its body temperature. Liu, et al. [14] established a regression model between surface and core body temperatures for an individual broiler. However, broiler surface temperature in the existing research was extracted manually by using the specific software provided by the thermal imager manufacturer [15]. This is time and labor consuming, especially when the number of images is large and the ROI (region of interesting) areas are small.

The broiler's head has few feathers, it is much easier to obtain the skin temperature in head than other parts of an individual broiler by using a top-view thermal camera. Inspired by the study done by Giloh, et al. [1] and Liu, et al. [14], an automatic head surface temperature extraction algorithm for top-view thermal image with individual broiler was developed in this study. At the same time, a method for head temperature time series smoothing was proposed, based on which, febrile broiler can be identified. This research lays a foundation for the development of an automatic system for febrile broiler identification.

2. Materials and Methods

2.1. Image System

The image system consisted of a portable scaffold and an infrared thermal camera (Fluke TI32, Avery, WA, USA), as shown in Figure 1.

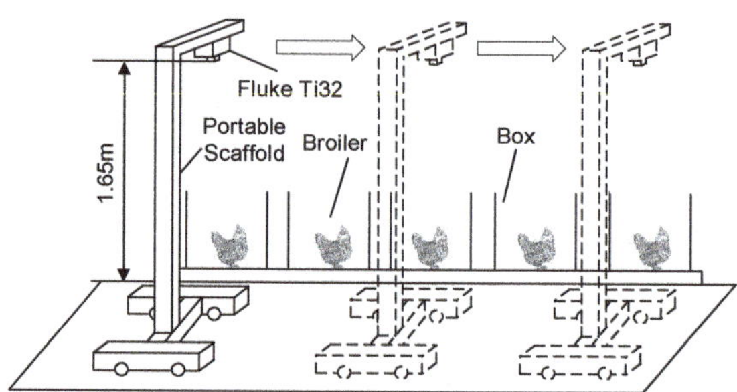

Figure 1. Image system.

The image resolution, thermal sensitivity, field of view (FoV) of the thermal camera was 320×240, ≤ 0.045 °C at 30 °C, and $23° \times 17°$, respectively. Five paper boxes were placed on the ground, each one with a broiler inside and with no cover. The camera was fixed on the portable scaffold at a height of 1.65 m above the ground. FoV of the camera at this height allows for coverage of the whole boiler in each paper box. According to the survey conducted by McManus, et al. [16] in 2016, emissivity value normally ranged from 0.98 to 0.93 for animal temperature monitoring. Therefore, the emission rate in this study was set to 0.95. The portable scaffold was pushed manually and slid from one box to another. Once the portable scaffold stopped beside a box, thermal imager was triggered manually to capture thermal images for the broiler inside the box. The thermal images were stored in the SD card of the thermal imager in IS2 format. Each IS2 file contained the radiometric data, infrared image,

IR-Fusion mode information, and so on, of the object in the FoV of the imager. At the end of thermal image acquisition, all the obtained IS2 files were transferred to a PC for further processing.

2.2. Broilers and Thermal Images Acquisition

20 QingJiaoMa broilers, with the age of around 45 days (45.3 days ± 1.23 days) and a body weight of around 1.5 kg (1.54 kg ± 0.12 kg) were randomly selected from Lihua Animal Husbandry Co. LTD, city of Changzhou, Jiangsu Province, China. All the selected broilers were confirmed to be healthy by a veterinarian. Thermal images of the broilers were acquired in an experimental animal house of Lihua Animal Husbandry Co. LTD from 19 September 2018 to 24 September 2018. Ambient temperature inside the experimental animal house was set to 27–30 °C by using an air conditioner. The environmental relative humidity was 55–65%. Three fluorescent lamps equipped on the ceiling were used to provide stable illumination conditions inside the house. There was no thermal heat from solar or other light sources during the whole course of the thermal images acquisition.

15 broilers were randomly chosen and served as the experimental group. Thermal images of the broilers in this group were captured in 3 batches (5 broilers in each batch). The remaining 5 broilers served as the control group in this study and were photographed in one batch. Thermal image acquisition began at 8:30 a.m. and ended at 11:00 p.m. each day. For each broiler in the experimental group, 0.1 mL of the pasteurella inoculum was injected into its breast at 12:00 a.m. For broilers in the control group, no treatment was given. A wireless wearable sensor based under-wing temperature monitoring system (CH-T2, Chero Technology Co. ltd, Zhejiang, China) was utilized to collect the under-wing temperature for each broiler. The temperature sensor of this system was positioned under each broiler's wing the night before the next day's thermal images acquisition. After the sensor installation was completed, each broiler was placed in one of the 5 boxes overnight to help it adapt to the experimental environment and the sensor under its wing. The under-wing temperature of each broiler was collected by the temperature sensor and sent to the application installed in a smart phone at a fixed time interval.

The inherent error between the actual and the monitoring temperature obtained by the thermal camera used in this study was ±2 °C. Averaging operation was employed to reduce the impact brought on by the inherent error. Therefore, five thermal images were captured for each broiler every 30 min. Each broiler of the control and experimental groups was photographed for 30 cycles and 15–21 cycles, respectively, where the quantity of the thermal image acquisition cycles of the broilers in experimental group depended on its dead time. At the end of thermal image acquisition, an image database with 2185 thermal images was procured.

2.3. Head Surface Temperature Extraction

An algorithm, named *HSTE* (Head Surface Temperature Extraction), was developed to extract head surface temperature from a top-view thermal image with individual broilers. *HSTE* consisted of 3 steps: thermal image pre-processing, head region locating, and representative head surface temperature extraction.

2.3.1. Thermal Image Pre-Processing

Each thermal image obtained in Section 2.2 was converted to a 240 × 320 matrix. The matrix contained the temperature value at each pixel. Taking a thermal image selected randomly from the image database as an example, its temperature matrix and thermal image are shown in Figure 2a,b, respectively.

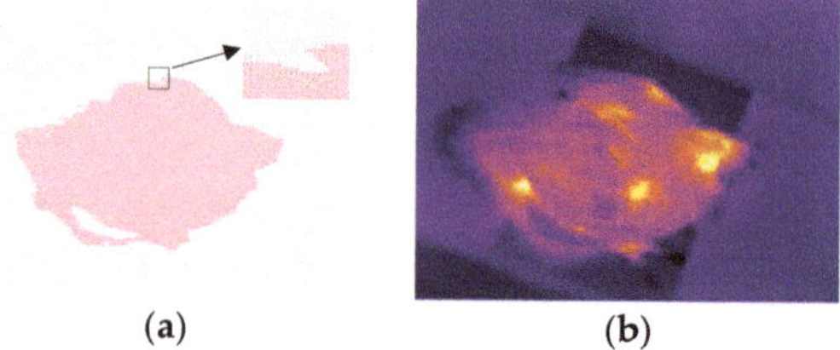

(a)   (b)

Figure 2. Temperature matrix and thermal image of an IS2 file: (**a**) Temperature matrix; (**b**) The corresponding thermal image.

Temperature matrix shown in Figure 2a was transformed into a grayscale image, as shown in Figure 3a, by using Formula (1):

$$Img\_gray[r,c] = \frac{Img\_m[r,c] - min(Img\_m)}{max(Img\_m) - min(Img\_m)} \; r \in [1,240], \; c \in [1,320]. \tag{1}$$

where $Img\_gray[r,c]$ was the intensity value of the pixel at the coordinate (r,c) in the grayscale image. $Img\_m[r,c]$ was the value at row r and column c of the temperature matrix, $max(Img\_m)$ and $min(Img\_m)$ were the maximum and minimum value of the temperature matrix, respectively.

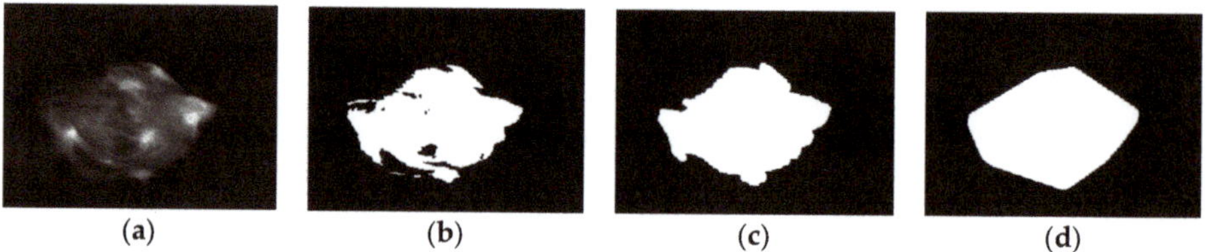

(a)   (b)   (c)   (d)

Figure 3. Image pre-processing: (**a**) Grayscale image; (**b**) Binary image; (**c**) Image after morphological processing; (**d**) Convex hull image.

The obtained grayscale image was converted to a binary one by using Otsu's method [17], as shown in Figure 3b. Then, morphological processing [18] was carried out to remove holes and spindly parts in the binary image. The morphological processing result of Figure 3b is shown in Figure 3c. Finally, the convex hull image was obtained by using the algorithm proposed by Barber, et al. [19] in 1996, as shown in Figure 3d.

2.3.2. Head Region Locating

The contour of a broiler can be approximately considered an ellipse, where head part is definitely located nearby one of the two endpoints of the major axis of the ellipse. At the same time, there are few feathers in a broiler's head, leading to a relatively high temperature region in the head part of a thermal image. Therefore, head region was located in a thermal image based on the features of surface temperature distribution and a rough elliptical body shape according to the following 3 sub-steps.

Sub-step 1; coordinates of the major axis endpoints extraction for the fitted ellipse. An ellipse was fitted for the convex hull image contour of an individual broiler by using direct least square method [20,21]. The fitted ellipse of the contour in Figure 4a is shown in Figure 4b, where the contour in Figure 4a was extracted from the convex hull image in Figure 3d by using the method proposed by Canny [22] in 1986. Once the fitted ellipse was obtained, the following parameters of the ellipse were extracted. The coordinate of the center point, which was denoted by $\left(x_0', y_0'\right)$ is shown in Figure 4b with a diamond point. Length of the major and minor axis were denoted by a and b, respectively. The inclined angle of the ellipse's major axis was denoted by α. Based on the aforementioned parameters, coordinates of the two endpoints of the major axis were calculated by Equations (2)–(5):

$$x_1 = \frac{a}{2}\cos\alpha + x_0'$$

(2)

$$y_1 = \frac{b}{2}\sin\alpha + y_0'$$

(3)

$$x_2 = -\frac{a}{2}\cos\alpha + x_0'$$

(4)

$$y_2 = -\frac{b}{2}\sin\alpha + y_0'$$

(5)

where x_1 and x_2, y_1, and y_2 were the horizontal and vertical coordinates of the two endpoints, respectively. The extracted endpoints are shown in Figure 4b with two star points.

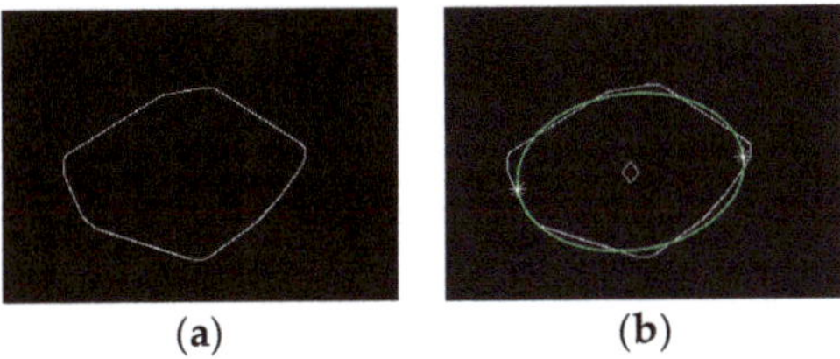

(a) **(b)**

Figure 4. Ellipse fitting for individual broiler body contour: (**a**) The contour of the convex hull image shown in Figure 3d; (**b**) The fitted ellipse.

Sub-step 2, candidate head regions extraction. Figure 2a indicates that there are several high intensity regions in a broiler's grayscale image. The higher the intensity of a region, the higher the temperature in this region. The head, which has no feather or just sparse feathers, is definitely among these high intensity regions. Adaptive K-means clustering method was carried out to locate these high intensity regions. Then, high intensity regions with the top half area were extracted as the candidate head regions. The main operation of candidate head regions extraction is indicated in the flowchart shown in Figure 5.

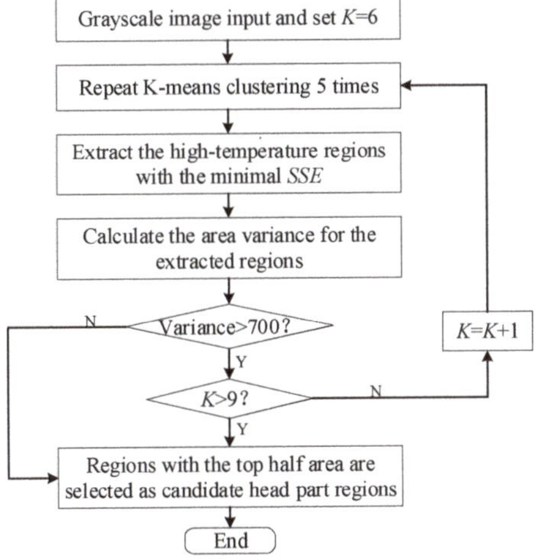

Figure 5. Flowchart of the candidate head regions extraction.

Parameter K of the adaptive K-means clustering [23] was initialized to be 6. That is, 6 pixels in the grayscale image shown in Figure 3a were randomly selected as the centroids. The Manhattan distance [24], calculated by Equation (6):

$$d_{pd} = \left| x_p - c_q \right| \ p \in [1, 76800], \ q \in [1, K],$$

(6)

was applied to measure the intensity distance from the rest of the pixels to the 6 centroids. x_p was the intensity of the pixel in row (p mod 240), column ($p/240 + 1$) of the grayscale image. "/" and "mod" were modulus and remainder operators on integer values, respectively. c_q was the intensity of the qth centroid. Operator $|w|$ referred to the operation of extracting the absolute value of w.

A maximum of 4 iterations of K-means clustering were carried out to obtain the high intensity regions, where each iteration had 5 clustering operations. That is, 5 sets of temporary high intensity regions were obtained for each iteration. The sum of square error, denoted by SSE and calculated by Equation (7):

$$SSE = \sum_{p=1}^{M} \sum_{q=1}^{K} d_{pq}^{2}$$

(7)

was utilized to choose the result high temperature regions for each iteration. Where, M was set to be 240×320 and K was adjustable.

Among the 5 sets of temporary high intensity regions, the one with the minimal SSE was selected as the result high intensity regions for the current iteration. The variance of the areas of the resulting high intensity regions was calculated. If the variance was smaller than 700, the resulting high intensity regions were chosen as the final high intensity regions. Otherwise, if K was smaller than 9, a new iteration of clustering was implemented with K increased by 1. Clustering terminated when either the variance was smaller than 700 or K was greater than 9. The final high intensity regions of the grayscale image shown in Figure 3a after the adaptive K-means clustering operation is shown in Figure 6a.

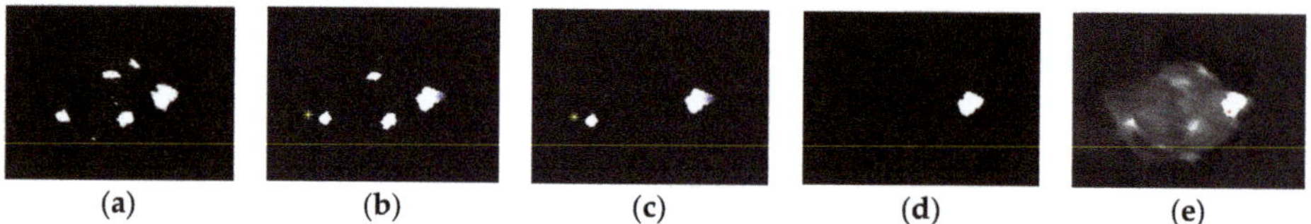

(a)   (b)   (c)   (d)   (e)

Figure 6. Head region locating: (**a**) The final high temperature regions extracted from the image in Figure 2a; (**b**) The candidate head regions; (**c**) The alternative head regions; (**d**) The extracted head region; (**e**) Relationship between the extracted head region and the gray-scale thermal image.

The area of each final high intensity region was calculated. Regions with the top half areas were extracted as the candidate head regions. The candidate head regions of Figure 6a are shown in Figure 6b, where the star points are the endpoints of the major axis of the fitted ellipse. The two endpoints were denoted by *EP1* and *EP2*, respectively.

Sub-step 3, head region locating. The broiler's head must be close to one of the endpoints of the major axis of the fitted ellipse of its top-view body contour. Therefore, two of the candidate head regions, each of which was closest to one of the endpoints of major axis of the fitted ellipse, were selected as the alternative head regions. The alternative head regions selected from the candidate head regions in Figure 6b are shown in Figure 6c. Head region was chosen from the two alternative head regions, denoted by *L1* and *L2*, respectively, according to the following 3 cases:

Case 1: *EP1* and *EP2* located inside *L1* and *L2*, one endpoint in one region. Then, the alternative head region with higher maximum temperature was extracted as the head region.

Case 2: Only one endpoint located inside *L1* or *L2*. Then, the alternative head region, in which the endpoint was located, was extracted as the head region.

Case 3: No endpoint located inside *L1* or *L2*. Then, the shorter distance of each endpoint to the center of *L1* and *L2*, denoted by *D_min1* and *D_min2*, respectively, were calculated and obtained. The alternative head region with a distance of *min{D_min1, D_min2}* s extracted as the head region. Where, *min{D_min1, D_min2}* was the smaller value of *D_min1* and *D_min2*.

Taking the alternative head regions in Figure 6c as an example, head region was located by using Case 2, as shown in Figure 6d. The relationship between the extracted head region and the original gray-scale thermal image is shown in Figure 6e. Examples of the head region located by using case 1 and case 3 are shown in Figure 7b,e, respectively.

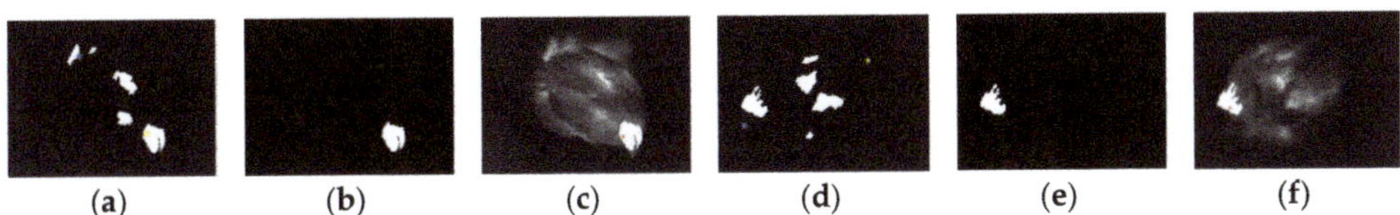

(a)   (b)   (c)   (d)   (e)   (f)

Figure 7. Examples of the head region locating: (a–c) Head region locating by using case 1; (d–f) Head region locating by using case 3.

Where the corresponding candidate head regions are shown in Figure 7a,d, respectively. The relationship between the extracted head region and the gray-scale thermal image is shown in Figure 7c,f. Operations described in Sections 2.3.1 and 2.3.2 were applied to all the 2185 thermal images in the database to locate the broiler's head region. As a result, the head region of the 2027 thermal images were located correctly.

Once the head region was located, the maximum temperature in the region was extracted as the head surface temperature, which was denoted by $HT_{i,j,k}$ ($i \in [1, 20]$). Where i was the number of the broiler and j was the number of the thermal image capture interval. For a broiler in the experimental group, the maximum value of j depended on its dead time. For broilers in the control group, the maximum value of j was 30. k was the number of the thermal image in each interval. It had a maximum value of 5.

According to the technical manual of the thermal imager used in this study, the maximum error between the temperatures recorded by the thermal imager and animal's actual superficial temperature is ±2 °C. In order to reduce the influence of this inherent error, representative head surface temperature (abbreviated as *RHT*) of the *i*th broiler in *j*th internal was calculated by Equation (8):

$$RHT_i^j = \left(\sum_{k=1}^{N_j} HT_{i,j,k}\right) / N_j \quad i \in [1, 20] \tag{8}$$

where N_j is the quantity of the thermal images in the *j*th interval.

2.3.3. Construction of RHT Time Series

Equation (8) was applied to all the 2027 thermal images obtained in Section 2.3.2. A *RHT* time series, which was denoted by *TSRHT*, could be constructed for each broiler. Taking the *i*th broiler as an example, its *TSRHT* was obtained by using expression (9):

$$TSRHT_i = \left\{RHT_i^1, RHT_i^2, \ldots, RHT_i^j, \ldots, RHT_i^{NIH_i}\right\} \quad i \in [1, 20]. \tag{9}$$

where NIH_i is the quantity of the intervals in the thermal images acquisition step for the *i*th broiler. Two broilers, one in the experimental group and another in the control group, were randomly selected, the corresponding *TSRHT* of which are shown in Figure 8.

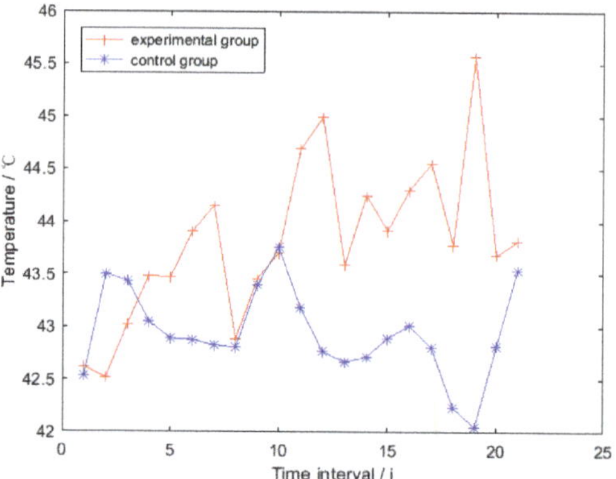

Figure 8. Representative head surface temperature (*RHT*) time series.

Data points in the *TSRHT* of the broiler in experimental and control groups are displayed with red cross and blue star points in Figure 8, respectively. It can be observed that sharp rising and drop exist in the whole *TSRHT*; even the average operation has been applied to reduce influence of the inherent error of the thermal imager. Therefore, the original *TSRHT* could not be utilized directly to identify febrile broiler. Smoothing operation was carried out to each *TSRHT* to obtain a gentle time series. Taking the *i*th broiler as an example, the *t*th data point in its smoothed $TSRHT_i$ ($i \in [1, 20]$) was obtained by calculating the mean of the first t ($t \in [1, NIH_i]$) elements in $TSRHT_i$ by using Equation (10):

$$m_i^t = \frac{\sum_{j=1}^{t} RHT_i^j}{t} \quad t \in [1, NIH_i] \, , \, i \in [1, 20].$$ (10)

where m_i^t is the smoothed value of the *t*th data point in the smoothed $TSRHT_i$. NIH_i and RHT_i^j are the quantity of the elements and the *j*th element in $TSRHT_i$, respectively. The smoothed *TSRHT* of the broilers in the experimental and control groups are shown in Figure 9a with red cross and blue star points, respectively. The corresponding under-wing temperature time series is shown in Figure 9b.

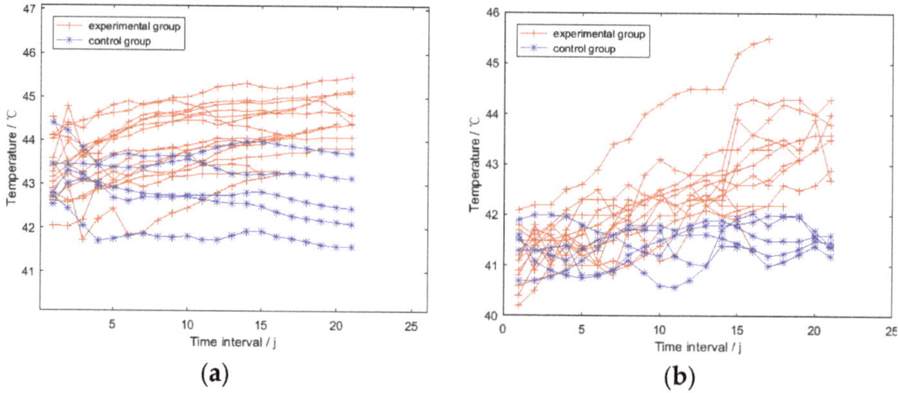

Figure 9. The smoothed representative head surface temperature time series (*TSRHT*) and under-wing temperature time series: (**a**) The smoothed *TSRHT* time series; (**b**) Under-wing temperature time series.

The maximum time intervals of the broilers in the experimental group was 21; data points of broilers in control group from the 22nd to 30th time interval are not plotted in Figure 9. It can be observed in Figure 9a that the smoothed *TSRHT* of broilers in the experimental and control groups had an overall slight increase and flat trends, respectively. Meanwhile, under-wing temperature time series in Figure 9b indicates that almost all the broilers in the experimental groups were feverish (under-wing temperature greater than 42 °C) from the 13rd time interval on. All the broilers in the control group were not feverish (under-wing temperatures less than 42 °C). This is consistent with the result found by Meltzer in 1983 and by Donkoh in 1989. Both studies suggested that most healthy broilers had an ordinary body temperature of 40–42 °C [25,26]. Therefore, all the broilers in the experimental group in this study were infected successfully by the pasteurella inoculum. It can be inferred from Figure 9 that different trend features existed in the smoothed *TSRHT* of febrile and non-febrile broilers, which can be utilized to identify individual febrile broilers.

3. Results

3.1. Testing of HSTE

Operations described in Sections 2.3.1 and 2.3.2 were applied to all the 2185 thermal images to locate a broiler's head region. Thermal images were classified into the following six categories: (i) broiler body was well covered by feathers, and both feet and uncooled manure appeared; (ii) broiler body was well covered by feathers, and only feet or only uncooled manure appeared, or no feet and uncooled manure appeared; (iii) broiler body had sparse feathers, no feet, and uncooled manure appeared; (iv) broiler body had sparse feathers, and feet appeared; (v) broiler body had sparse feathers, an uncooled manure appeared; and (vi) broiler body had sparse feathers, and both feet and uncooled manure appeared. An example image for each category is shown in Figure 10.

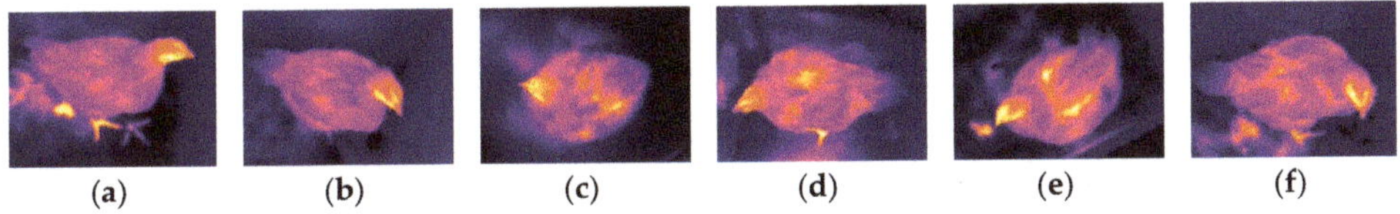

| (a) | (b) | (c) | (d) | (e) | (f) |

Figure 10. Example image of each category: (**a**–**f**) Example images of category (i)–(vi).

Algorithm testing results indicated that the head region of the thermal images in category (a)–(f) could be correctly located with a ratio of 78.05%, 94.90%, 94.94%, 91.53%, 84.89%, and 73.68%, respectively, as shown in Table 1.

Table 1. Ratio of correct locating of head region for images in different categories.

Category	(i)	(ii)	(iii)	(iv)	(v)	(vi)
Number of images	41	785	849	295	139	76
Number of correct locating	32	745	806	270	118	56
Ratio of correct locating	78.05%	94.90%	94.94%	91.53%	84.89%	73.68%

As a result, the average ratio of the correct locating of the head region was 92.77%. That is, the head region of 2027 thermal images were located correctly. These images were picked out for the *TSRHT* construction described in Section 2.3.3.

In order to evaluate the accuracy of the head temperature obtained automatically by extracting the maximum value in the head region, 100 thermal images were randomly selected from the 2027 images whose broiler head regions were located correctly. Head temperatures of the selected thermal images were obtained automatically and manually by using *HSTE* and Fluke Smartview 4.3, respectively. In the latter case, head region was selected by drawing a circle in the head for each selected thermal image. SmartView 4.3 provided the maximum, mean, and minimum temperatures of the selected region, as shown in Figure 11a.

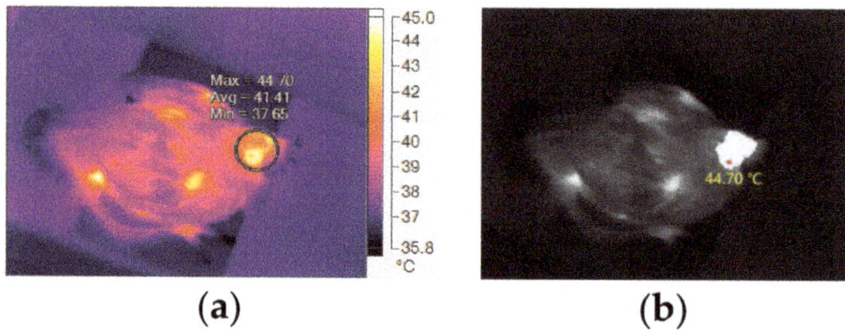

(a)   (b)

Figure 11. Head temperature extraction: (**a**) Head temperature extracted manually by using Smartview; (**b**) Head temperature extracted automatically by *HSTE*.

Head temperature extracted automatically for the thermal image in Figure 11a is shown in Figure 11b. Using the maximum temperature in the selected region provided by the SmartView 4.3 as the gold standard, the accuracy of the head surface temperature extracted by *HSTE* was evaluated. For the thermal images whose broiler head part was correctly located, correlation coefficient between the maximum temperatures extracted by *HSTE* (38.86 °C ± 3.77 °C) and by Smartview (38.9 °C ± 3.76 °C) was 99.99%. The standard deviation of the maximum temperatures was close to 4 °C; this is because the thermal images were randomly selected from different broilers in different groups (including experimental and control groups). The absolute value of the error of the head surface temperature was extracted automatically and by SmartView 4.3 was denoted by *ESA*. For all the selected 100 thermal images, *ESA* is shown in Figure 12.

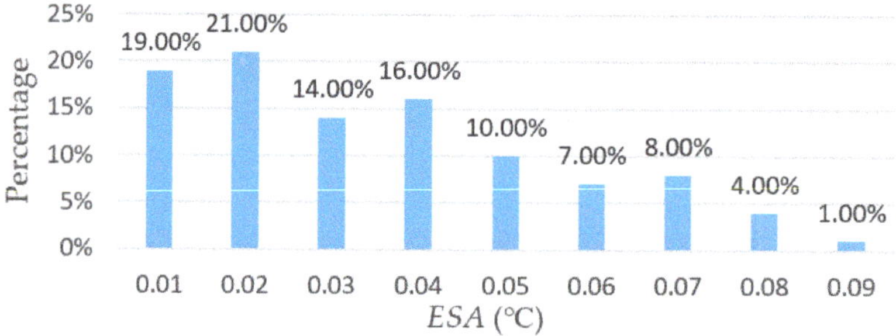

Figure 12. The errors between head temperatures extracted automatically by *HSTE* and by using Smartview.

Figure 12 indicates that the maximum *ESA* of the head temperature, extracted automatically and manually, is less than 0.1 °C. Where 95% of the *ESA* are distributed in the range of 0.01–0.07 °C, it can be concluded that once the head region is located correctly, the head temperature can be extracted automatically with an acceptable accuracy.

3.2. Overall Trend Analysis for the Smoothed TSRHT

It was observed in Figure 9a that overall trends of slight increase and flat existed in the smoothed *TSRHT* of broilers in experimental and control groups, respectively. Slope of the straight line fitted by the first t' ($t' \in [5, NIH_i]$, $i \in [1, 20]$) elements in a smoothed $TSRHT_i$, which was denoted by $Slope\_TSRHT_i^{t'}$, was used to describe these two different overall trend features. For the *i*th broiler, all of its $Slope\_TSRHT_i^{t'}$ formed a slope time series. This series was denoted by $Slope\_TSRHT_i$ and expressed in expression (11).

$$Slope\_TSRHT_i = \left\{ Slope\_TSRHT_i^5, \ldots, Slope\_TSRHT_i^{t'}, \ldots, Slope\_TSRHT_i^{NIH_i} \right\} t' \in [5,\, NIH_i],\ i \in [1, 20] \qquad (11)$$

As the maximum and minimum value of NIH_i respectively was 15 and 21, quantity of the elements in each $Slope\_TSRHT_i$ was between 11 and 17. $Slope\_TSRHT_i$ of all the 20 broilers are plotted in Figure 13.

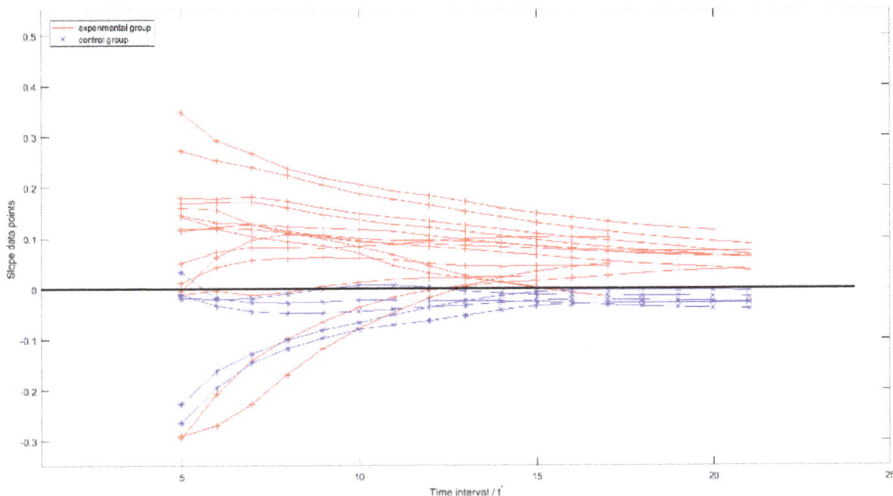

Figure 13. Slope series of each smoothed $TSRHT_i$.

Where data points of the broilers in the experimental and control groups are displayed with red cross and blue star points, respectively, the black line is a boundary, which separates slope data points into positive and non-positive groups. For each given t', the percentage of all the $Slope\_TSRHT_i^{t'}$ ($i \in [1, 20]$) in the experimental and control groups that respectively had positive and non-positive values were listed in Table 2.

Table 2. Percentage of positive and non-positive values for different t'.

t'	5	6	7	8	9	10	11	12	13	14	15	16	17	18	19	20	21
$E(\%)$	73	80	80	80	86.7	86.7	86.7	86.7	100	100	100	92.9	92.3	100	100	100	100
$C(\%)$	80	100	100	100	100	80	80	80	100	100	100	100	100	100	100	100	100

E: experimental group; C: control group.

Preliminary statistical results in Table 2 indicated that from the 13th time interval on, almost all the $Slope\_TSRHT_i^{t'}$ of the smoothed $TSRHT$ at each t' in control and experimental groups had non-positive and positive values, respectively. It could be concluded that the positive or non-positive of the $Slope\_TSRHT_i^{t'}$ can be used to identify whether a broiler is febrile at a given time interval.

4. Discussion

An algorithm was developed to extract head surface temperature from top-view thermal image with an individual broiler. Algorithm testing result indicated that head regions in 7.23% of the thermal images were wrongly located. This would bring about a wrong head temperature, which would in turn influence the trend feature of the smoothed $TSRHT$. Therefore, thermal images, whose head region were wrongly located, were picked out manually before *RHT* was calculated. This set a barrier against the automatic identification of the febrile broiler. Aiming at this insufficiency, the following two possible solutions are worthy of further study in the future.

The first possible solution is to introduce more image information for head region locating. RGB and thermal cameras can be integrated together to capture RGB and thermal images for a broiler at the same time and from the same angle. Head region can be located based on the fused RGB and thermal images. Where the color of the comb, the shape, and context characters of the boiler head, combined with the fitted ellipse and the surface temperature distribution of the broiler body, can be integrated to improve the ratio of the correct locating of head region. The second possible solution is to identify febrile broiler based on the average temperature time series of all the high temperature regions in a broiler's body surface. High temperature regions can be located much easier than the head region. However, the relationship between the under-wing temperature and the feature of the average temperature time series of the high temperature regions is still unknown and deserving further study.

It is important to note that the ambient temperature and relative humidity were respectively 27-30 °C and 55–65% in this study. Both of the environmental parameters were stable during the whole process of the thermal images collection. A great deal of studies indicated that ambient temperature and relative humidity had impacts on the animal's superficial temperature [11]. That is to say, head surface temperature should be considered a dependent variable on several independent variables, such as animal body temperature, ambient temperature, relative humidity, and so on. However, for a different ambient temperature and relative humidity range, as long as the environmental parameters are stable, the same impact would be brought to all the data points in a smoothed *TSRHT*. That is to say, a different ambient temperature and relative humidity range would not influence the overall trend of the smoothed *TSRHT*. Therefore, head surface temperature extracted by *HSTE* should be suitable for febrile broiler identification in different ambient temperatures, as long as the variation range of the environment parameters are small.

Furthermore, it can be observed from Figure 13 that some non-positive data points appeared in the smoothed *TSRHT* of the broilers in the experimental group in the first several time intervals. If the positive or non-positive of $Slope\_TSRHT_i^{t'}$ is used to identify whether a broiler is febrile, a method to determine the start time of the febrile broiler identification should be developed in further study. Finally, different trend features were found in the smoothed *TSRHT* of febrile and non-febrile broilers. It can be inferred that a different proportion of febrile broilers in a broiler flock could bring different features of the population mean superficial temperature time series. Combined with the adhesive animal image segmentation methods [21,27,28], *HSTE* could be employed to extract population mean head surface temperature for broiler flocks. Therefore, the algorithm developed in this research can be regarded as a first step of the development of an automatic warning system for febrile broiler flocks.

5. Conclusions

Ellipse fitting and adaptive K-means clustering were integrated to extract broiler head surface temperature from top-view thermal image with individual broilers in this study. Algorithm testing results indicated that 92.77% of the thermal images could be processed correctly, and the maximum error of the extracted head temperatures was less than 0.1 °C. For the thermal images whose broiler head part was correctly located, correlation coefficient between the maximum temperatures extracted by the developed algorithm and by Smartview was 99.99%. Different overall trend features were observed in the smoothed representative head surface temperature time series (*TSRHT*) of febrile and non-febrile broilers. A slope time series was constructed based on the slopes of the straight lines fitted by the first several data points in each smoothed *TSRHT*. Preliminary statistical analyses were carried out for the signs (positive and non-positive) feature of these slope time series, and the result indicated that the signs feature of the data points in a slope time series can be used to identify whether a broiler is febrile or not. The presented method lays a foundation for the development of an automatic system for febrile broiler identification.

Author Contributions: The main contributions of each author of the manuscript were as follows, X.X.: data collection and analysis, algorithm development, programing of the algorithm, writing—original draft preparation; M.L.: funding acquisition, project administration, algorithm development, design of field experiments, writing—review and editing; W.Y.: algorithm testing, data curation; G.D.: setup of field experiment and data collection, data analysis; Q.Y.: setup of field experiments, selection and management of the experimental broilers; M.S.: funding acquisition, supervision of the whole study; T.N.: supervision of the whole study, writing—review and editing; D.B.: supervision of the whole study, writing—review and editing. All authors discussed the results and implications, and everyone provided helpful feedback. The manuscript was written by X.X. and M.L., and revised by all co-authors.

Funding: This work was funded by the National Natural Science Foundation of China (Nr. 31972615), the Key projects of intergovernmental cooperation in international scientific and technological innovation (Nr. 2017YFE0114400), and the "13th Five-Year" National Key R&D Plan Project of China (Nr. 2017YFD0701600). This work was also supported by the Key Project of R&D of Jiangsu Science and Technology Department (Nr. BE2018433), the Natural Science Foundation of Jiangsu Province (Nr. BK20191315), and the Key Project of R&D of Changzhou Science and Technology Department (Nr. CE20172005).

Acknowledgments: All experiments were carried out in compliance with and using protocols approved by the biosafety committee of Nanjing Agricultural University. The handling of the boilers was performed in accordance with the guidelines approved by the experimental animal administration and ethics committee of Nanjing Agricultural University. All the authors would like to acknowledge the work done by Xianchen Meng (animal experimenter of Jiangsu Lihua Animal Husbandry CO., LTD.) in the setup and implementation of the wireless wearable sensor based broiler under-wing temperature monitoring system.

Conflicts of Interest: The authors declare no conflict of interest.

References

1. Giloh, M.; Shinder, D.; Yahav, S. Skin surface temperature of broiler chickens is correlated to body core temperature and is indicative of their thermoregulatory status. *Poult. Sci.* **2012**, *91*, 175–188. [CrossRef] [PubMed]

2. Moller, A.P. Body temperature and fever in a free-living bird. *Comp. Biochem. Physiol. B Biochem. Mol. Biol.* **2010**, *156*, 68–74. [CrossRef] [PubMed]

3. Cooper, M.A.; Washburn, K.W. The relationships of body temperature to weight gain, feed consumption, and feed utilization in broilers under heat stress. *Poult. Sci.* **1998**, *77*, 237–242. [CrossRef] [PubMed]

4. Iwasaki, W.; Ishida, S.; Kondo, D.; Ito, Y.; Tateno, J.; Tomioka, M. Monitoring of the core body temperature of cows using implantable wireless thermometers. *Comp. Electron. Agricult.* **2019**, *163*, 104849. [CrossRef]

5. Nogami, H.; Arai, S.; Okada, H.; Zhan, L.; Itoh, T. Minimized bolus-type wireless sensor node with a built-in three-axis acceleration meter for monitoring a Cow's Rumen conditions. *Sensors* **2017**, *17*, 687. [CrossRef]

6. Andersson, L.M.; Okada, H.; Miura, R.; Zhang, Y.; Yoshioka, K.; Aso, H.; Itoh, T. Wearable wireless estrus detection sensor for cows. *Comp. Electron. Agricult.* **2016**, *127*, 101–108. [CrossRef]

7. Reid, E.; Fried, K.; Velasco, J.; Dahl, G. Correlation of rectal temperature and peripheral temperature from implantable radio-frequency microchips in Holstein steers challenged with lipopolysaccharide under thermoneutral and high ambient temperatures. *J. Anim. Science* **2012**, *90*, 4788–4794. [CrossRef]

8. Li, L.; Chen, H.; Yu, H.; Huang, R.; Huo, L. Dynamic monitoring device of hens temperature based on wireless transmission. *Trans. Chin. Soc. Agricult. Mach.* **2013**, *6*, 242–245.

9. Zaninelli, M.; Redaelli, V.; Luzi, F.; Bronzo, V.; Mitchell, M.; Dell'Orto, V.; Bontempo, V.; Cattaneo, D.; Savoini, G. First evaluation of infrared thermography as a tool for the monitoring of udder health status in farms of dairy cows. *Sensors* **2018**, *18*, 862. [CrossRef]

10. Alsaaod, M.; Schaefer, A.; Büscher, W.; Steiner, A. The role of infrared thermography as a non-invasive tool for the detection of lameness in cattle. *Sensors* **2015**, *15*, 14513–14525. [CrossRef]

11. Soerensen, D.D.; Pedersen, L.J. Infrared skin temperature measurements for monitoring health in pigs: A review. *Acta Veterinaria Scandinavica* **2015**, *57*, 5. [CrossRef] [PubMed]

12. LokeshBabu, D.S.; Jeyakumar, S.; Vasant, P.J.; Sathiyabarathi, M.; Manimaran, A.; Kumaresan, A.; Pushpadass, H.A.; Sivaram, M.; Ramesha, K.P.; Kataktalware, M.A. Siddaramanna, Monitoring foot surface temperature using infrared thermal imaging for assessment of hoof health status in cattle: A review. *J. Therm. Biol.* **2018**, *78*, 10–21.

13. Schaefer, A.L.; Cook, N.; Tessaro, S.V.; Deregt, D.; Desroches, G.; Dubeski, P.L.; Tong, A.K.W.; Godson, D.L. Early detection and prediction of infection using infrared thermography. *Can. J. Anim. Sci.* **2004**, *84*, 73–80. [CrossRef]

14. Liu, X.; Liu, Y.; Liu, X.; Cao, C.; Lu, H.; Wang, F.; Ge, Y.; Lu, Y. Feasibility study for the use of surface-temperature discriminant poultry health. In *2017 ASABE Annual International Meeting*; ASABE: St. Joseph, MI, USA, 2017; p. 1.

15. Naeaes, I.d.A.; Bites Romanini, C.E.; Neves, D.P.; do Nascimento, G.R.; Vercellino, R.d.A. Broiler surface temperature distribution of 42 day old chickens. *Scientia Agricola* **2010**, *67*, 497–502. [CrossRef]

16. McManus, C.; Tanure, C.B.; Peripolli, V.; Seixas, L.; Fischer, V.; Gabbi, A.M.; Menegassi, S.R.; Stumpf, M.T.; Kolling, G.J.; Dias, E. Infrared thermography in animal production: An overview. *Comp. Electron. Agricult.* **2016**, *123*, 10–16. [CrossRef]

17. Otsu, N. A threshold selection method from gray-level histograms. *IEEE Trans. Syst. Man Cybernet.* **1979**, *9*, 62–66. [CrossRef]

18. Vincent, L. Morphological area openings and closings for grey-scale images. In *Shape in Picture*; Springer: Cham, Switzerland, 1994; pp. 197–208.

19. Barber, C.B.; Dobkin, D.P.; Huhdanpaa, H. The Quickhull algorithm for convex hulls. *ACM Trans. Math. Software* **1996**, *22*, 469–483. [CrossRef]

20. Halir, R.; Flusser, J. *Numerically Stable Direct Least Squares Fitting of Ellipses*; University of West Bohemia Press: Plzen, Czech Republic, 1998; pp. 125–132.

21. Lu, M.; Xiong, Y.; Li, K.; Liu, L.; Yan, L.; Ding, Y.; Lin, X.; Yang, X.; Shen, M. An automatic splitting method for the adhesive piglets' gray scale image based on the ellipse shape feature. *Comp. Electron. Agricult.* **2016**, *120*, 53–62. [CrossRef]

22. Canny, J. A computational approach to edge detection. *IEEE Trans. Pattern Anal. Mach. Intell.* **1986**, *8*, 679–698. [CrossRef]

23. Arthur, D.; Vassilvitskii, S. *Siam/Acm, k-Means Plus Plus: The Advantages of Careful Seeding*; IEEE: Piscataway, NJ, USA, 2007; pp. 1027–1035.

24. Leisch, F. A toolbox for -centroids cluster analysis. *Comput. Stat. Data Anal.* **2006**, *51*, 526–544. [CrossRef]

25. Meltzer, A. The effect of body temperature on the growth rate of broilers. *Br. Poult. Sci.* **1983**, *24*, 489–495. [CrossRef] [PubMed]

26. Donkoh, A. Ambient-temperature—A factor affecting performance and physiological-response of broiler-chickens. *Int. J. Biometeorol.* **1989**, *33*, 259–265. [CrossRef] [PubMed]

27. Nasirahmadi, A.; Richter, U.; Hensel, O.; Edwards, S.; Sturm, B. Using machine vision for investigation of changes in pig group lying patterns. *Comp. Electron. Agricult.* **2015**, *119*, 184–190. [CrossRef]

28. Psota, E.T.; Mittek, M.; Pérez, L.C.; Schmidt, T.; Mote, B. Multi-pig part detection and association with a fully-convolutional network. *Sensors* **2019**, *19*, 852. [CrossRef]

MDPI

St. Alban-Anlage 66

4052 Basel

Switzerland

Tel. +41 61 683 77 34

Fax +41 61 302 89 18

www.mdpi.com

Sensors Editorial Office

E-mail: sensors@mdpi.com

www.mdpi.com/journal/sensors

www.ingramcontent.com/pod-product-compliance
Lightning Source LLC
LaVergne TN
LVHW071305170726
843515LV00006BA/1488